Geometry – von Staudt's Point of View

# NATO ADVANCED STUDY INSTITUTES SERIES

*Proceedings of the Advanced Study Institute Programme, which aims at the dissemination of advanced knowledge and the formation of contacts among scientists from different countries*

The series is published by an international board of publishers in conjunction with NATO Scientific Affairs Division

| | |
|---|---|
| A Life Sciences<br>B Physics | Plenum Publishing Corporation<br>London and New York |
| C Mathematical and Physical Sciences | D. Reidel Publishing Company<br>Dordrecht, Boston and London |
| D Behavioural and Social Sciences<br>E Applied Sciences | Sijthoff & Noordhoff International Publishers<br>Alphen aan den Rijn and Germantown U.S.A. |

**Series C – Mathematical and Physical Sciences**

*Volume 70 – Geometry - von Staudt's Point of View*

# Geometry - von Staudt's Point of View

*Proceedings of the NATO Advanced Study Institute held at Bad Windsheim, West Germany, July 21-August 1, 1980*

edited by

PETER PLAUMANN

and

KARL STRAMBACH

*Mathematisches Institut, Universität Erlangen-Nürnberg, Erlangen, West Germany*

D. Reidel Publishing Company

Dordrecht : Holland / Boston : U.S.A. / London : England

Published in cooperation with NATO Scientific Affairs Division

**Library of Congress Cataloging in Publication Data**

CIP

NATO Advanced Study Institute (1980: Bad Windsheim, Germany)
Geometry–von Staudt's point of view.

(NATO advanced study institutes series. Series C, Mathematical and physical sciences ; v. 70)
Includes index.
1. Geometry, Projective–Congresses. 2. Staudt, Karl Georg Christian von, 1798–1867. I. Plaumann, P. (Peter) II. Strambach, K. (Karl) III. Title. IV. Series.
QA471.N37 1980 516.5 81-5843
ISBN-13: 978-94-009-8491-2 e-ISBN-13: 978-94-009-8489-9
DOI: 10.1007/978-94-009-8489-9

Published by D. Reidel Publishing Company
P.O. Box 17, 3300 AA Dordrecht, Holland

Sold and distributed in the U.S.A. and Canada
by Kluwer Boston Inc.,
190 Old Derby Street, Hingham, MA 02043, U.S.A.

In all other countries, sold and distributed
by Kluwer Academic Publishers Group,
P.O. Box 322, 3300 AH Dordrecht, Holland

D. Reidel Publishing Company is a member of the Kluwer Group

TABLE OF CONTENTS

# PREFACE

Ever since F. Klein designed his "Erlanger Programm", geometries have been studied in close connection with their groups of automorphisms. It must be admitted that the presence of a large automorphismgroup does not always have strong implications for the incidence-theoretical behaviour of a geometry. For example, O. H. Kegel and A. Schleiermacher [Geometriae Dedicata 2, 379 - 395 (1974)] constructed a projective plane with a transitive action of its collineation group on quadrangles, in which, nevertheless every four points generate a free subplane. However, there are several important special classes of geometries, in which strong implications are present. For instance, every finite projective plane with a doubly transitive collineation group is pappian (Theorem of Ostrom-Wagner), and every compact connected projective plane with a flag-transitive group of continuous collineations is a Moufang plane [H. Salzmann, Pac. J. Math. 60, 217 - 234 (1975)]. Klein's point of view has been very useful for numerous incidence structures and has established an intimate connection between group theory and geometry



*P. Plaumann and K. Strambach (eds.), Geometry – von Staudt's Point of View, vii–xi.*

which is a guidepost for every modern treatment of geometry.

A few decades earlier than Klein's proposal, K. G. Ch. von Staudt stated a theorem which indicates a different point of view and is nowadays sometimes called the "Fundamental Theorem of Projective Geometry". This theorem shows clearly that it is important to consider the manner in which the blocks are embedded in order to get information on the surrounding geometrical structure. It turns out that many incidence structures permit an adequate definition of the group of projectivities of a block, so that using the insights of von Staudt and Hessenberg a kind of "Fundamental Theorem" can be proven. For all those geometries of the same type the group of projectivities operates n-fold transitively on the points of a block, where $n$ is characteristic for the respective type.

The fundamental theorems then characterize the classical models by sharply n-transitive action; in all these cases the group of projectivities can be exactly determined. Generally it is altogether useless to try to compute the group of projectivities of a non-classical geometry. For finite projective planes, however, or for projective planes which are manifolds some results are known; the reader can get information on the state of the art in Part III of this book.

In spite of the apparently weak link between the group of projectivities and the structure of geometry, regularity of the action on the points of a block has decisive consequences. This was first realized by A. Barlotti and A. Schleiermacher for projective and affine planes. A geometry is said to be m-regular or to have the property $P_m$, if the stabilizer of any $m$ points on a block within the group of projectivities consists only of the identity. Surprisingly, for every suitable type of geometry there is a small natural number $m$ such that the following holds: the free geometries of this type satisfy $P_m$, while for every $k > m$ one can construct geometries of this type satisfying $P_k$ but not $P_{k-1}$. For projective planes and for Benz geometries $m = 6$, while for affine planes and k-nets $m = 4$. Further, it seems to be true that $P_k$ for $n < k < m$ implies $P_n$, where $n$ is the minimal possible value for this type; by the fundamental theorem the geometry then is classical. Since for projective planes this result is due to A. Schleiermacher, we call a proposition of this kind a "von Staudt-Schleiermacher Theorem". Until now von Staudt-Schleiermacher theorems have been proven only for projective planes, Minkowski planes and 3-nets. Considering only geometries in special categories, for example, finite geometries or locally compact, connected geometries, weaker assumptions of regularity are sufficient to characterize the classical models.

Of course, the picture we sketched in the last section is only a very rough one. To transform it into concrete mathematics with definitions, theorems and proofs takes the larger part of Part I of this book. As in every well developed mathematical theory, there are various aspects which do not fit into a short linear description such as the one we gave above. We do not mind if the reader has perhaps more interest in these aspects than in the mathematical ideology we have stressed in this preface. Already in von Staudt's work the group of projectivities is intimately interwoven with cross ratios and conic sections; these matters are considered by some contributors to this book.

For readers who are interested in the history of mathematical ideas we have included an article by H. Freudenthal which gives a thorough survey of von Staudt's work and its importance for mathematics.

We are very grateful to the NATO Science Division

for providing the funds for an Advanced Study Institute and to the participants in this summer school who were a receptive audience for the lectures. But primarily we wish to thank the authors who adapted themselves willingly to a given concept without forgetting their personal preferences. A book written by one hand surely would be more uniform, but it could never encompass so much experience and knowledge, never show so many facets of one principle. Certainly this book can serve as a substitute during the next few years for the missing monograph in this area of geometry.

Erlangen, December 31st, 1980

Peter Plaumann and Karl Strambach.

# PROJECTIVITIES IN PROJECTIVE PLANES

Günter Pickert

Universität Gießen

Abstract.
In section A the question, under what conditions a projectivity of a line - defined as a product of perspectivities - with 3 fixed points is the identity (condition $P_3$), leads to the condition of Desargues and Pappos, the Hessenberg Theorem ("Pappos" implies "Desargues") and the Fundamental Theorem of Projective Geometry, giving several conditions equivalent to $P_3$; historical notes to these developments are provided in B. Section C proves the Schleiermacher Theorem: If every projectivity of a line with 5 fixed points is the identity, then the Pappos condition holds. Also consequences of the similar condition with 6 fixed points are considered. In D the Moufang planes are characterized by the existence of a permutation group on a line, sharply transitive on $l \setminus \{P\}$ for a point P on l, and normalized by those projectivities of l onto itself with fixed point P (Generalized Lüneburg-Yaqub-Theorem).

## A. The fundamental theorem of projective geometry

In this section a *projective plane* is a pair $(\mathcal{P}, \mathcal{L})$, where $\mathcal{L}$ is a set of subsets, called *lines*, of the set



*P. Plaumann and K. Strambach (eds.), Geometry - von Staudt's Point of View, 1–49.*

$\mathcal{P}$, the elements of which are called *points*, with the three properties:

(J) *For every two points* P,Q *there is exactly one line* l *with* $P,Q \in l$.

(I) *For every two lines* k,l *there is a point* $P \in k,l$.

(Q) *There are four points, no three of which are collinear.*

Here "collinear" means, that there is a line containing the points. The line determined by P,Q with $P \neq Q$ according to (J) is denoted by PQ and called the *join* of P,Q. It follows from

$$P \neq Q,\ P,Q \in k,l \Longrightarrow k = PQ = l$$

and (I), that the meet of two lines k,l contains exactly one point, the *intersection* of k,l, denoted by $k \cap l$ (misusing the set-theoretical notation). A set of 4 points with the property in (Q) is called a (non-degenerate) *quadrangle*.

For two lines a,b and a point $C \notin a,b$ the mapping $\pi : a \to b$ determined by

$$CX = CX^{\pi}\ ,\quad X^{\pi} \in b \quad \text{for all } X \in a$$

is called the *perspectivity of* a *to* b *from center* C. A product $\pi_1 \ldots \pi_m$ of perspectivities $\pi_i : a_i \to a_{i+1}$

($i=1,\dots,m$; $m \in N$) is called a *projectivity* (of $a_1$ to $a_{m+1}$). Of course the projectivities of a line l to itself form a permutation group, the *projective group* $\Pi_l$ of the line l. It is easily seen, that this group operates transitively on the triples of (different) points of l and that the groups $\Pi_l$ (for all lines l of the plane) are isomorphic (as permutation groups). Now the question arises very naturally, under what conditions $\Pi_l$ is even sharply transitive on point triples, that is:

($P_3$) *A projectivity of a line* l *to itself with* 3 *fixed points is the identity* $1_l$.

Since the $\Pi_l$ are isomorphic, the property ($P_3$) for one line already implies ($P_3$) for all lines. Specializing ($P_3$) to ($P_{3,m}$) by restricting the projectivities to products of (at most) m perspectivities, we see, that ($P_{3,2}$) holds in every projective plane: If the perspectivities $\pi_1 : a_1 \to a_2$, $\pi_2 : a_2 \to a_1$ are not inverses of one another, their centers $C_1, C_2$ are different and $\pi_1\ \pi_2$ has only the fixed points $a_1 \frown a_2$, $C_1C_2 \frown a_1$. Therefore ($P_{3,3}$) would be valid too, if one could show every product of two perspectivities to be also a perspectivity. This raises the question: Under what conditions is the product of the perspectivities

$\pi_{12} : a_1 \to a_2$ from $C_{12}$ and $\pi_{23} : a_2 \to a_3$ from $C_{23}$

also a perspectivity? In the cases $a_1=a_2$, $a_2=a_3$, $a_3=a_1$, $C_{12}=C_{23}$ the answer is trivial; thus they will be excluded in the sequel. Then $\pi_{12}\,\pi_{23}$ can only be a perspectivity, if it fixes $a_1 \cap a_3$, and this gives the condition

$$a_1 \cap a_2 \neq a_2 \cap a_3 \Longrightarrow a_1 \cap a_3 \in C_{12}C_{23} \ .$$

Thus we have to consider the two cases

$(F_1)$ $a_1 \cap a_2 \neq a_1 \cap a_3 \in C_{12}C_{23}$ ;

$(F_2)$ $a_1 \cap a_2 = a_2 \cap a_3$

(i.e. the $a_i$ are concurrent).

Beginning with $(F_1)$ we put

$$P_1 = a_2 \cap a_3 \ , \ P_3 = a_1 \cap a_2 \ , \ Q_3 = a_1 \cap a_3, \ Q_1 = C_{12},$$
$$Q_2 = C_{23} \ .$$

If $\pi_{12}\,\pi_{23}$ is a perspectivity $\pi_{13}$, then its center $C_{13}$ must lie on $Q_2P_3$ as well as on $P_1Q_1$ :

$$C_{13} = P_1Q_1 \cap Q_2P_3 \ .$$

On the other hand, without assumption on $\pi_{12}\,\pi_{23}$, this equation defines a point $C_{13} \notin a_1, a_3$, so that we can introduce the perspectivity $\pi_{13}$ of $a_1$ to $a_3$ from $C_{13}$. Now for $Q_3, P_3$ and $P_1^{\pi_{31}}$ (with $\pi_{31} = \pi_{13}^{-1}$) $\pi_{13}$ gives the same images as $\pi_{12}\,\pi_{23}$. Thus $\pi_{12}\,\pi_{23}$ is a perspectivity if and only if

$$(*) \quad X^{\pi_{12}\,\pi_{23}} = X^{\pi_{13}} \text{ for all } X \in a_1 \setminus \{Q_3, P_3, P_1^{\pi_{31}}\} \ .$$

Putting $X^{\pi_{12}} = P_2$ $(\in a_2 \setminus \{a_2 \cap Q_1Q_2, P_3, P_1\})$, $P_2^{\pi_{23}} = X'$ we have $X = P_3Q_3 \cap Q_1P_2$ , $X' = P_2Q_2 \cap Q_3P_1$ , and (*) simply means the collinearity of $X, X', C_{13}$ (see fig. 1).

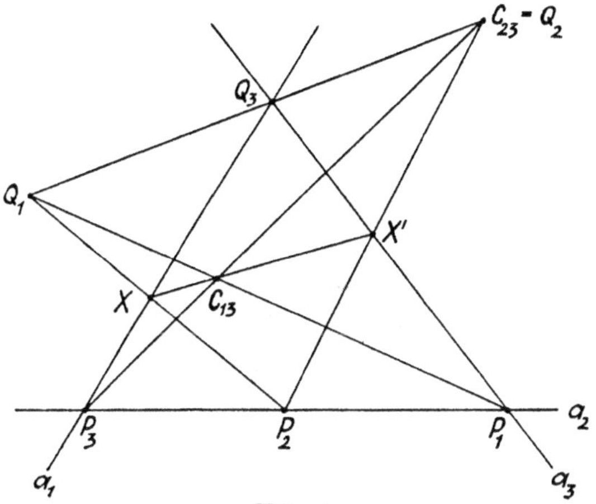


Fig. 1

Thus we get the *Pappos Condition:*

(P) *If the three points* $P_1, P_2, P_3$ *are on a line and the three points* $Q_1, Q_2, Q_3$ *are on another line but not on the first, then the points*

$$P_1Q_1 \cap Q_2P_3, \quad P_2Q_2 \cap Q_3P_1, \quad P_3Q_3 \cap Q_1P_2$$

*are collinear.*

Since under the assumptions in (P) we may put $a_1 = P_3Q_3$, $a_2 = P_1P_3$, $a_3 = Q_3P_1$ and introduce the perspectivities $\pi_{12}: a_1 \to a_2$ from $C_{12} = Q_1$, $\pi_{23}: a_2 \to a_3$ from $C_{23} = Q_2$, our derivations prove

<u>Proposition 1.</u> *A projective plane fulfils the Pappos Condition* (P) *if and only if every product of two perspectivities with* $(F_1)$

*is a perspectivity.*

We get further

Corollary 1. $(P_3) \Longrightarrow (P_{3,3}) \Longrightarrow (P)$ .

The first implication here being trivial, we have, according to Prop. 1, only to show $\pi_{12}\,\pi_{23} = \pi_{13}$, assuming $(P_{3,3})$ and the notations introduced above. But this results from the fact, that $\pi_{12}\pi_{23}\pi_{31}$ has the three fixed points $P_3, Q_3$, $a_1 \cap P_1Q_1$ .

In the case $(F_2)$ $\pi_{12}\pi_{23}$ can only be the perspectivity $\pi_{13}: a_1 \rightarrow a_3$ from $C_{13}$, if $C_{13} \in C_{12}C_{23}$; this is obvious, if the common point C of the $a_i$ is not on $C_{12}C_{23}$, and if C, but not $C_{13}$ would be on $C_{12}C_{23}$, then $C \notin C_{12}C_{13}$, which from $\pi_{12}\pi_{23} = \pi_{13}$ and therefore $\pi_{21}\pi_{13} = \pi_{23}$ would give $C_{23} \in C_{12}C_{13}$, contradicting $C_{13} \notin C_{12}C_{23}$ . Choosing $A_1 \in a_1 \setminus \{C, a_1 \cap c\}$ with $c = C_{12}C_{23}$ then we get $C_{13} = A_1A_3 \cap c$ with $A_2 = A_1^{\pi_{12}}$, $A_3 = A_2^{\pi_{23}}$ .

From this we see, that $\pi_{12}\pi_{23} = \pi_{13}$ simply means the *Desargues Condition* (see fig. 2).

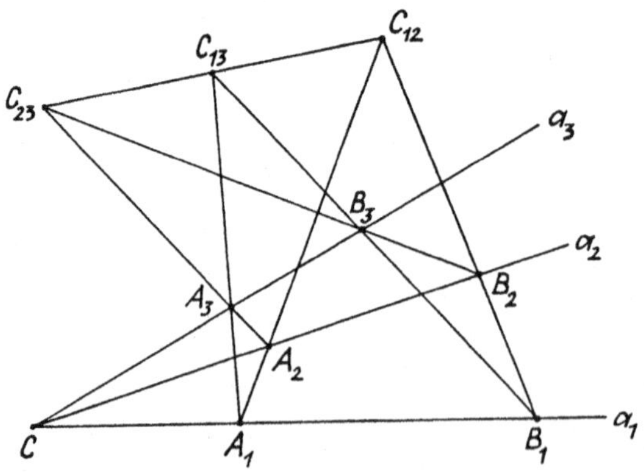


Fig. 2

(D) $C, A_i, B_i \in a_i$, $|\{C, A_i, B_i\}| = 3$ $(i=1,2,3)$,
$|\{a_1, a_2, a_3\}| = 3$, $A_3 \notin A_1A_2$, $B_3 \notin B_1B_2$, $C_{ij} \in A_iA_j$, $B_iB_j$ $(i < j;\ i,j=1,2,3)$, $C_{12} \neq C_{23} \Longrightarrow C_{13} \in C_{12}C_{23}$.

Thus we have

Proposition 2. *A projective plane satisfies the Desargues Condition* (D) *if and only if every product of two perspectivities with* ($F_2$) *is a perspectivity.*

A projective plane satisfying (D) is called *desarguesian.*

Using the reformulation of Pappos and Desargues Condition given by the Propositions 1,2 we get the famous

Hessenberg Theorem. (P) $\Longrightarrow$ (D).

To prove it (in a way essentially due to Hessenberg) we start according to Proposition 2 with perspectivities $\pi_{12}: a_1 \to a_2$ from $C_{12}$ and $\pi_{23}: a_2 \to a_3$ from $C_{23}$ with $C = a_1 \cap a_2 = a_2 \cap a_3 = a_3 \cap a_1$, $C_{12} \neq C_{23}$, $C_{12} \notin a_1, a_2$, $C_{23} \notin a_2, a_3$. Since the plane of order 2 fulfils (D) trivially, we may suppose, that every line contains at least 4 points. We choose (see fig. 3)

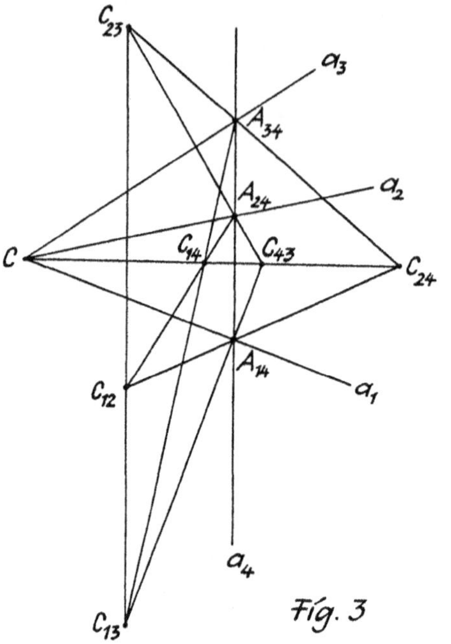


Fig. 3

$A_{14} \in a_1 \setminus \{C_{12}C_{23} \cap a_1,\ C\}$ , $A_{34} \in a_3 \setminus \{A_{14}{}^{\pi_{12}\pi_{23}},\ C,\ A_{14}C_{23} \cap a_3\}$ .

Then $C_{24} = C_{12}A_{14} \cap C_{23}A_{34}$ is neither on $a_2$ nor on $a_4 = A_{14}A_{34}$, so that we can introduce the perspectivity $\pi_{24}: a_2 \to a_4$ from $C_{24}$ and with $\pi_{42} = \pi_{24}^{-1}$ arrive at

$$\pi_{12}\pi_{23} = (\pi_{12}\pi_{24})(\pi_{42}\pi_{23}).$$

If $A_{34} \in C_{12}C_{23}$ , then $C_{24} = C_{12}$ and $\pi_{12}\pi_{24}$ is the perspectivity $\pi_{14}$ of $a_1$ to $a_4$ from $C_{14} = C_{24}$. Otherwise we have $C_{24} \neq C_{12}$ and

$$a_1 \cap a_2 = C \neq a_1 \cap a_4 = A_{14} \in C_{12}C_{24},$$

so that $(F_1)$ with 4 instead of 3 is valid. It thus follows from (P) and Proposition 1 that $\pi_{12}\pi_{24}$ is the perspectivity of $a_1$ to $a_4$ from

$$C_{14} = C_{12}A_{24} \cap C_{24}C \quad \text{with} \quad A_{24} = a_2 \cap a_4 \ .$$

In the same way we see, that $\pi_{42}\pi_{23}$ is the perspectivity $\pi_{43}$ of $a_4$ to $a_3$ from

$$C_{43} = C_{24}C \cap C_{23}A_{24} \ ;$$

here we have $C_{24} \neq C_{23}$, since $A_{14} \notin C_{12}C_{23}$. Now $C_{24} \notin a_2$ implies $C_{14} \neq C_{43}$, and $(F_1)$ with 4 instead of 2 is valid. Thus

$$\pi_{14}\pi_{43} = \pi_{12}\pi_{24}\pi_{42}\pi_{23} = \pi_{12}\pi_{23}$$

is a perspectivity, so that according to Proposition 2 (D) is proved.

Now Cor.1 and the Hessenberg Theorem give the so-called

<u>Fundamental Theorem of Projective Geometry.</u> *For a projective plane the following conditions are equivalent:* (P), $(P_3)$, $(P_{3,3})$,

$(P_3')$ *a projectivity is already determined by the images of* 3 *points,*

(PF) *a projectivity of a line to itself with a fixed point is a product of two perspectivities,*

(PF') *a projectivity of a line onto another line with a fixed point is a perspectivity.*

Proof. By standard argument (see e.g. [6], p.139/40) we get

$$(P_3) \Longrightarrow (P_3') \Longrightarrow (PF') \Longrightarrow (PF) \Longrightarrow (P_3) \ .$$

Since we have Cor.1, only

$$(P) \Longrightarrow (P_3)$$

must be proved. This can be done using the Hessenberg Theorem and the fact, that the points and lines of a desarguesian plane (i.e. a projective plane, in which (D) is valid) can be represented by the 1- resp. 2-dimensional subspaces of a 3-dimensional vector space over a sfield K, which is a field (i.e. commutative) since we have (P): To 4 collinear points A,B,C,D there exists exactly one $u \in K$, such that A,B,C,D are the subspaces generated resp. by the vectors $\vec{a}$, $\vec{b}$, $\vec{a}+\vec{b}$, $\vec{a}u+\vec{b}$ ; this u, the crossratio of A,B,C,D is invariant under perspectivities, therefore also under projectivities, and $(P_3)$ now follows since D is determined by A,B,C,u. A conceptually simpler proof (see [3],§34) consists in deriving from (D) alone, that for $a \neq b$ every product $\pi: a \to b$ of 3 perspectivities can be written as product of 2 perspectivities, so that the proof of (PF') can be reduced by mathematical induction to the Propositions 1,2; together with the Hessenberg Theorem this gives

$$(P) \Longrightarrow (PF') .$$

## B. Historical Notes.

Originally the expression "Fundamental Theorem of Projective Geometry" was only used for the statement, that $(P_3)$ is valid in the real projective plane. In a "purely projective" way this was first proved 1847 by von Staudt (see [10], §9). He defined a projectivity as a mapping conserving harmonic quadruples without stating explicitly the continuity used afterwards in his proof (A proof without continuity assumption was given by Darboux in 1880). In modern language his proof runs as follows: The set $\mathcal{F}$ of fixed points of a projectivity of a line l onto itself is a closed subset of this (real projective) line provided with the usual topology (1-sphere); for a point $P \in l \setminus \mathcal{F}$ the connectedness of l would give 2 points $A,B \in \mathcal{F}$ separating every point of $\mathcal{F}$ from P; but since $|\mathcal{F}| \geq 3$, there is a point $C \in \mathcal{F} \setminus \{A,B\}$, and then the fourth harmonic point D to A,B,C also belongs to $\mathcal{F}$, but is not separated from P by A,B. Von Staudt derived also $(P_3')$,(PF') and showed, that every projectivity (in his sense) is a product of at most 3 perspectivities.

In 1891 H. Wiener formulated the statement

$$(P),(D) \Longrightarrow (P_3) ,$$

which was proved 1898 by F. Schur. The Hessenberg Theorem of 1905 then gave

$$(P) \Longrightarrow (P_3) .$$

It follows from results of G. Ancochea (1941, 1947) and L.-K. Hua (1949), that with von Staudt's definition of "projectivity" $(P_3)$ is valid in a desarguesian projective plane of characteristic $\neq 2$ (i.e. the diagonal points of a quadrangle are not collinear) if and only if the sfield, used in the vector space representation of the plane, has no other automorphism or anti-automorphism than the identity; thus, considering the inner automorphisms of a sfield, we have for desarguesian planes $(P_3) \Longrightarrow (P)$, but the converse is not true, since there exist fields with more than one automorphism. The theorem of von Staudt follows then from the fact, that in the (ordered) field of real numbers the axiom of Archimedes holds and every positive element is a square ("sum of squares" is already sufficient).

## C. The Schleiermacher Theorem

Barlotti [1] was the first to consider the following generalization of $(P_3)$:

$(P_n)$ *A projectivity of a line* $l$ *to itself with* $n$ *fixed points is the identity* $1_l$.

Of course $(P_1)$ is false in every projective plane, and $(P_2)$ is valid only in the projective plane of order 2 (with exactly 3 points on every line). Barlotti proved 1964 $(P_6)$ to be valid in every free plane; thus, loosely speaking, no interesting theorem can be deduced from $(P_6)$ alone. But already with $(P_5)$ the situation is totally different. Introducing the spezialisation

$(P_{n,m})$ of $(P_n)$ as in the case n=3 before we have the

<u>Schleiermacher Theorem</u> $(P_{5,5}) \Longrightarrow (P)$ .

We modify the original proof given by Schleiermacher 1967 [7] using an idea of Fritsch [2]. From $(P_{5,5})$ we deduce at first (D) in the equivalent form given in Proposition 2, using the notations in fig. 2 with different lines $a_1, a_2, a_3$ through C, different points $C_{12}, C_{23}, C_{31} (=C_{13})$ on c, $\pi_{ij}$ as the perspectivity of $a_i$ to $a_j$ from $C_{ij}$ and $A_1 \in a_1 \setminus \{C, a_1 \cap c\}$, $A_2 = A_1^{\pi_{12}}$, $A_3 = A_2^{\pi_{23}} \mp A_1^{\pi_{13}}$ :

(1) $\pi_{12}\pi_{23}\pi_{31} = 1_{a_1}$

This will be done at first under the restrictions

(2) $C \notin c, \quad C_{12} \notin a_3, \quad C_{23} \notin a_1$ .

It is easy to see, that $C, A_1, a_1 \cap c$ are 3 fixed points of $\pi = \pi_{12}\pi_{23}\pi_{31}$ .

Tō prove (1) with the help of $(P_{5,5})$, which of course implies $(P_{5,3})$, we need therefore two more fixed points of $\pi$. For

$$B_1 \in a_1 \setminus \{C, A_1, a_1 \cap c\}, \quad B_2 = B_1^{\pi_{12}}, \quad B_3 = B_2^{\pi_{23}}$$

we try to construct $B_1^{\pi}$ as image of $B_1$ under a product of 5 perspectivities:

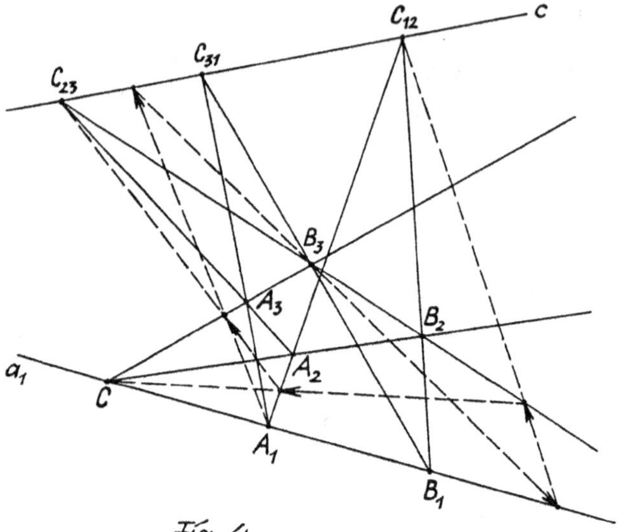


Fig. 4

Forming the product $\pi'$ (see the arrows in fig. 4) of the perspectivities of $a_1$ to $B_2B_3$ from $C_{12}$, of $B_2B_3$ to $A_1A_2$ from C, of $A_1A_2$ to $a_3$ from $C_{23}$, of $a_3$ to c from $A_1$, of c to $a_1$ from $B_3$; their existence is guaranteed by (2), and we have $B_1^{\pi'} = B_1^{\pi}$. Now it is easy to see, that $\pi'$ has the fixed points $C, A_1$, $a_1 \cap c$, $a_1 \cap B_2B_3$, $a_1 \cap B_3C_{12}$ and, using (2), that these are different, if

$$B_3 \notin A_1A_2 \ , \ A_1C_{23} \ .$$

Under this condition we get $\pi' = 1_{a_1}$ from $(P_{5,5})$. Therefore every point of

$$a_1 \setminus \{(A_1A_2 \cap a_3)^{\pi_{32}\pi_{21}} \ , \ (A_1C_{23} \cap a_2)^{\pi_{21}}\}$$

is a fixed point of $\pi$. Since (D) is valid in every projective plane with less than 7 points on a line (see e.g. [6],p.302), we may assume that there are at least

7 points on $a_1$. Thus $\pi$ has indeed at least 5 fixed points and is therefore the identity on $a_1$.

Now we have to remove the restrictions (2). Beginning with the second of these we suppose $C_{12} \in a_3$, but $C \notin c$. Since we may assume, that there are at least 5 points on every line, we can choose (see fig. 5)

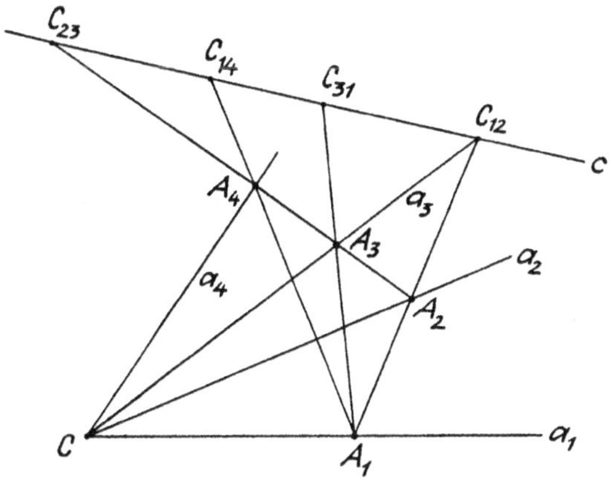


Fig. 5

$$A_4 \in A_2A_3 \setminus \{A_2, A_3, C_{23}, a_1 \cap A_2A_3\} .$$

Then the point

$$C_{14} = c \cap A_1A_4$$

is neither on $a_1$ nor on $a_4 = CA_4$ and $C_{23}$ not on $a_4$, so that the perspectivities $\pi_{14}: a_1 \to a_4$ from $C_{14}$, $\pi_{41} = \pi_{14}^{-1}$ and $\pi_{24}: a_2 \to a_4$, $\pi_{43}: a_4 \to a_3$ from $C_{23}$ can be introduced. Applying the already proved restricted result and observing $C_{12} \notin a_4$, $C_{14} \notin a_3$ (since $A_4 \neq A_2A_3$), we get

(3) $\quad \pi_{12}\pi_{24}\pi_{41} = 1_{a_1} = \pi_{14}\pi_{43}\pi_{31}$ , if $C_{23} \notin a_1$

and thus, since $\pi_{24}\pi_{43} = \pi_{23}$ ,

(4) $\quad \pi_{12}\pi_{23}\pi_{31} = \pi_{12}\pi_{24}\pi_{41}\pi_{14}\pi_{43}\pi_{31} = 1_{a_1}$ , if $C_{23} \notin a_1$.

Since (1) is equivalent to $\pi_{31}\pi_{12}\pi_{23} = 1_{a_3}$, this also proves

(5) $\quad \pi_{12}\pi_{23}\pi_{31} = 1_{a_1}$ , if $C_{12} \notin a_3$ .

Now using (5) with 4 instead of 3 resp. 2 and observing $C_{12} \notin a_4$, $C_{14} \notin a_3$ we see, that the condition $C_{23} \notin a_1$ in (3) and therefore in (4) becomes superfluous. Thus only $C \notin c$ remains as restriction for the validity of (1). But this can be removed in the well known way, reducing the proof of (D) with $C \in c$ to the case $C' \notin c$ (see fig. 6, where c is the line "at infinity",

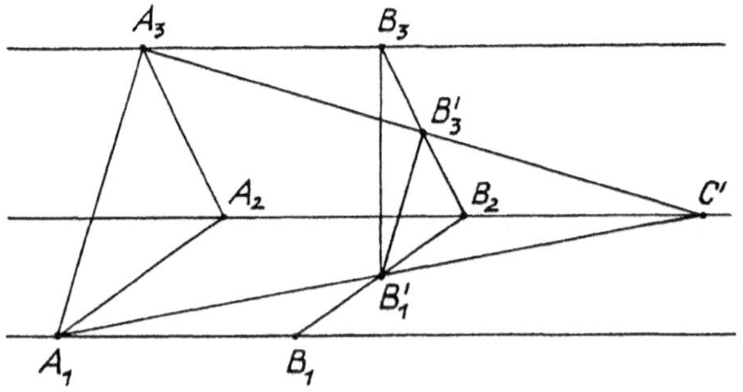


Fig. 6

and compare with fig. 2): For every point $B_1' \in B_1B_2 \setminus \{B_1, B_2, C_{12}\}$ we construct $C' = a_2 \cap A_1B_1'$ with $A_3, A_2, B_2 \neq C' \notin c$ and $B_3' = B_2B_3 \cap A_3C'$. Then (D) with C' instead of C gives $C_{13} \in B'_1B'_3$ and therefore $C_{12} \in B'_1B_3$, so that $C_{13}B_3$, which does not contain the points $B_2, C_{12}$ of $B_1B_2$, must go through $B_1$. This finishes the proof of $(P_{5,5}) \Longrightarrow (D)$.

The proof of the theorem will now be completed by

(6) $(P_{5,5}),(D) \Longrightarrow (P)$.

We prove instead

(6') $(P_{5,4}),(D) \Longrightarrow (P)$,

which of course implies (6) and on the other hand is implied by (6), considering

(6") $(P_{n,4}),(D) \Longrightarrow (P_{n,5})$,

which follows from the fact (see [3], § 34), that with the help of (D) a product of 5 perspectivities can be written as a product of 4. For the proof of (6') we use the (already mentioned) consequence of (D), that the plane can be represented by a 3-dimensional vector space V over a sfield K. Let $(\vec{e}_0, \vec{e}_1, \vec{e}_2)$ be a basis of V and $E_i = \vec{e}_iK$ (i=0,1,2). Then the points not on $E_1E_2$ can be uniquely written as $(\vec{e}_0 + \vec{e}_1x_1 + \vec{e}_2x_2)K$ and thus represented by coordinate pairs $(x_1,x_2)$ $(\in K \times K)$. Let E,A be the points represented in this way by (1,1),(a,a); then $A' = E_2A \cap E_1E$ has the coordinate pair (a,1) (see fig. 7).

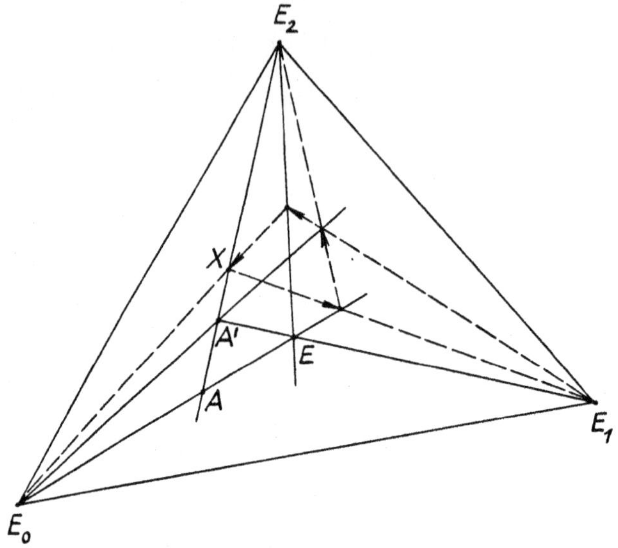


Fig. 7

Now we consider the product $\pi$ of the 4 perspectivities of $E_2A$ to $E_oE$ from $E_1$, of $E_oE$ to $E_oA'$ from $E_2$, of $E_oA'$ to $E_2E$ from $E_1$, of $E_2E$ to $E_2A$ from $E_o$. Of course $E_2$ is a fixed point of $\pi$. The other points X of $E_2A$ have the coordinate pairs (a,x) and the 4 perspectivities applied successively give the points with the coordinate pairs $(x,x)$, $(x,a^{-1}x)$, $(1,a^{-1}x)$, $(a,a^{-1}xa)$. Thus we have for all $X \in E_2A \setminus \{E_2\}$:

(7) $\quad X^\pi = X \iff x = a^{-1}xa$ .

If $a \neq 1,-1$, $\pi$ has according to (7) the other 4 fixed points given by $x = 0,1,a,a^{-1}$ and therefore, according to $(P_{5,4})$ must be the identity on $E_2A$. Using (7) this gives the commutativity of K and thus (P).

The original proof of Schleiermacher for

$$(P_{5,5}) \Longrightarrow (D)$$

uses the concept of a *desarguesian quintuple*. This is a non-degenerate quintuple $(E_1,\dots,E_5)$ of points (i.e. no three of the five points are collinear) such that with $d_i = E_{i+1}E_{i+3}$ (addition of indices mod. 5) the product $\pi$ of the perspectivities of $d_i$ to $d_{i+1}$ from $E_i$ $(i=1,\dots,5)$ is the identity $1_{d_1}$. Since $\pi$ has the 5 fixed points $E_2, E_4$, $d_1 \cap E_5E_1$, $d_1 \cap d_2, d_1 \cap d_5$ (see fig. 8),

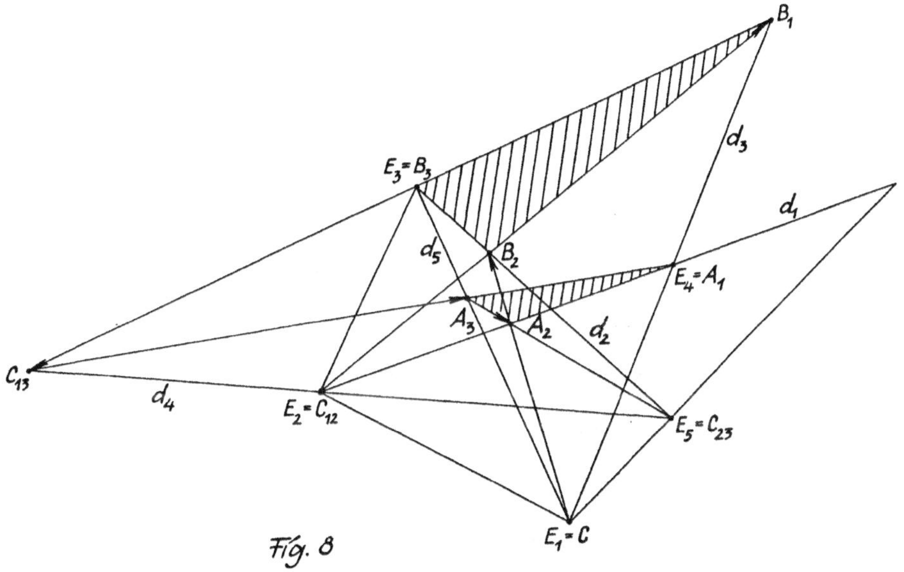

Fig. 8

it follows from $(P_{5,5})$, that every non-degenerate quintuple of points is desarguesian. So one needs

Proposition 3. *A projective plane is desarguesian if and only if every non-degenerate quintuple of points is desarguesian.*

Proof. For a non-degenerate quintuple $(E_1,\dots,E_5)$ and a point $A_2 \in d_1$, different from the 5 fixed points of $\pi$ already recognized, we see (fig. 8) that the points $C = E_1$, $A_1 = E_4$, $A_2$, $A_3 = d_5 \cap E_5A_2$, $B_3 = E_3$, $B_2 = d_2 \cap CA_2$, $B_1 = d_3 \cap E_2B_2$, $C_{12} = E_2$, $C_{23} = E_5$, $C_{13} = A_1A_3 \cap B_1B_3$ satisfy the hypothesis in (D), so that in a desarguesian plane $C_{13} \in C_{12}C_{23}$ follows. This means (see fig. 8) $A_2^{\pi} = A_2$, so that $(E_1,\dots,E_5)$ is desarguesian. Vice versa we introduce, under the hypothesis in (D) the quintuple $(C,C_{12},B_3,A_1,C_{23})$. To guarantee that it is non-degenerate, one needs besides (2) the restrictions

(2') $\qquad B_3 \notin A_1C_{12} \quad , \quad A_1 \notin B_3C_{23}$ .

Now from fig. 8 one can see, how (D) follows, of course with the restrictions (2),(2'). Since in the proof of Proposition 5 as well as in Ch. III Proposition 3 is needed only under the further assumption $(P_6)$, we use $(P_6)$ to remove the restrictions more easily, than it had been done by Schleiermacher without the help of $(P_6)$ ([7], Lemma 2). With the notations in (1) we see from the result just proved, that under the restrictions (2) the product $\pi_{12}\pi_{23}\pi_{31}$ leaves fixed all points of $a_1$ with the possible exeption, due to (2'), of

$$B = ((A_1A_2 \cap a_3)C_{23} \cap a_2)C_{12} \cap a_1 \quad ,$$

$$B' = (A_1C_{23} \cap a_2)C_{12} \cap a_1$$

(see fig. 9 with c as the line "at infinity"). Using $(P_6)$ and the not altogether trivial result (e.g. [6], p. 302), that (D) holds in every projective plane with less than 8 points on a line,

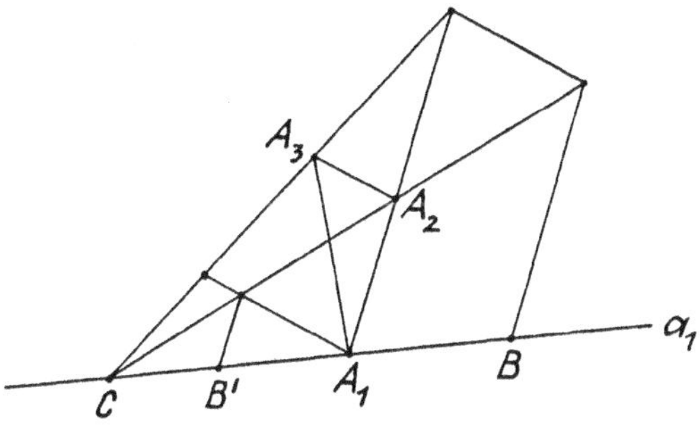


Fig. 9

one gets (1) under the restrictions (2); these can finally be removed as it had been done in (3-5).

Concerning the conditions $(P_n)$ with $n \geqq 6$ Schleiermacher [9] proved the following two propositions:

<u>Proposition 4.</u> $(P_{n,4}),(D) \Longrightarrow (P)$ .

Proof. Of course the proposition is trivially true for $n \leqq 2$, is a consequence of the Fundamental Theorem for n=3 and of the Schleiermacher Theorem together with (6") for $n \in \{4,5\}$. Now (7) in the proof of (6') shows that for the centralizer

$$C(a) \quad ( =_{def} \{x \mid xa=ax,\ x \in K\} )$$

and the center $K_o$ of the sfield $K(P_{n,4})$ has the consequence

$$|C(a)| \geqq n-1 \Longrightarrow a \in K_o$$

and thus in particular

(8) $C(a)$ infinite $\Longrightarrow a \in K_o$.

Since $C(a) \supseteqq K_o$ we get from (8) the commutativity of K and thus (P), if $K_o$ is infinite. Therefore from now on we suppose the finiteness of $K_o$. Let L be a (commutative) subfield of K and thus $C(a) \supseteqq L$ for every $a \in L$. Then, due to (8)

$$L \text{ infinite}, a \in L \Longrightarrow a \in K_o$$

and thus

$$L \text{ infinite} \Longrightarrow L \subseteqq K_o,$$

so that, $K_o$ being finite, every L must be finite. Therefore the subfield $K_o(x)$ is finite for every $x \in K$ and thus according to an extension due to Jacobson ([4], p. 183, Th. 2) of the Wedderburn Theorem (every finite sfield is commutative) K is commutative and therefore (P) valid. Slightly modifying the proof given in [4] one can also go back to the Wedderburn Theorem and the conclusion from Cor. 1 in [4], p. 162 (see also E), that for every finite subfield $L \supseteqq K_o$ of K every automorphism of L leaving the elements of $K_o$ fixed can be extended to an inner automorphism of K. With $|K_o| = q$ the automorphism $x \to x^q$ of L satisfies this condition, and thus there is an element $a \in K \setminus \{0\}$ with

(9) $ax = x^q a$ for all $x \in L$.

Since $K_o(a)$ is finite, there exists n such that the $a^i$ $(i=0,\ldots,n-1)$ form a basis of $K_o(a)|K_o$, and thus $\sum_{i=o}^{n-1} La^i$ is a finite subring of K and thus commutative. Therefore (9) gives $x^q=x$ for all $x \in L$, that is $L = K_o$. But since for every $x \in K$ we can use $K_o(x)$ as L, this gives $K = K_o$.

For the next proposition the concept of a *desarguesian configuration* is needed. This is a set $\mathcal{D}$ of 10 points and 10 lines of a projective plane, for which a labelling as $P_I$, $l_J$ with $I,J \subset \{1,2,3,4,5\}$, $|I| = 2$, $|J| = 3$ exists with

(10) $I \subset J \Longrightarrow P_I \in l_J$ ;

such a labelling is then called a *representation* of $\mathcal{D}$. If $P_I \in l_J$, $|I \cap J| = 1$, there exists $J' \subset \{1,\ldots,5\}$ with $|J'| = 3$, $I \subset J'$, $|J \cap J'| = 2$ and because of (10)

$$P_I , P_{J \cap J'} \in l_J, l_{J'} ,$$

which is a contradiction since $I \not\subset J \cap J'$, $J \not\subset J'$.

Thus we have also a "partial converse" of (10):

(10') $P_I \in l_J \Longrightarrow I \subset J$ or $I \cap J = \emptyset$.

$\mathcal{D}$ is called n-*fold degenerate*, if there exists exactly n incident point-line pairs in $\mathcal{D}$ besides the 30 pairs following from (10). According to (10') in every labelling these pairs must have the form $(P_I, l_{\bar{I}})$ with $\bar{I} = \{1,...,5\} \setminus I$. It follows from $I \cap I' = \emptyset$, $P_I \in l_{\bar{I}}$, $P_{I'} \in l_{\bar{I}'}$ that $I' \subset \bar{I}$, $I \subset \bar{I}'$ and thus

$$P_I, P_{I'} \in l_{\bar{I}}, l_{\bar{I}'} \quad ,$$

which contradicts $I \neq I'$. Therefore the degree n of degeneracy is at most 4 (see [6], p. 88). It is easy to see, that the 10 points fulfilling the hypothesis as well as the conclusion in (D) are the points of a desarguesian configuration with the representation given by

(11) $$P_{\{1,2\}} = C, \; P_{\{1,6-i\}} = A_i, \; P_{\{2,6-i\}} = B_i \; (i=1,2,3),$$
$$P_{\{4,5\}} = C_{12}, \; P_{\{3,4\}} = C_{23}, \; P_{\{3,5\}} = C_{13} \; .$$

Thus a projective plane is desarguesian, if and only if for every set of 10 points $P_I$ and 10 lines $l_J$ $(I,J \subset \{1,...,5\}, |I| = 2, |J| = 3)$, fulfilling 29 of the 30 incidences following from (10), also the remaining incidence is valid.

In fig. 8 we have a desarguesian configuration with the points (11). To connect the concept "desarguesian configuration" and "desarguesian quintuple" we allow therefore from now on only those representation of desarguesian configurations for which

$$(P_{\{1,2\}}, P_{\{4,5\}}, P_{\{2,3\}}, P_{\{1,5\}}, P_{\{3,4\}})$$

is a non-degenerate quintuple, called the *adjoint quintuple* of the representation. Such a representation always exists, if the desarguesian configuration has a degree of degeneracy at most 2. To prove this the representation must be chosen in such a way that none of the points $P_{\{1,2\}}$, $P_{\{2,3\}}$, $P_{\{4,5\}}$ belongs to an incident pair $(P_I, P_{\bar{I}})$. But since $n \leq 2$, there are at most two points $P_I$ in such pairs and then, according to the proof of $n \leq 4$ above, of the form $P_{\{i,j\}}$, $P_{\{j,k\}}$ in an arbitrarily chosen representation, and a permutation of $\{1,\ldots,5\}$ sending $i,j,k$ resp. to $1,3,4$ leads to the desired representation.

Proposition 5. If in a projective plane there exists a desarguesian configuration, which is at most 2-fold degenerate, then

$$(P_6) \Longrightarrow (P) .$$

We cannot give the full proof here, which is somewhat involved (see [9]). The following two lemmata are essential.

Lemma 1. In a projective plane with $(P_{6,5})$ the adjoint quintuple for every allowed representation of a desarguesian configuration is desarguesian.

Proof. With the notations of fig. 8 one sees that the product of 5 perspectivities introduced in the definition of "desarguesian quintuple" has the 6 fixed points $C_{12}, A_1, A_2$, $d_1 \cap d_2$, $d_1 \cap d_5$, $d_1 \cap E_1E_5$ and is therefore, according to $(P_{6,5})$ the identity $1_{d_1}$.

Lemma 2. If in a projective plane with $(P_{6,5})$ the non-degenerate quintuples $(E_1', E_2,\ldots,E_5)$, $(E_1'',E_2,\ldots,E_5)$ with $E_1' \neq E_1''$, $E_1'E_4 = E_1''E_4 = d_3$ are desarguesian, then also every non-degenerate quintuple $(E_1,E_2,\ldots,E_5)$ with $E_1E_4 = d_3$.

Proof. The essential idea consists in keeping the points $A_2(=R),E_2,\ldots,E_5$ of fig. 8 fixed and moving $E_1$ on a line $d_3$ through $E_4$ with $E_2,E_3,E_5 \notin d_3$. Of course R on $d_1(= E_2E_4)$ must be different from (see the proof of proposition 3)

(12) $E_2,\ E_4,\ d_1 \cap d_2,\ d_1 \cap d_5,\ d_1 \cap E_1E_5$ ,

with excludes for $E_1$ the possibilities (see fig. 10)

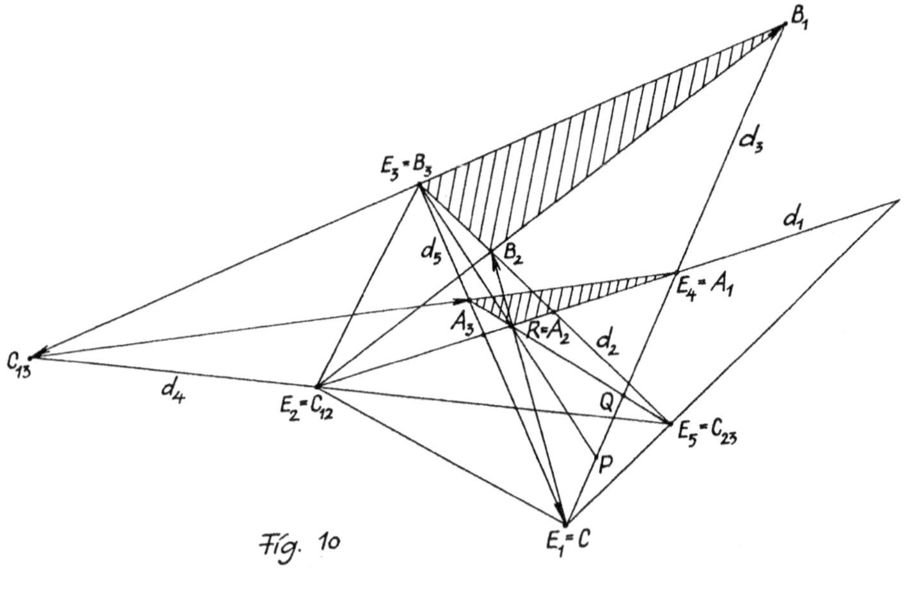


Fig. 10

(12') $P = d_3 \cap E_3R$ , $Q = d_3 \cap E_5R$ .

Then the perspectivities $\pi_1:d_3 \to d_2$ from R, $\pi_2:d_2 \to d_3$ from $E_2$, $\pi_3:d_3 \to d_4$ from $E_3$ , $\pi_4:d_4 \to E_5R$ from $E_4$, $\pi_5:E_5R \to d_3$ from $E_3$ (see the arrows in fig. 10) exist and their product $\pi$ has the 4 fixed points

(12") $E_4$, $d_2 \cap d_3$ , P , Q .

Further we have $E_1^{\pi} = E_1$ if and only if we get a desarguesian configuration with (11) by introducing the notation (see fig. 10)

(13) $C=E$, $B_3=E_3$, $B_2=E^{\pi_1}$, $B_1=B_2^{\pi_2}$, $C_{13}=B_1^{\pi_3}$, $A_3=C_{13}^{\pi_4}$,

$A_2=R$, $A_1=E_4$, $C_{12}=E_2$, $C_{23}=E_5$.

Now comparing with fig. 8 this turns out to be true if $(E_1,\ldots,E_5)$ is desarguesian, and vice versa lemma 1 shows, that $(E_1,\ldots,E_5)$, being the adjoint quintuple of the representation described by (13), (11), is desarguesian if the configuration with the points (13) is desarguesian. Thus we have

(14) $E_1^{\pi} = E_1 \Longleftrightarrow (E_1,\ldots,E_5)$ desarguesian,

provided R differs from the 5 points (12). In order to apply (14) to the desarguesian quintuples $(E_1',E_2,\ldots,E_5)$, $(E_1'',E_2,\ldots,E_5)$ we must therefore choose R on $d_1$ different from the 7 points resulting of (12) by substituting $E_1'$ resp. $E_1''$ for $E_1$ . Then $\pi$ has also the fixed points $E_1',E_2''$ , and these are different from the 4 fixed points (12"), since the quintuples $(E_1',E_2,\ldots,E_5)$,

$(E_1'', E_2, \ldots, E_5)$ are non-degenerate and $E_1', E_1''$ are different from the points (12'). Therefore according to $(P_{6,5})$ $\pi$ is the identity $1_{d_3}$ , and (14) shows, that $(E_1, \ldots, E_5)$ is desarguesian, if it is non-degenerate and $E_1 \in d_3 \setminus \{P,Q\}$ with (12'). To take care of the exceptions P,Q , overlooked in the proof of Schleiermacher ([9], p. 327), we use not only one point R on $d_1$, different from the 7 points described above, but 3 of them: R,R',R" with the exceptional pairs (P,Q), (P',Q'),(P",Q"). Now no point of $d_3$ can be exceptional relative to all 3 points R,R',R", because this would mean e.g. P=Q'=Q" and therefore R'=R". This proves the assertion of lemma 2 provided there are at least 10 points on $d_1$. We have used already (in the proof of Proposition 3) the fact, that a projective plane with less than 8 points on a line is desarguesian. Fortunately enough M. Hall together with J.D. Swift and R.J. Walker (1953 - 56; see the references in [6], p. 302) proved, that all projective planes with exactly 8 or 9 points on a line are desarguesian (in the case of 9 points a computer was needed). But in a desarguesian plane all non-degenerate quintuples of points are desarguesian according to Proposition 3. This finishes the proof of lemma 2.

Now the proof of Proposition 5 proceeds from the given

desarguesian configuration with the help of Lemma 1 to a desarguesian quintuple, from there in several steps, essentially using Lemma 2 to the desarguesian property of all non-degenerate quintuples in the case of an infinite plane; Proposition 3 then gives (D) and Proposition 4 (with n=6) finally (P). For finite planes Schleiermacher ([9], p. 332/3) proved $(P_6) \Longrightarrow (P)$ even without the assumption of a desarguesian configuration (see Ch. III, <u>3.1</u>).

If we assume the condition (d), resulting from (D) by the restriction $C \in C_{12}C_{23}$ - in this case the projective plane is called a *Moufang plane* - there exist desarguesian configurations with degree of degeneracy at most 2, provided the plane has at least 5 points on a line. But since (P) holds in the projective planes with less than 5 points on a line, we get from Proposition 5

$$(P_6),(d) \Longrightarrow (P) .$$

This result can be proved independently from and simpler than Proposition 5 in the following slightly stronger form:

<u>Proposition 6.</u> $(P_{6,4}),(d) \Longrightarrow (P)$ .

Proof. In a Moufang plane the points not on an arbitrarily chosen line ("at infinity") can be represented in the same way by pairs of coordinates in an *alternative division ring* K (see e.g. [6], 6,7) and the lines given by equations of the same form as in the desarguesian case with a sfield K. Since in an alternative division ring one is allowed to use the associative law of multiplication, if only 2 elements are concerned (exact formulation and proof see e.g. [6], p. 162), (7) remains valid. Now for $a \in K \setminus \{0,1,-1\}$ and $a^3 \neq 1$ we get from (7), that $\pi$ has besides $E_2$ the 5 fixed points given by $x=0,1,a,a^{-1},a^2$, and therefore, according to $(P_{6,4})$, $\pi$ is the identity. With

$$K_o = \{a \mid ax=xa \text{ for all } x \in K\}$$

and since $0,1,-1 \in K_o$, this gives

$$(15) \qquad a^3 \neq 1 \Longrightarrow a \in K_o$$

for all $a \in K$. Since every commutative alternative division ring is a field (see e.g. [6], p. 162), we get (P) from the commutativity of K. To derive this property from (15), it is sufficient to prove ab=ba in the case

$$(15_o) \qquad a^3 = 1 = b^3 \ , \ 1 \neq a,b,ab \ ,$$

and since $\quad -a \in K_o \Longrightarrow a \in K_o$,

$$a(ab) = (ab)a \Longrightarrow ab = ba \ ,$$

we can also suppose

$$(15_1) \qquad (-a)^3 = 1,$$

$$(15_2) \qquad (ab)^3 = 1;$$

from $(15_{o,1})$ we get $-a = a \ (\neq 0)$, so that K has

characteristic 2. Therefore

$(15_{o,2})$ give

$$a^2 = a + 1 \ , \ b^2 = b + 1 \ , \ (ab)^2 = ab + 1 \ ,$$

$$b^2a^2 = b^{-1}a^{-1} = (ab)^{-1} = (ab)^2 = ab + 1 \ ,$$

thus, since $(b+1)(a+1) = ba + b + a + 1$ ,

$$ab + ba = a + b$$

and therefore $(a+b)^2 = a + 1 + a + b + b + 1 = 0$,

$a=b$, $ab=ba$ .

## D. The generalized Lüneburg-Yaqub-theorem

Lüneburg [5] and Yaqub [11] proved for a finite projective plane of order q (i.e. q+1 points on a line; Lüneburg: q odd ; Yaqub: q even), that the plane is desarguesian, if and only if the stabilizer $\Pi_{l,P}$ of the point $P \in l$ in the projective group $\Pi_l$ (of a line l) contains a normal subgroup $\Delta$ of order q . Thus we have a property of the permutation group $\Pi_l$, totally different in outlook compared with $(P_5)$, which implies also (due to the finiteness of the plane) (P). To get rid of the finiteness assumption one remarks, that $\Delta$ must be transitive on $l \setminus \{P\}$: Since $|\Delta| > 1$ there exist $\delta \in \Delta$ and $A,B \in l \setminus \{P\}$ with $A^{\delta} = B \neq A$ ; for every 2 points $Q,R \in l \setminus \{P\}$ there exists $\pi \in \Pi_{l,P}$ with $A^{\pi} = Q$ , $B^{\pi} = R$ , since $\Pi_l$ is 3-fold transitive on l, and with $\delta' = \pi^{-1}\delta\pi \in \Delta$ we get $Q^{\delta'} = A^{\delta\pi} = B^{\pi} = R$ . Thus it seems appropriate to assume instead of $|\Delta| = q$,

that $\Delta$ is sharply transitive (regular) on $l \setminus \{P\}$.
Since Lüneburg and Yaqub use the theorem of Levi, that every finite Moufang plane satisfies (P), one can expect only a Moufang plane in the infinite case.
These considerations led Schleiermacher [8] to the Generalized Lüneburg-Yaqub-Theorem. *The Moufang planes are characterized by the existence of a permutation group* $\Delta$ *on a line* $l$ *, sharply transitive on* $l \setminus \{P\}$ *for some* $P \in l$ *with*

(1) $$\pi^{-1}\Delta\pi = \Delta \quad \textit{for all} \quad \pi \in \Pi_{l,P} \, .$$

$\Delta$ *is then a normal abelian subgroup of* $\Pi_{l,P}$, *induced by the group of elations with center* $P$.
Of course if such a group $\Delta$ exists for one incident pair $(P,l)$, this is true for every such pair. We prove at first in a Moufang plane for $P \in l$ the existence of a permutation group $\Delta$ on $l$, sharply transitive on $l \setminus \{P\}$ with (1). $\Gamma(C,a)$ is defined as the group of all collineations (of the plane) with *center* $C$ and *axis* $a$ (i.e. every point on $a$ and every line through $C$ are kept fixed); for $C \in a$ the elements of $\Gamma(C,a)$ are called *elations*. Then the Moufang planes can be characterized also by the transitivity of $\Gamma(C,a)$ on $l \setminus \{C\}$, where $l$ is a line $\neq a$ through $C$, for all incident pairs $(C,a)$. Thus every perspectivity of $b$ to $c$ from $C$ is induced by an elation with center $C$

and axis $a = (b \cap c)C$ (see fig. 11). Therefore the stabilizer $\Gamma_{l,P}$ of $l,P$ in the group $\Gamma$, generated by all elations, induces $\Pi_{l,P}$. Now the group $\Gamma(P)$ of all elations with center P is a normal subgroup of $\Gamma_{l,P}$, since

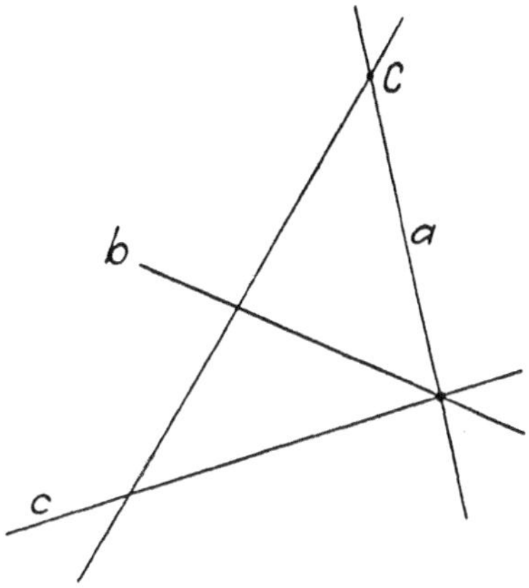


Fig. 11

$$\gamma^{-1}\,\Gamma(P)\gamma = \Gamma(P^{\gamma}) \quad \text{for all collineations } \gamma.$$

Further for points A,B on $l \setminus \{P\}$ and lines $a_1, a_2 \neq l$ through P there are $\gamma_i \in \Gamma(P,a_i)$ with $A^{\gamma_i} = B$ $(i=1,2)$; but then $\gamma = \gamma_1 \gamma_2^{-1}$ belongs to $\Gamma(P,a)$ for some a through P , keeps A fixed and induces therefore the identity on l. Thus the normal subgroup $\Delta$ of $\Pi_{l,P}$ induced by $\Gamma(P)$ is sharply transitive on $l \setminus \{P\}$.

For the proof of the converse we go over to an *affine* plane by deleting a line $u \neq l$ through P (together of course with all its points) as line "at infinity". This plane is called a *translation plane* (and the elations with axis u, restricted to the affine plane, its *translations*), if for all $C \in u$ the group $\Gamma(C,u)$ is transitive on every affine line with C as point "at infinity"; this means the transitivity of the translation group on the point set of the affine plane.

Those elements of $\Pi_1$, which are products of perspectivities with centers on u, restricted to the affine plane, are called *affine projectivities*; they form a group $\Pi_1^*$, induced by the stabilizer of 1 in the group generated by the elations with center on u. Now there is an affine variant of the theorem, where the affine line $1 \setminus \{P\}$ is denoted by 1':

<u>Proposition 7.</u> *The translation planes are characterized by the existence of a sharply transitive group* $\mathrm{T}$ *on a line* 1' *with*

(1') $\alpha^{-1}\mathrm{T}\alpha = \mathrm{T}$ *for all* $\alpha \in \Pi_1^*$ ;

$\mathrm{T}$ *is then a normal abelian subgroup of* $\Pi_1^*$, *induced by the group of translations with fixed line* 1'.

Before proving this, we will use Proposition 7 to complete the proof of the theorem. Let u be any line, $P \in u$ and 1 another line through P. By the remark following the theorem we have a sharply transitive permutation group $\Delta$ on $1 \setminus \{P\}$ with (1). This implies, that the group $\mathrm{T}$, induced by $\Delta$ on $1' = 1 \setminus \{P\}$ is sharply transitive on 1' and satisfies (1'). Thus according to Proposition 7 the affine plane with u as line at infinity is a translation plane and $\mathrm{T}$ induced by $\Gamma(P,u)$. But u being arbitrary, the projective plane is a Moufang plane, $\Delta$ is induced by $\Gamma(P)$ (as well as by $\Gamma(P,u)$) and thus a normal abelian subgroup of $\Pi_{1,P}$. Thus the theorem is proved.

Proof of Proposition 7. First given a translation plane and a line $l'$ in it we shall determine a permutation group $T$ on $l'$ with the properties mentioned in the proposition (this part is not needed in the proof of the theorem). With $u$ as the line at infinity and $P \in u$ completing $l'$ to the projective line $l$, we define $T$ as the group induced by $\Gamma(P,u)$, which is of course abelian and sharply transitive. It only remains to show, that $T$ satisfies (1'). Considering the definition of $\Pi^*_{1,}$ (1') follows, if for every perspectivity $\pi: a \to b$ from $C \in u$ with $a \nparallel b$, $A = u \cap a$, $B = u \cap b$ and every $\tau \in \Gamma(A,u)$ there exists $\tau' \in \Gamma(B,u)$ with

(2) $$X^{\tau\pi} = X^{\pi\tau'} \quad \text{for all } X \in a .$$

If $D = a \cap b$ is on $u$, (2) holds with $\tau' = \tau$, since $C, X, X^{\pi}$ are collinear and therefore also $C$ $(=C^{\tau})$, $X^{\tau}$, $X^{\pi\tau}$ (see fig. 12), which together with $X^{\pi\tau} \in b$ gives (2). If $D \notin u$, we use $\tau_1 \in \Gamma(C,u)$ with $D^{\tau\tau_1} = D^{\tau\pi}$. Then $\tau' = \tau\tau_1 = \tau_1\tau$ as an elation with axis $u$ belongs to $\Gamma(B,u)$, since $D, D^{\tau'} \in b$. The

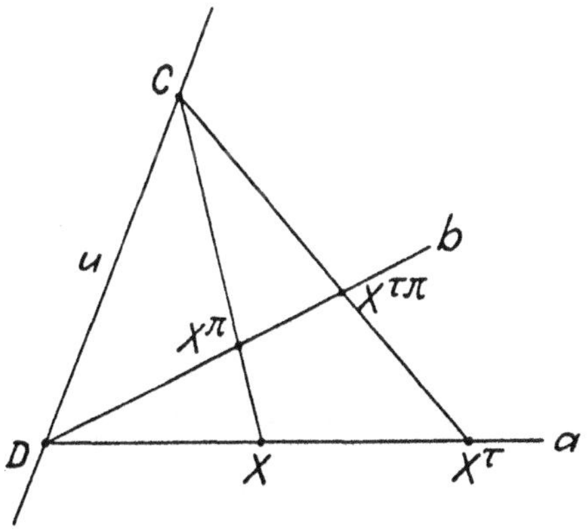


Fig. 12

points $C, X, X^{\pi\tau_1}$ (for $X \in a$) are collinear and therefore also $C\ (=C^{\tau}), X^{\tau}$, $X^{\pi\tau_1\tau}$, which together with $X^{\pi\tau_1\tau} = X^{\pi\tau'} \in b$ gives (2).

To prove the converse we start with a sharply transitive permutation group $T$ on a line $l'$ with (1'). Then for every other line $l_1'$ and a parallel projection (i.e. the affine restriction of a perspectivity with center on u) $\pi$ from $l'$ to $l_1'$

$$(3) \qquad T_1 = \pi^{-1} T \pi$$

is a sharply transitive permutation group on $l_1'$; for $\alpha_1 \in \Pi^*_{l_1}$ one gets

$$\pi\alpha_1\pi^{-1} = \alpha \in \Pi^*_l$$

and from (3),(1') therefore

$$\alpha_1^{-1} T_1 \alpha_1 = \pi^{-1}\alpha^{-1}\pi T_1 \pi^{-1}\alpha\pi = \pi^{-1}\alpha^{-1} T \alpha\pi = T_1 ,$$

that is the analogue of (1') for $T_1$. Since two parallel projections $\pi, \pi' : l \to l_1$ give $\pi'\pi^{-1} \in \Pi^*_l$ and therefore by (1') $\pi^{-1} T \pi = \pi'^{-1} T \pi'$, $T_1$ in (3) is uniquely determined by $l_1'$, independent of $\pi$. Thus to every line $l'$ there exists a sharply transitive permutation group $T_l$ on $l'$ with

$$(1'') \qquad \alpha^{-1} T_l \alpha = T_l \qquad \text{for } \alpha \in \Pi^*_l ,$$

and for two lines $l, l_1$ and every product $\pi : l \to l_1$ of parallel projections we have (3) with $T = T_l$, $T_1 = T_{l_1}$. Because of (3) the $T_l$ are isomorphic even as permutation groups. For two elements $\tau$, $\tau_1$ of $T_l$ and a point $O \in l'$

there exists $\alpha \in \Pi_1^*$ with (see fig. 13)

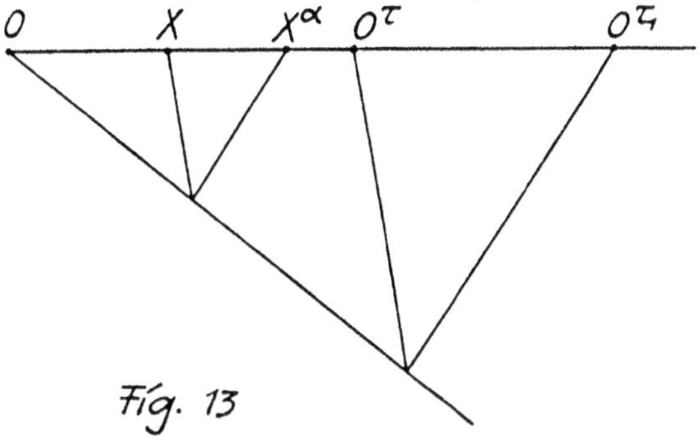


Fig. 13

$$O^\alpha = O \ , \ O^{\tau\alpha} = O^{\tau_1}$$

and thus

$$O^{\alpha^{-1}\tau\alpha} = O^{\tau_1} \ .$$

Since according to (1") $\alpha^{-1}\tau\alpha \in T_1$ and since $T_1$ is sharply transitive, this gives

$$\alpha^{-1}\tau\alpha = \tau_1 \ ,$$

i.e. the elements of $T_1$ are conjugates of each other under $\Pi_1^*$ (we have not yet proved, that $T_1$ is a subgroup of $\Pi_1^*$!). So there are only the following cases:

I. $\tau^2 \neq 1_{1'}$ for all $\tau \in T_1 \setminus \{1_{1'}\}$ ;

II. $\tau^2 = 1_{1'}$ for all $\tau \in T_1$ .

In case I we introduce for every point O of the affine plane the mapping $\delta_O$ by

(4) $O^{\delta_O} = O \ ; \ X^{\delta_O} = O^\tau$ , if $X \neq O$, $X^\tau = O$, $\tau \in T_{OX}$ .

Since I,(4) and

$$X^\tau = O, \ O^\tau = X' \Longleftrightarrow X'^{\tau^{-1}} = O \ , \ O^{\tau^{-1}} = X$$

$\delta_O$ is an involution (i.e. $\delta_O^2$, but not $\delta_O$ is the identity) and leaves all lines through O fixed. For two (affine) points X,Y not collinear with O and the parallel projection $\pi$ of OX to OY with $X^{\pi} = Y$ (3) gives

$$\pi^{-1} T_{OX} \pi = T_{OY} ,$$

therefore

$$\tau \in T_{OX}, \quad X^{\tau} = O \Longrightarrow \pi^{-1}\tau\pi \in T_{OY}, \quad Y^{\pi^{-1}\tau\pi} = O, \quad O^{\pi^{-1}\tau\pi} = O^{\tau\pi}$$

and thus according to (4) $Y^{\delta_O} = X^{\delta_O \pi}$ :
$\delta_O$ maps every line onto a parallel one and is therefore induced by an involutory element of $\Gamma(O,u)$, i.e. $\delta_O$ is the *point reflection* with center O. From a result of Baer (see e.g. [6], p. 213) it follows now, that the affine plane is a translation plane. We shorten the proof a little bit, using the definition (4). To construct for two points A,B the translation $\delta$ with $A^{\delta}=B$

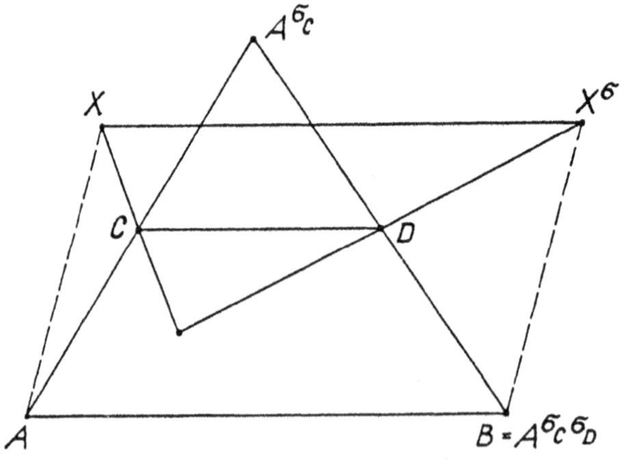


Fig. 14

one chooses a point $C \notin AB$ and determines the point $D \in A^{\delta_C}B$ such that CD is parallel to AB (see fig. 14). Then $\delta = \delta_C \delta_D$ maps every line onto a parallel one; it is therefore either a translation or a dilatation.
It follows from $\tau \in T_{AC}$, $A^\tau = C$, $C^\tau = A^{\delta_C}$, that for the parallel projection $\pi: AC \to BD$ with $A^\pi = B$ (and therefore $C^\pi = D$) and $\tau' = \pi^{-1}\tau^{-1}\pi \in T_{BD}$ we have

$$A^{\delta_C \tau'} = C^{\tau\tau'} = C^{\tau\pi\tau'} = C^\pi = D \ ,$$

$$D^{\tau'} = C^{\pi\tau'} = C^{\tau^{-1}\pi} = A^\pi = B.$$

According to (4) this gives $A^\delta = B$ and thus $(AB)^\delta = AB$, since this must be a line parallel to AB. If $\delta$ would be a dilatation, the fixed point (center) of $\delta$ must therefore lie on AB. On the other hand one has

$$X^\delta = X \Longrightarrow X^{\delta_C} = X^{\delta_D} \Longrightarrow X \in CD \ ,$$

so that, since $AB \nparallel CD$, $\delta$ is no dilatation and thus a translation. This proof shows also, that the group of translations with fixed line $l'$ consists of the $\delta_A \delta_B$ with $A, B \in l'$. Their restrictions to $l'$ form of course a sharply transitive normal abelian subgroup of $\Pi_1^*$; thus if we can show, that they all belong to $T_1$, then $T_1$ is this subgroup. Now for $O \in l'$ the restriction $\delta_O^*$ of $\delta_O$ to $l'$ belongs to $\Pi_1^*$: If $l_1'$, $l_2'$ are two lines $\nparallel l'$ through O and $\pi_1$, $\pi_2$, $\pi_3$ resp. the projections of $l'$ to $l_1'$ parallel to $l_2'$, of $l_1'$ to $l_2'$ parallel to $l'$, of $l_2'$ to $l'$ parallel to $l_1'$, then $\delta_O^* = \pi_1\pi_2\pi_3$ (see fig. 15), since we are in a translation plane. For a

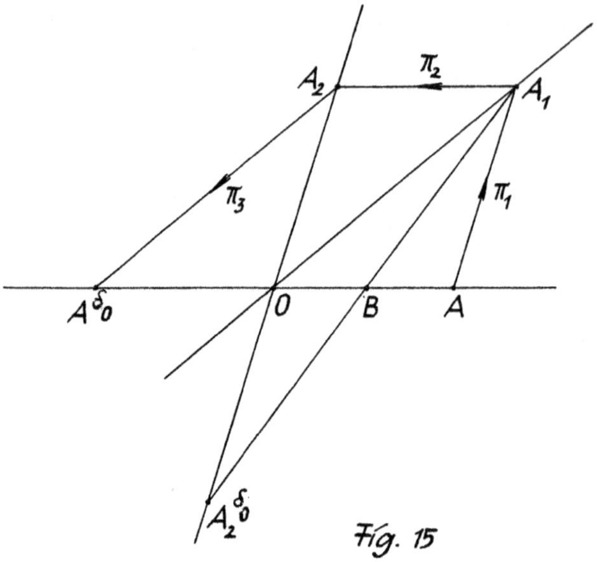


Fig. 15

simplification of the proof for case I, hinted at later on in case II, we prove here this equation, using only the fact, that $\delta_0$ defined in (4) is a point reflection. For $A \in l'$, $A_1 = A^{\pi_1}$, $A_2 = A_1^{\pi_2}$, $B = A_2^{\delta_0} A_1 \cap OA$ (see fig. 15) we have (as in the proof to fig. 14) $A_1 = A_2^{\delta_0 \delta_B}$ and thus $O^{\delta_B} = A$, $A_1^{\delta_B \delta_0} = A_2$. Therefore $OA_1 = l_1'$ is parallel to $A^{\delta_0} A_2$ and thus $A^{\delta_0} = A_2^{\pi_3} = A^{\pi_1 \pi_2 \pi_3}$. For $\tau \in T_1$ it follows now from $\delta_0^* \in \Pi_1^*$ and (1"), that

$$\delta_0^* \tau \delta_0^* \in T_1 \quad ,$$

and since (4) implies

$$O^{\delta_0^* \tau \delta_0^*} = X = O^{\tau^{-1}} \quad ,$$

this gives

(5) $\delta_0^* \tau \delta_0^* = \tau^{-1}$ for all $\tau \in T_1$, $0 \in l'$

because $T_1$ is sharply transitive. (5) applied to $\tau = \tau_1 \tau_2$ gives the commutativity of $T_1$. The case $A=B$ being trivial, we consider now for $A,B \in l'$, $A \neq B$ the restriction $\delta_A^* \delta_B^*$ of $\delta_A \delta_B$. Let X be any point of $l'$ and $\tau, \tau' \in T_1$ determined by $A^\tau = B$, $A^{\tau'} = X$. Then by (4), (5) and the commutativity of $T_1$ we obtain $X^{\delta_A^* \delta_B^*} = A^{\tau' \delta_A^* \delta_B^*} = A^{\delta_A^* \tau' \delta_A^* \delta_B^*} = A^{\tau'^{-1} \delta_B^*} = B^{\tau' \tau} = B^{\tau \tau'} = A^{\tau^2 \tau'} = X^{\tau^2}$ and thus $\delta_A^* \delta_B^* = \tau^2 \in T_1$. This finishes the proof in case I.

In case II the commutativity of $T_1$ is a well known fact. For the following proof only this consequence of II is needed and not the full information, that $T_1$ is a vector space over GF(2); thus the result in case II could be applied to shorten the proof of case I: we need only (5), showing that $T_1$ is abelian. Now given two points A,B we only have to construct a translation, which induces on AB the Element $\tau \in T_{AB}$ with $A^\tau = B$. To do this a coordinate system is introduced with AB (here all lines are considered as affine lines) as second axis $a_2$, a line $\neq AB$ through A as first axis $a_1$, defining $P(\tau_1, \tau_2)$ with $\tau_i \in T_i = T_{a_i}$ $(i=1,2)$ as the intersection of the parallel to $a_2$ through $A^{\tau_1}$ with the parallel to $a_1$ through $A^{\tau_2}$

(see fig. 16).

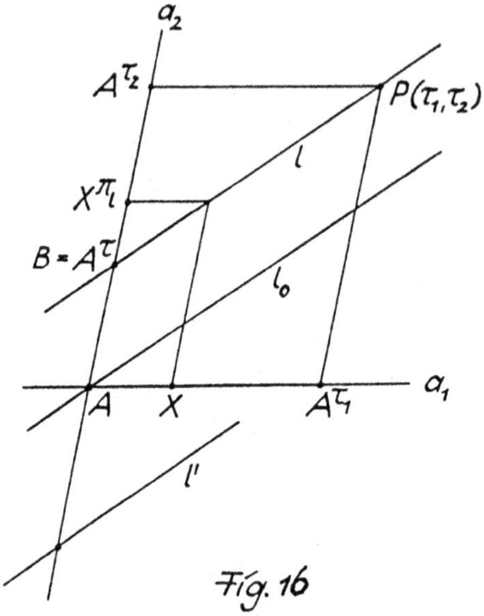


Fig. 16

Then the permutation $\tau^*$ of the affine plane defined by

(6) $$P(\tau_1, \tau_2)^{\tau^*} = P(\tau_1, \tau_2\tau)$$

induces $\tau$ on AB since

$$A^{\tau_2\tau^*} = P(1_{a_1}, \tau_2)^{\tau^*} = A^{\tau_2\tau}.$$

Of course $\tau^*$ has no fixed point; thus it only remains to prove, that $\tau^*$ maps every line onto a parallel one. This is trivial for the lines parallel to $a_1$ or $a_2$. Now let $l$ be a line through B, neither parallel to $a_1$ nor to $a_2$, further $l_o$ the parallel to $l$ through A and $\pi_l$ the product of the parallel projections of $a_1$ to $l$ parallel to $a_2$ and of $l$ to $a_2$ parallel to $a_1$ (see fig. 16), so that

(7) $$P(\tau_1, \tau_2) \in l \iff A^{\tau_1\pi_l} = A^{\tau_2} .$$

According to (1"),(3)

$$\tau_1 \longrightarrow \pi_1^{-1}\tau_1\,\pi_1$$

is an isomorphism of $T_1$ to $T_2$, which will be, as usual, denoted also by $\pi_1$:

$$\tau_1^{\pi_1} = \pi_1^{-1}\,\tau_1\,\pi_1 \quad .$$

In the sequel $\pi_1$, $\pi_{1_i}$ will be abbreviated by $\pi$, $\pi_i$.

For a given $\tau_o \in T_1 \setminus \{1_{a_1}\}$the line

(8) $\quad l_1 = P(\tau_o, \tau_o^{\pi})\,A$

is introduced. From the definitions and the commutativity of $T_2$ we get the equations

$$A^{\tau_1\pi} = A^{\pi\tau_1^{\pi}} = A^{\tau\tau_1^{\pi}} = A^{\tau_1^{\pi}\tau} \qquad ,$$

$$A^{\tau_1\pi_i} = A^{\tau_1^{\pi_i}} \qquad (i=0,1)$$

and thus from (7), since $T_2$ is sharply transitive:

(9) $\quad P(\tau_1, \tau_2) \in l \Longleftrightarrow \tau_1^{\pi}\tau = \tau_2 \quad ,$

(10) $\quad P(\tau_1, \tau_2) \in l_i \Longleftrightarrow \tau_1^{\pi_i} = \tau_2 \quad (i=0,1)$ .

Now, since $A \in l_1$, $A \notin l$, the lines $l, l_1$ are different and it follows from (9),(10), that they have as many points in common as there are permutations $\tau_1 \in T_1$ with

(11) $\quad \tau_1^{\pi}\tau = \tau_1^{\pi_1} \quad ;$

thus this equation cannot have two solutions for $\tau_1$.

But (8),(10) give

(12) $\quad \tau_o^{\pi_1} = \tau_o^{\pi}$

and (11) therefore

$$(\tau_o\tau_1)^{\pi}\tau = (\tau_o\tau_1)^{\pi_1} \quad ,$$

so that (11) is impossible since $\tau_o\tau_1 \neq \tau_1$. Thus $l, l_1$

are parallel and therefore $l_1 = l_o$, $\pi_1 = \pi_o$. Now (12) with $\pi_1 = \pi_o$ is trivially true in the excluded case $\tau_o = 1_{a_1}$; thus

$$\tau_1^{\pi_o} = \tau_1^{\pi}$$

is valid for all $\tau_1 \in T_1$ and gives together with (6)

$$P(\tau_1, \tau_1^{\pi_o})^{\tau^*} = P(\tau_1, \tau_1^{\pi}\tau) .$$

Because of (9),(10) this means

(13) $$l_o^{\tau^*} = l.$$

For another line $l'$ parallel to $l_o$ with $l' \cap a_2 = A^{\tau'}$ (see fig. 15) this result (13) (with $\tau'$ instead of $\tau$) gives

$$l_o^{\tau'^*} = l'$$

and shows further (with $\tau'\tau$ instead of $\tau$), that

$$l'^{\tau^*} = l_o^{\tau'^*\tau^*} = l_o^{(\tau'\tau)^*}$$

is a line parallel to $l_o$ and therefore to $l'$.

## E. A lemma on sfields

The proof of Corollary 1 in [4], p. 162, needed in the proof of Proposition 4 for the derivation of (9), uses a lot of the general theory on sfields. Therefore it might be convenient for the reader to see here, how this proof can be developed from elementary facts of linear algebra alone.

Lemma. *Let* $K$ *be a sfield with center* $K_o$ *and* $L$ *a subsfield* $\supset K_o$ *of* $K$ *having finite dimension over* $K_o$. *Then every isomorphism of* $L$ *into* $K$ *leaving the elements of* $K_o$ *invariant is induced by an inner automorphism of* $K$.

Proof. Let $\varphi$ be an isomorphism of $L$ into $K$ with $x^\varphi = x$ for all $x \in K_o$. Our first aim is to prove, that $\varphi$ is induced by an element of the ring $\mathcal{L}$ consisting of those linear mappings of $K$ as vector space over $K_o$, which are sums of products of left and right multiplications with elements of $K$. For $S \subseteq K$, $\mathcal{S} \subseteq \mathcal{L}$ we define

$$S^{\perp} = \{\lambda \in \quad \mid x^\lambda = 0 \text{ for all } x \in S\},$$

$$\mathcal{S}^{\perp} = \{x \in K \mid x^\lambda = 0 \text{ for all } \lambda \in \mathcal{S}\}$$

and prove by induction on $m$ $(=0,1,\ldots)$ for all $a_1,\ldots,a_m \in K$

$$(1) \qquad \{a_1,\ldots,a_m\}^{\perp\perp} \subseteq \sum_{i=1}^{m} K_o a_i$$

(the other inclusion " $\supseteq$ " is trivial). For m= 0 this follows from $\emptyset^{\perp} = \mathcal{L}$, $\mathcal{L}^{\perp} = \{0\}$. For m > 0 and with the right ideal

$$\mathcal{L}_o = \{a_1, \ldots, a_{m-1}\}^{\perp}$$

of $\mathcal{L}$ we write the induction hypothesis (m-1 instead of m) as

$$(1') \qquad \mathcal{L}_o^{\perp} \subseteq \sum_{i=1}^{m-1} K_o a_i \ .$$

Then (1) follows from (1') with the help of

$$(2) \qquad (\mathcal{L}_o \cap \{a_m\}^{\perp})^{\perp} \subseteq \mathcal{L}_o^{\perp} + K_o a_m \ ,$$

since

$$\mathcal{L}_o \cap \{a_m\}^{\perp} = \{a_1, \ldots, a_m\}^{\perp} \ .$$

Thus only (2) is needed. For its proof we put $a = a_m$ and consider $b \in (\mathcal{L}_o \cap \{a\}^{\perp})^{\perp}$, so that for all $\lambda \in \mathcal{L}_o$:

$$(3) \qquad a^{\lambda} = 0 \Longrightarrow b^{\lambda} = 0 \ .$$

Therefore we have a homomorphism (relative to addition) $\psi$ of $a^{\mathcal{L}_o}$ onto $b^{\mathcal{L}_o}$ determined by

$$(4) \qquad (a^{\lambda})^{\psi} = b^{\lambda} \qquad \text{for all } \lambda \in \mathcal{L}_o \ .$$

Since (2) is trivially true for $\mathcal{L}_o \subseteq \{a_m\}^{\perp}$, we can suppose $a^{\mathcal{L}_o} \neq \{0\}$ and choose $\lambda_o \in \mathcal{L}_o$ with $a^{\lambda_o} \neq 0$, putting $a_o = a^{\lambda_o}$, $b_o = b^{\lambda_o}$. Now every $x \in K$ can be written as

$$x = xa_o^{-1}a^{\lambda_o} = a^{\lambda_o}a_o^{-1}x,$$

and since, $\mathcal{L}_o$ being a right ideal of $\mathcal{L}$, the mappings

$$y \to xa_o^{-1}y^{\lambda_o} \ , \quad y \to y^{\lambda_o}a_o^{-1}x$$

of K both belong to $\mathcal{L}_o$, we have $x \in a^{\mathcal{L}_o}$ (for every

$x \in K$) and by (4) therefore for all $x \in K$

(5) $$xa_o^{-1}b_o = x^\psi = b_o a_o^{-1}x \, .$$

For x=1 this gives $a_o^{-1}b_o = b_o a_o^{-1}$ and therefore, using (5) for every $x \in K$, $b_o a_o^{-1} \in K_o$. Thus (4),(5) show, that there exists $c \in K_o$ with $ca^\lambda = b^\lambda$ and therefore $(b - ca)^\lambda = 0$ for all $\lambda \in \mathcal{L}_o$. This means $b - ca \in \mathcal{L}_o^\perp$ and thus $b \in \mathcal{L}_o^\perp + K_o a$ , which proves (2).

For a basis $(e_1,\ldots,e_n)$ of L over $K_o$ we know from (1) (with $a_i=e_i$, $m=n-1$), that there exists $\lambda \in \{e_1,\ldots,e_{n-1}\}^\perp$ with $e_n^\lambda \neq 0$. Thus we can choose $\lambda_i \in \mathcal{L}$ $(i=1,\ldots,n)$ with

(6) $$e_i^{\lambda_j} = 0 \quad \text{for } i \neq j \ , \ e_i^{\lambda_i} \neq 0 \quad (i,j=1,\ldots,n) \ ;$$

then for $\lambda \in \mathcal{L}$ determined by

(7) $$x^\lambda = \sum_{j=1}^{n} e_j^\varphi (e_j^{\lambda_j})^{-1} x^{\lambda_j} \quad \text{for all } x \in K$$

we have, due to (6),(7)

$$e_i^\lambda = e_i^\varphi \quad (i=1,\ldots,n),$$

which gives the desired property $x^\lambda = x^\varphi$ for all $x \in L$. Since every product of a left with a right multiplication can be written as a product of an inner automorphism of K with a left multiplication, there are restrictions $\delta_i$ to L of such inner automorphisms and $a_i \in K$ $(i=1,\ldots,n)$ such that in the left K-vector space $\mathcal{L}_L$ of the $K_o$-linear mappings of L into K

(8) $$\varphi = \sum_{i=1}^{m} a_i \cdot \delta_i \ ,$$

where the K-multiplication is defined by

(9) $a . \delta = \delta \overline{a}$ for all $a \in K$, $\delta \in \mathcal{L}_L$ ,

$\overline{a}$ denoting the left multiplication with a. Of course we can suppose the $\delta_i$ in (8) being linearly independent (over K) and $a_i \neq 0$ ($i=1,\ldots,m$); since $\varphi$ is an isomorphism, we have $m \geq 1$. For $c \in L$, $\delta = \varphi$, $\delta_i$ (9) gives

$$\overline{c}\,\delta = c^{\delta} . \delta$$

and (8),(9) therefore

$$\sum_{i=1}^{m} c^{\varphi} a_i . \delta_i = c^{\varphi} . \varphi = \overline{c}\varphi = \sum_{i=1}^{m} \overline{c}\,\delta_i \overline{a}_i = \sum_{i=1}^{m} (c^{\delta_i} . \delta_i)\overline{a}_i$$
$$= \sum_{i=1}^{m} a_i c^{\delta_i} . \delta_i \quad .$$

Since the $\delta_i$ are linearly independent, we have therefore

$$c^{\varphi} a_1 = a_1 c^{\delta_1}$$

and thus

$$c^{\varphi} = a_1 c^{\delta_1} a_1^{-1} \quad \text{for all } c \in L ,$$

so that $\varphi$ is indeed the restriction of an inner automorphism of K.

## Literaturverzeichnis

1 BARLOTTI, A.: Sul gruppo delle proiettività di una retta in sè nei piani liberi e nei piani aperti. Rend. Sem. Mat. Univ. Padova 34, 135 - 159 (1964).

2 FRITSCH, R.: Ein affiner Beweis des Satzes von Staudt-Schleiermacher. Monatshefte Math. 86, 177 - 184 (1978).

3 HESSENBERG, G., DILLER, J.: Grundlagen der Geometrie. Berlin 1967.

4 JACOBSON, N.: Structure of rings. Amer. Math. Soc. Publ., vol.37, Providence 1968².

5 LÜNEBURG, H.: An axiomatic treatment of ratios in an affine plane. Arch. d. Math. 18, 444 - 448 (1967),

6 PICKERT, G.: Projektive Ebenen. Berlin, Heidelberg, New York 1975².

7 SCHLEIERMACHER, A.: Bemerkungen zum Fundamentalsatz der projektiven Geometrie. MZ 99, 299 - 304 (1967).

8 - - : Reguläre Normalteiler in der Gruppe der Projektivitäten bei projektiven und affinen Ebenen. MZ 114, 313 - 320 (1970).

9 - - : Über projektive Ebenen, in denen jede Projektivität mit sechs Fixpunkten die Identität ist. MZ 123, 325 - 339 (1971).

10 STAUDT, Ch. von : Geometrie der Lage. Nürnberg 1847.

11 YAQUB, J. C. D. S. : On the group of projectivities on a line in a finite projective plane. MZ 104, 247 - 248 (1968).

# PERSPECTIVITIES IN CIRCLE GEOMETRIES

Helmut Karzel and Hans-Joachim Kroll

Technische Universität München

During recent years research in the theory of circle geometries has made enormous progress. This development is most impressively reflected by the bibliography of BENZ's book [4]. Circle geometries, as the term is used in our investigations, were first considered in the last century in the works of MÖBIUS, LIE and LAGUERRE. It was not until the paper of B. L. van der WAERDEN and L. J. SMID that modern methods were introduced in the research of circle geometries. Also PETKANTSCHIN investigated circle geometries in 1940. Both papers received no attention for quite a while due to difficult times. In the fifties R. FURCH suggested the study of circle geometries and two of his advanced students (EWALD, BENZ) took up this suggestion to do work for a doctorate. Especially the beautiful results obtained by BENZ stimulated research in this field so that an increasing number of mathematicians started work in the field of circle geometries. Among those who made important contributions we would like to mention R. ARTZY, P. DEMBOWSKI, W. HEISE, H. MÄURER, P. QUATTROCCHI.

Three types of circle geometries are known, the Möbius, Laguerre, and Minkowskian planes. These geometries can be defined by a unified method (cf.[15]). For this purpose we shall start with the concepts of I-nets and I-chain-nets (§2). Nets are also called webs (German: Gewebe, Wabe) by W. BLASCHKE and G. THOMSEN. The theory of nets has been mainly developed by W. BLASCHKE [6], G. THOMSEN [33], K. REIDEMEISTER [31], G. BOL [8], R. ARTZY [1] and others. From I-chain-nets we obtain by specialization the following structures: I-structure,



*P. Plaumann and K. Strambach (eds.), Geometry - von Staudt's Point of View, 51–99.*

2-structure, affine plane, hyperbola structure, Möbius, Laguerre and Minkowskian planes. Some of these geometric structures can be described by algebraic structures which are introduced in §1. In these geometries one can define several types of perspectivities, hence projectivities and groups formed by projectivities mapping a generator or a chain onto itself. These groups will be called here von Staudt groups (§2, §7 and §8). The von Staudt group of a 2-chain-net has some nice properties if the so called rectangle axiom is valid (§3). A special case of the rectangle axiom is the REIDEMEISTER condition, used by REIDEMEISTER [32] to characterize the 3-nets of groups. In §4 we shall discuss only briefly the von Staudt groups of 3-nets with Reidemeister condition, since A. BARLOTTI and K. STRAMBACH are planning to publish a longer report on 3-nets.

In a (3.1)-net we can define essentially two von Staudt groups $\Gamma_{12}$ and $\Gamma_{123}$, which are both 2-transitive (§5). A (3.1)-net can be embedded in a 2-structure if $\Gamma_{12}$ is sharply 2-transitive, and in a desarguesian affine plane if $\Gamma_{123}$ is sharply 2-transitive (theorem (5.3) and (5.5)).

Hyperbola structures with a sharply 3-transitive von Staudt group $\Gamma_{12}$ are characterized in theorem (6.2). These geometries can also be described algebraically by KT-fields.

For the proper circle geometries such as Möbius, Laguerre, and Minkowskian planes we define different types of perspectivities, where circles are projecting rays. In this way one obtains different types of von Staudt groups (§7). H. FREUDENTHAL and K. STRAMBACH [13] studied von Staudt groups in Möbius and Laguerre planes. Their work was continued by H.-J. KROLL [28],[29],[30] who under considerably weakened conditions gave a unified theory, which also included the Minkowskian planes. Recently M. FUNK [12] generalized KROLL's results by assuming only conditions $(P_4)$ and $(P_5)$ (cf. §1). In §7 we give a report on this subject.

A local von Staudt group $\Gamma$ in circle planes is introduced in §8 by using the basic perspectivities of type 4 from §7. This group $\Gamma$ is also 3-transitive. If we assume sharp 3-transitivity we obtain a theorem (cf. (8.6)) of the same type as the BUEKENHOUT theorem on pascalian ovals [10]. In the proof we use results of W. KERBY, H. WEFELSCHEID [26] and K. SÖRENSEN [23].

Recently a new kind of circle geometries of higher

dimension, the so-called Burau geometries were axiomatically defined by G. KIST and the authors [21]. The notion of the Burau geometry is rooted in the concept of a chain which was originally defined by C. von STAUDT for the projective line over the complex numbers $\mathbb{C}$ and then extended to the plane and higher dimensional case by C. JUEL and W. BURAU respectively. The perspectivities of the underlying projective space $(P,\mathfrak{G})$ of the Burau geometry (cf. §9) are circle preserving and generate a sharply 3-transitive von Staudt group of $(P,\mathfrak{G})$ (theorem (9.1)).

The treatment of the various von Staudt groups in circle geometries is closely connected to sharply 2- and 3-transitive permutation groups. The main theorems of this topic will be used in this paper. In this connection we would like to point out that G. P. KIST [27] has just extended the theory of these permutation groups to sharply 2- and 3-transitive permutation sets. One can expect that the results of G. P. KIST will find applications in circle geometries.

## § 1 BASIC ALGEBRAIC CONCEPTS

Permutation sets. A set $E$ together with a subset $\Sigma$ of the symmetric group $S_E$ is called a permutation set $(E,\Sigma)$. For $n \in N$ a permutation set $(E,\Sigma)$ is called n-transitive if for any two n-tuples $(a_1,a_2,\dots,a_n)$, $(b_1,b_2,\dots,b_n) \in E^n$ with $|\{a_1,\dots,a_n\}| = |\{b_1,\dots,b_n\}| = n$ there is a $\sigma \in \Sigma$ with $\sigma(a_i) = b_i$ for $i \in \{1,2,\dots,n\}$. $(E,\Sigma)$ has the property $(P_n)$ if only the identity fixes $n$ distinct points of $E$, $(E,\Sigma)$ is called sharply n-transitive if $(E,\Sigma)$ is n-transitive with the property $(P_n)$, and regular if $(E,\Sigma)$ is sharply 1-transitive.

From any quasigroup $(E,+)$ one can derive the regular permutation set $(E,E^+)$ with

$E^+ := \{a^+ : E \longrightarrow E;\ x \longrightarrow a+x \mid a \in E\}$.

Conversely let $(E,\Sigma)$ be a regular permutation set, let $0 \in E$ be distinguished and for $a \in E$ let $a^+$ be the uniquely determined permutation of $\Sigma$ with

$a^+(0) = a$. Then $(E,+)$ with $a+b := a^+(b)$ is a quasigroup. A quasigroup is a loop if and only if the corresponding regular permutation set contains the identity.

Near domains, KT-fields. A set $F$ together with two binary operations "+" and "·" is called a near-domain if $(F,+)$ is a loop (the neutral element will be denoted by 0), if $(F^*,\cdot)$ with $F^* := F\setminus\{0\}$ is a group (1 denotes the neutral element) and if the following conditions hold:

F1 For $a,b \in F$ with $a+b=0$ we have $b+a=0$; $b$ is then denoted by $-a$.

F2 For $a,b,c \in F$ we have $a(b+c) = ab+ac$ and $0\cdot a = 0$

F3 For any $a,b \in F$ there is exactly one $d_{a,b} \in F$ such that for all $c \in F$ we have $a+(b+c) = (a+b)+d_{a,b}\cdot c$

A near-domain $(F,+,\cdot)$ is called a KT-field $(F,+,\cdot,\sigma)$ if there is an involutory automorphism $\sigma$ of the multiplicative group $(F^*,\cdot)$ which satisfies the functional equation

KT For all $x \in F\setminus\{0,-1\}$ we have $\sigma(1+\sigma(x)) = 1-\sigma(1+x)$.

Sharply 2-transitive and sharply 3-transitive permutation groups. For a near-domain $(F,+,\cdot)$ the set $A(F)$ of all maps

$[c,m] : F \longrightarrow F;\ x \longrightarrow c+mx$ with $c,m \in F$, $m \neq 0$

forms a group which acts sharply 2-transitive on $F$. We call this group the affine group of the near-domain. If $(F,+,\cdot,\sigma)$ is a KT-field we enlarge $F$ by a new element $\infty$ to $\overline{F} := F \cup \{\infty\}$. If we extend the maps $[c,m]$ and $\sigma$ by

$[c,m](\infty) := \infty$ and $\sigma(0) = \infty$, $\sigma(\infty) = 0$

then the set $\{[c,m] \mid c,m \in F,\ m \neq 0\} \cup \{\sigma\}$

generates a group which acts sharply 3-transitive on $\overline{F}$. This group is called the <u>projectivity group</u> of the KT-field.

We can now state the representation theorem [19],[26]:

<u>(1.1)</u> Any sharply 2-transitive or sharply 3-transitive permutation group is the affine group of a near-domain or the projectivity group of a KT-field respectively.

Remarks concerning the representation theorem:

<u>(1.2)</u> For a sharply 2-transitive permutation group $(E,\Sigma)$ we obtain the addition and the multiplication of the corresponding near-domain $(E,+,\cdot)$ in the following way: let $0,1$ be two distinct elements of $E$; let $\Sigma_0 := \{\gamma \in \Sigma \mid \gamma(0) = 0\}$ be the stabilizer of $0$; for any $a \in E$ let $\tilde{a}$ be the involution interchanging $0$ and $a$ if $a \neq 0$; and let $\tilde{0}$ be the involution fixing $0$, if there is such an involution or otherwise the identity. For $a \in E$ let $a^+ := \tilde{a}\tilde{0}$, and for $a \in E^* := E\setminus\{0\}$ let $a^\cdot \in \Sigma_0$ with $a^\cdot(1) = a$. Then $(E,E^+)$ with $E^+ := \{a^+ \mid a \in E\}$ and $(E^*,\Sigma_0)$ are regular permutation sets and we have $a+b := a^+(b)$, $a \cdot b := a^\cdot(b)$ if $a \neq 0$ and $a \cdot b = 0$ if $a = 0$. The characteristic of the near-domain $(E,+,\cdot)$ is $2$ if and only if the involutions have no fixed points.

<u>(1.3)</u> In the case of a sharply 3-transitive permutation group $(E,\Sigma)$ we have to distinguish three distinct points, $0,1,\infty \in E$. Then $(E\setminus\{\infty\}, \Sigma_\infty := \{\gamma \in \Sigma \mid \gamma(\infty) = \infty\})$ is a sharply 2-transitive permutation group and hence $F := E\setminus\{\infty\}$ can be provided with an addition and a multiplication according to (1.2). The restriction of the involution, fixing $1$ and interchanging $0$ and $\infty$,

onto the set $E\setminus\{0,\infty\}$ is the involutory automorphism $\sigma$ of $(F^*,\cdot)$ satisfying the functional equation KT.

(1.4) In a near-domain $(F,+,\cdot)$ for any $a,b \in F$ we have $a\cdot(-b) = (-a)\cdot b = -a\cdot b$.

Proof. By [18] (12.23) $a\cdot(-b) = -ab$ and $-1 \in Z(F) := \{z \in F \mid \forall x \in F : zx = xz\}$.
Hence $(-a)\cdot b = (a\cdot(-1))\cdot b = (a\cdot b)(-1) = -ab$.

We shall need the following theorem of W. KERBY and H. WEFELSCHEID [26]:

(1.5) A near-domain $(F,+,\cdot)$ is distributive if and only if $(F,+,\cdot)$ is a skewfield.

(1.6) If $(F,+,\cdot,\sigma)$ is a KT-field with commutative multiplication, then $(F,+,\cdot)$ is a commutative field and $\sigma(x) = x^{-1}$ for all $x \in F\setminus\{0\}$. Every commutative field $(K,+,\cdot)$ is, with respect to the mapping $\sigma : x \longrightarrow x^{-1}$ for $x \in K\setminus\{0\}$, a KT-field (cf. [5] or [26]).

## § 2 BASIC GEOMETRIC CONCEPTS

Nets and chain nets. Let $P$ be a set and let $\mathfrak{G}$ and $\mathfrak{K}$ be subsets of the power set of $P$; the elements of $P,\mathfrak{G}$ and $\mathfrak{K}$ will be called points, generators and chains respectively. The pair $(P,\mathfrak{G})$ is called an I-net, if there is a partition $\mathfrak{G} = \bigcup_{i\in I} \mathfrak{G}_i$ of $\mathfrak{G}$ such that the following two conditions are valid:

N1 For each point $p \in P$ and each $i \in I$ there is exactly one generator $G \in \mathfrak{G}_i$ with $p \in G$; this generator will be denoted by $[p]_i$.

N2 Any two generators of distinct classes $\mathfrak{G}_i \neq \mathfrak{G}_j$ intersect in exactly one point.

The triple $(P,\mathfrak{G},\mathfrak{K})$ is called an <u>I-chain net</u> if $(P,\mathfrak{G})$ is an I-net and if the following condition holds:

<u>N3</u> Every element $K$ of $\mathfrak{K}$ intersects every generator in exactly one point.

<u>Examples</u>. Let $(P,\mathfrak{G})$ be an affine plane. If we select a set $\{\mathfrak{G}_i \mid i \in I\}$ of parallel pencils then $(P, \bigcup_{i\in I} \mathfrak{G}_i)$ is an I-net. For any subset $\mathfrak{K} \subset \mathfrak{G} \setminus (\bigcup_{i\in I} \mathfrak{G}_i)$ the triple $(P, \bigcup_{i\in I} \mathfrak{G}_i, \mathfrak{K})$ is an I-chain net.

<u>(2.1)</u> If $(E,\Sigma)$ is a permutation set then $(P,\mathfrak{G}_1 \cup \mathfrak{G}_2,\mathfrak{K})$ with $P := E \times E$, $\mathfrak{G}_1 := \{\{x\} \times E \mid x \in E\}$, $\mathfrak{G}_2 := \{E \times \{y\} \mid y \in E\}$ and $\mathfrak{K} := \{\{(x,\sigma(x)) \mid x \in E\} \mid \sigma \in \Sigma\}$ is a 2-chain net. This 2-chain net will be denoted by $\gamma(E,\Sigma) := (P,\mathfrak{G}_1 \cup \mathfrak{G}_2,\mathfrak{K})$ and will be called the <u>associated 2-chain net</u> of the permutation set $(E,\Sigma)$.

<u>Derivations</u>. If $|I| \geq n$ let $n$ distinguished elements of $I$ be denoted by $1,2,\ldots,n$. If $(P,\mathfrak{G})$ is an I-net and if $(P,\mathfrak{G},\mathfrak{K})$ is an I-chain net with $n \geq 2$ then $(P,\mathfrak{G}_1 \cup \mathfrak{G}_2, \bigcup_{i\in I\setminus\{1,2\}} \mathfrak{G}_i)$ and $(P,\mathfrak{G}_1 \cup \mathfrak{G}_2,\mathfrak{K} \cup \bigcup_{i\in I\setminus\{1,2\}} \mathfrak{G}_i)$ are 2-chain nets.

For any point $q \in P$ we need the following notations:

$\mathfrak{K}(q) := \{X \in \mathfrak{K} \mid q \in X\}$, $\mathfrak{K}^q := \{X \setminus \{q\} \mid X \in \mathfrak{K}(q)\}$.

For $I = \emptyset$: $P^q := P \setminus \{q\}$, $(P,\mathfrak{G},\mathfrak{K})^q := (P^q,\mathfrak{K}^q)$.

For $|I| = 1$: $P^q := P\setminus[q]_1$, $\mathfrak{G}_1^q := \mathfrak{G}_1 \setminus\{[q]_1\}$, $(P,\mathfrak{G},\mathfrak{K})^q := (P^q,\mathfrak{G}^q,\mathfrak{K}^q)$.

For $|I| = 2$: $P^q := P\setminus([q]_1 \cup [q]_2)$, $\mathfrak{G}_i^q := \{X \cap P^q \mid X \in \mathfrak{G}_i, X \neq [q]_i\}$, $i \in I$, $(P,\mathfrak{G},\mathfrak{K})^q := (P^q,\mathfrak{G}_1^q \cup \mathfrak{G}_2^q,\mathfrak{K}^q)$.

For $|I| \geq 2$: $(P,\mathfrak{G},\mathfrak{K})^q := (P,\mathfrak{G}_1 \cup \mathfrak{G}_1,\mathfrak{K} \cup \bigcup_{i\in I\setminus\{1,2\}} \mathfrak{G}_i)^q$.

In each of these cases $(P,\mathfrak{G},\mathfrak{K})^q$ is called the <u>derivation in the point q</u>.

<u>Particular structures</u>. An I-chain net $(P,\mathfrak{G},\mathfrak{K})$ is called an <u>(I,J)-net</u> if there is a subset $J \subset P$ with $\mathfrak{K} = \bigcup_{j\in J} \mathfrak{K}(j)$ and such that for every $j \in J$ the derivation $(P,\mathfrak{G},\mathfrak{K})^j$ in the point $j$ is a $\{j\}$-net for $I = \emptyset$, a $\{1,j\}$-net for $I = \{1\}$ or a $\{1,2,j\}$-net for $|I| \geqslant 2$. $(P,\mathfrak{G},\mathfrak{K})$ is called <u>non-trivial</u> if $|\mathfrak{G} \cup \mathfrak{K}| \geqslant 2$ and if $|X| \geqslant 2$ for all $X \in \mathfrak{G} \cup \mathfrak{K}$. A non-trivial (I,P)-net (i.e. $J = P$) is also called an <u>I-structure</u> (cf. [19]).

For the case $|I| =: n \in \mathbb{N}$ or $|J| =: m \in \mathbb{N}$ an I-net, I-chain net, (I,J)-net or I-structure is also called an <u>n-net</u>, <u>n-chain net</u>, (n,m)-net or <u>n-structure</u> respectively.

A 2-chain net $(P,\mathfrak{G},\mathfrak{K})$ is called an <u>hyperbola structure</u> if the derivation $(P,\mathfrak{G},\mathfrak{K})^p$ in any point $p \in P$ is a 2-structure.

An I-chain net $(P,\mathfrak{G},\mathfrak{K})$ with the property that for any point $p \in P$ the derivation $(P,\mathfrak{G},\mathfrak{K})^p$ with respect to $(P^p, \bigcup_{i\in I} \mathfrak{G}_i^p \cup \mathfrak{K}^p)$ is an affine plane will be called a <u>Möbius plane</u> if $I = \emptyset$, a <u>Laguerre plane</u> if $|I| = 1$ and a <u>Minkowskian plane</u> if $|I| = 2$. All Möbius, Laguerre and Minkowskian planes are called <u>circle planes</u> or <u>Benz planes</u>, and the chains of a circle geometry are called <u>circles</u>.

Every circle plane $(P,\mathfrak{G},\mathfrak{K})$ has the properties:

<u>C</u> For any three distinct points $a,b,c$ with $[a]_i \neq [b]_i \neq [c]_i \neq [a]_i$ for $i \in I$ there is exactly one circle $C \in \mathfrak{K}$ such that $a,b,c \in C$.

T (Touching axiom) For any circle $C \in \mathfrak{K}$, any $a \in C$ and any $b \in P \setminus K$ with $[a]_i \neq [b]_i$ and $i \in I$ there is exactly one $T \in \mathfrak{K}$ with $b \in T$ and $C \cap T = \{a\}$.

These two properties are characteristic of the circle planes.

(2.2) Let $(P, \mathfrak{G}, \mathfrak{K})$ be a non-trivial I-chain net with $|I| \leq 2$. Then:

(1) $(P, \mathfrak{G}, \mathfrak{K})$ is a hyperbola structure if and only if $|I| = 2$ and C holds.

(2) $(P, \mathfrak{G}, \mathfrak{K})$ is a circle plane if and only if C and T hold.

Examples and remarks. 1. Let $(P, \mathfrak{G})$ be an affine plane, let $J \subset P$ be a subset of the point set $P$, let $\{\mathfrak{G}_i \mid i \in I\}$ be a set of parallel pencils and let $\mathfrak{K} := \{X \in \mathfrak{G} \setminus \bigcup_{i \in I} \mathfrak{G}_i \mid J \cap X \neq \emptyset\}$. Then $(P, \bigcup_{i \in I} \mathfrak{G}_i, \mathfrak{K})$ is an (I,J)-net. For $I = \{1, 2\}$ and $J = P$ the triple $(P, \mathfrak{G}_1 \cup \mathfrak{G}_2, \mathfrak{K})$ is a 2-structure and $(P, \mathfrak{G}_1 \cup \mathfrak{G}_2 \cup \mathfrak{K})$ is the given affine plane. For $I = \emptyset$ and $J = P$ we have $\mathfrak{K} = \mathfrak{G}$ and the derivation in a point $q \in P$ is the 1-net $(P, \mathfrak{K})^p = (P^p, \mathfrak{K}^p)$.

2. For every I-structure $(P, \mathfrak{G}, \mathfrak{K})$ the pair $(P, \mathfrak{G} \cup \mathfrak{K})$ is an incidence space, i.e. any two distinct points a,b have exactly one "line" $X \in \mathfrak{G} \cup \mathfrak{K}$ in common: $a, b \in X$. A non-trivial (I,J)-net $(P, \mathfrak{G}, \mathfrak{K})$ is an I-structure if and only if $(P, \mathfrak{G} \cup \mathfrak{K})$ is an incidence space.

3. Every Minkowskian plane is a hyperbola structure.

4. Let Q be a quadric in a 3-dimensional projective space over a commutative field K, let $\mathfrak{G}$ be the set of all lines on the quadric and let $\mathfrak{K}$ be the set of

all plane sections of the quadric $Q$ with more than one point and not containing a line. Then $(Q,\mathfrak{S},\mathfrak{K})$ is a Möbius, Laguerre, or Minkowskian plane if $Q$ is an ellipsoid, a cone, or a ruled quadric respectively. In the case of a cone the vertex is not considered as a point of $Q$. Any circle plane, which can be represented in this way is called a miquelian circle plane.

Perspectivities and projectivities. Let $(P,\mathfrak{S},\mathfrak{K})$ be an I-chain net, let $i\in I$ and let $A,B\in(\mathfrak{K}\cup\mathfrak{S})\setminus\mathfrak{S}_i$. Then the bijection

$$[A \xrightarrow{i} B] : A \longrightarrow B;\ a \longrightarrow [a]_i \cap B$$

is called an i-perspectivity and every bijection $A \longrightarrow B$, which can be decomposed into a product of i-perspectivities $(i\in I)$ is named an I-projectivity. For $[A_n \xrightarrow{i_n} A_{n+1}]\circ \dots \circ[A_2 \xrightarrow{i_2} A_3]\circ[A_1 \xrightarrow{i_1} A_2]$ we use the notation $[A_1 \xrightarrow{i_1} A_2 \xrightarrow{i_2} A_3 \xrightarrow{i_3} \dots \xrightarrow{i_n} A_{n+1}]$.

The set of all I-projectivities $A \longrightarrow A$ with the identity forms a group $\Gamma_I(A)$, which we call the von STAUDT group of the structure $(P,\mathfrak{S},\mathfrak{K})$.

(2.3) For $A,B\in\mathfrak{K}\cup\mathfrak{S}$ the groups $\Gamma_I(A)$ and $\Gamma_I(B)$ are isomorphic.

Proof. If $|I|\leq 1$ then $\Gamma_I=\{1_A\}$. Let $|I|\geq 2$.
a) If there is an $i\in I$ with $A,B\notin\mathfrak{S}_i$ then the i-perspectivity $\varphi=[A \xrightarrow{i} B]$ gives us the isomorphism $\Gamma(A)\longrightarrow\Gamma(B);\ \alpha\longrightarrow\varphi\alpha\varphi^{-1}$.
b) In the other case we have $|I|=2$ (hence $I=\{1,2\}$) and $A\in\mathfrak{S}_1$, $B\in\mathfrak{S}_2$ or $A\in\mathfrak{S}_2$, $B\in\mathfrak{S}_1$. If $\mathfrak{K}=\emptyset$, then $\Gamma(A)=\{1_A\}$ and $\Gamma(B)=\{1_B\}$. For $E\in\mathfrak{K}$ the map

$\varphi = [A \xrightarrow{2} E \xrightarrow{1} B]$ determines an isomorphism as in a).

Representation by permutation sets. Let $(P,\mathfrak{G},\mathfrak{K})$ be a 2-chain net with $\mathfrak{K} \neq \emptyset$ and $E$ a distinguished element of $\mathfrak{K}$. We associate to every $K \in \mathfrak{K}$ the projectivity $\tilde{K} := [E \xrightarrow{1} K \xrightarrow{2} E]$. The set $\tilde{\mathfrak{K}} := \{\tilde{K} \mid K \in \mathfrak{K}\}$ of permutations of $E$ is called the associated permutation set $(E,\tilde{\mathfrak{K}})$ of the 2-chain net $(P,\mathfrak{G},\mathfrak{K})$ and is denoted by $\Pi_E(P,\mathfrak{G},\mathfrak{K})$. $\tilde{\mathfrak{K}}$ contains the identity $\tilde{E}$ and we have $\gamma(E,\tilde{\mathfrak{K}}) \cong (P,\mathfrak{G},\mathfrak{K})$ (cf. (2.1)).

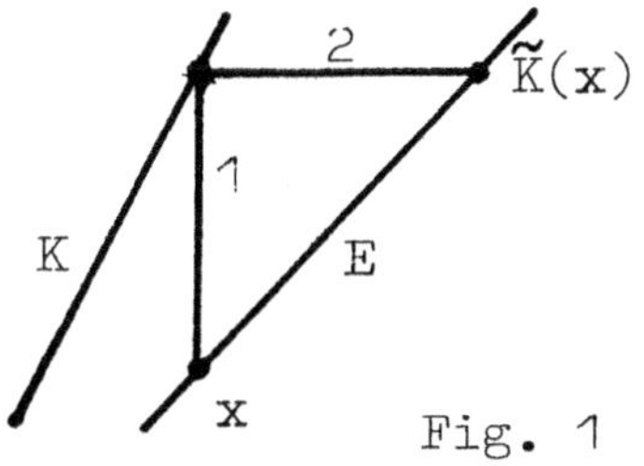


Fig. 1

For a 2-chain net $(P,\mathfrak{G}_1 \cup \mathfrak{G}_2,\mathfrak{K})$ with $\mathfrak{K} \neq \emptyset$ and $E \in \mathfrak{K}$ we have the following lemmas:

(2.4) $\tilde{\mathfrak{K}}$ generates the von STAUDT group $\Gamma_{12}(E)$.

(2.5) The following statements are equivalent:
(1) $\mathfrak{K}$ is a partition of $P$.
(2) $(P,\mathfrak{G}_1 \cup \mathfrak{G}_2 \cup \mathfrak{K})$ is a 3-net.
(3) $(E,\tilde{\mathfrak{K}})$ is a regular permutation set.

Since to every loop $(E,+)$ there exists the regular permutation set $(E,E^+)$ every loop can be represented by a 3-net $(P,\mathfrak{G}_1 \cup \mathfrak{G}_2 \cup \mathfrak{K}) = \gamma(E,E^+)$ according to (2.5).

(2.6) $(P,\mathfrak{G}_1 \cup \mathfrak{G}_2,\mathfrak{K})$ is a 2-structure if and only if $(E,\tilde{\mathfrak{K}})$ is a sharply 2-transitive permutation set.

(2.7) $(P,\mathfrak{G}_1 \cup \mathfrak{G}_2,\mathfrak{K})$ is a hyperbola structure if and only if $(E,\tilde{\mathfrak{K}})$ is a sharply 3-transitive permutation set.

In (2.5),(2.6),(2.7) we have $\mathfrak{R} = \{\{(x,\gamma(x)) \mid x \in E\} \mid \gamma \in \tilde{\mathfrak{R}}\}$ .

Since the proofs of (2.5),(2.6),(2.7) are similar, we shall give only the proof of (2.7). We may identify $(P,\mathfrak{G}_1 \cup \mathfrak{G}_2,\mathfrak{R})$ with $\gamma(E,\tilde{\mathfrak{R}})$ . Then for $(a_1,a_2),(b_1,b_2) \in P = E \times E$ we have $[(a_1,a_2)]_1 \nparallel [(b_1,b_2)]_1$ if and only if $a_1 \nparallel b_1$ , and $[(a_1,a_2)]_2 \nparallel [(b_1,b_2)]_2$ if and only if $a_2 \nparallel b_2$ . Therefore, (2.7) follows by (2.2.1).

<u>Affine central perspectivities</u>. Let $(P,\mathfrak{G},\mathfrak{R})$ be an I-structure. Since $(P,\mathfrak{G} \cup \mathfrak{R})$ is an incidence space any two distinct points $a,b \in P$ have exactly one element $X \in \mathfrak{G} \cup \mathfrak{R}$ with $a,b \in X$ ; we denote this element by $\overline{a,b}$ . For $i \in I$ , $A,B \in \mathfrak{G}_i$ and $z \in P\setminus(A \cup B)$ the prescription $[A \xrightarrow{z} B] : A \longrightarrow B;\ x \longrightarrow \overline{x,z} \cap B$ defines a bijection called an <u>affine central perspectivity</u>.

<u>(2.8)</u> Let $|I| \geq 2$ and $|G| \geq 3$ for $G \in \mathfrak{G}$ . For $\Sigma := \{[B \xrightarrow{y} A] \circ [A \xrightarrow{x} B] \mid x,y \in P\setminus(A \cup B)\}$ the pair $(A,\Sigma)$ is a 2-transitive permutation set.

Proof. Let $a_1,a_2,a_1',a_2' \in A$ with $a_1 \nparallel a_2$ , $a_1' \nparallel a_2'$ . For $j \in I$ , $j \neq i$ , $b_1 := [a_1']_j \cap B$ , $x \in \overline{a_1,b_1}\setminus\{a_1,b_1\}$ , $b_2 := [A \xrightarrow{x} B](a_2)$ the point $y := [a_1']_j \cap \overline{b_2,a_2'}$ exists, and the bijection $[B \xrightarrow{y} A] \circ [A \xrightarrow{x} B]$ maps $a_1$ onto $a_1'$ and $a_2$ onto $a_2'$ .

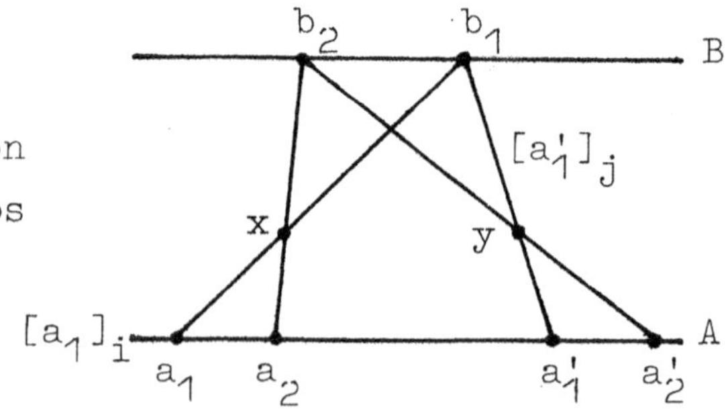


Fig. 2

(2.9) Let $|I| \geq 6$ and $A \in \mathfrak{R} \cup \mathfrak{G}$, $A \notin \mathfrak{G}_i$ for $i = 1,2,\ldots,6$ and let $\Sigma$ denote the set of all projectivities $[Y \xrightarrow{3} A][A \xrightarrow{j} Y][X \xrightarrow{3} A][A \xrightarrow{i} X]$ with $X,Y \in \mathfrak{R} \cup \mathfrak{G}$, $X,Y \notin \mathfrak{G}_1 \cup \mathfrak{G}_2$, $i,j \in \{4,5,6\}$. Then the pair $(A,\Sigma)$ is a 2-transitive permutation set.

Proof. Let $a_1,a_2,a_1',a_2' \in A$, $a_1 \neq a_2$, $a_1' \neq a_2'$. We may assume $a_1 = a_1'$. There exists an $i \in \{4,5,6\}$ such that for $x := [a_2]_i \cap [a_2']_3$ the line $X = \overline{x,a_1} \notin \mathfrak{G}_1 \cup \mathfrak{G}_2$ and we have $[X \xrightarrow{3} A][A \xrightarrow{i} X](a_1) = a_1$ and $[X \xrightarrow{3} A][A \xrightarrow{i} X](a_2) = a_2'$.

(2.10) Let $(P,\mathfrak{G},\mathfrak{R})$ be an I-structure with $|I| \leq 2$ such that $(P,\mathfrak{G} \cup \mathfrak{R})$ is an affine plane of order not less than $|I| + 3$. Then $(P,\mathfrak{R},\mathfrak{G})$ is an I'-structure with $I' = \mathfrak{R}/_{\|}$. Let $A \in \mathfrak{R}$ and let $\Sigma$ denote the set of all projectivities $[Y \xrightarrow{i} A][A \xrightarrow{j} Y][X \xrightarrow{k} A][A \xrightarrow{l} X]$ with $X,Y \in \mathfrak{R}$, $i,j,k,l \in I'$. Then the pair $(A,\Sigma)$ is a 2-transitive permutation set.

Proof. For $|I'| \geq 5$ (2.10) is a consequence of (2.9). Let $|I'| = 4$. Then the order of the affine plane is $|I| + 3$. It is sufficient to show that for $a \in A$ the set $\Sigma_a = \{\sigma \in \Sigma \mid \sigma(a) = a\}$ is transitive on $A \backslash \{a\}$. Let $b,c \in A \backslash \{a\}$ with $b \neq c$. We shall show that there is an $s \in P \backslash A$ such that $B := \overline{a,s}$, $\overline{b,s}, \overline{c,s} \in \mathfrak{R}$. Then the bijection $\sigma = [B \xrightarrow{k} A][A \xrightarrow{l} B] \in \Sigma$, where $k$ and $l$ denote the directions given by $\overline{c,s}$ and $\overline{b,s}$ respectively, fixes the point $a$ and maps $b$ onto $c$.

Case 1: $I = \emptyset$. For $s \in P \backslash A$ we have $B := \overline{a,s}, \overline{b,s}, \overline{c,s} \in \mathfrak{R}$.

Case 2: $|I| = 1$. There is a $C \in \mathfrak{R}$ with $c \in C$, $C \neq A$. Since $|C| = |I| + 3 = 4$ there is a point $s \in C$ with $s \neq c$, $s \neq [a]_1 \cap C$, $s \neq [b]_1 \cap C$ and we have $\overline{a,s}, \overline{b,s}, \overline{c,s} \in \mathfrak{R}$ (Fig.3).

Case 3: $|I| = 2$. Let $d := [a]_1 \cap [b]_2$, $C := \overline{c,d}$, $e := [a]_2 \cap C$, $f := [b]_1 \cap C$. Then $C \in \mathfrak{K}$. Since $|C| = |I| + 3 = 5$ there is a point $s \in C \backslash \{c,d,e,f\}$ and $\overline{s,a}, \overline{b,s}, \overline{c,s} \in \mathfrak{K}$ (Fig. 4).

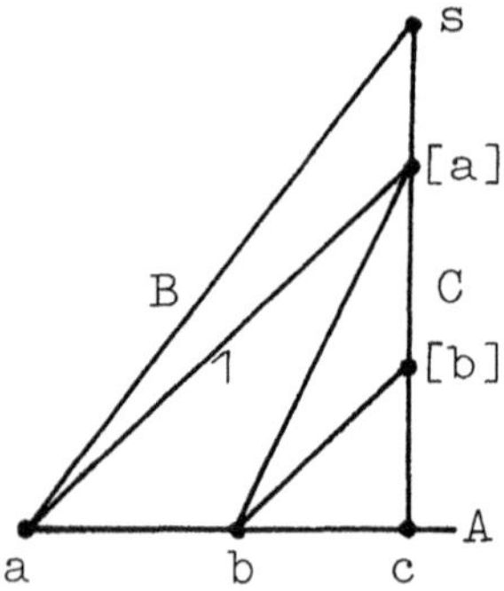


Fig. 3

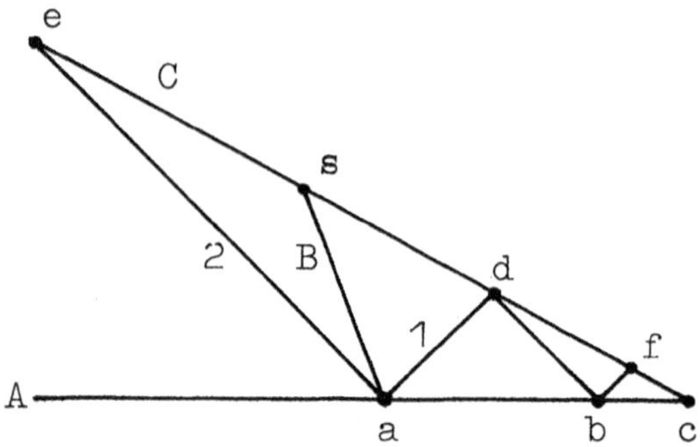


Fig. 4

§ 3 THE RECTANGLE AXIOM

In this section let $(P, \mathfrak{G}_1 \cup \mathfrak{G}_2, \mathfrak{K})$ be a 2-chain net with $\mathfrak{K} \neq \emptyset$ and $E \in \mathfrak{K}$. We ask for conditions that the associated permutation set $(E, \tilde{\mathfrak{K}}) = \Pi_E(P, \mathfrak{G}, \mathfrak{K})$ coincides with the von STAUDT group $\Gamma(E) = \Gamma_{12}(E)$. For this purpose we shall need the following rectangle axioms:

$\underline{R}_s$ For any $A, B, C \in \mathfrak{K}$ the set $\{[[a]_1 \cap B]_2 \cap [[a]_2 \cap C]_1 \mid a \in A\}$ is a chain of $\mathfrak{K}$.

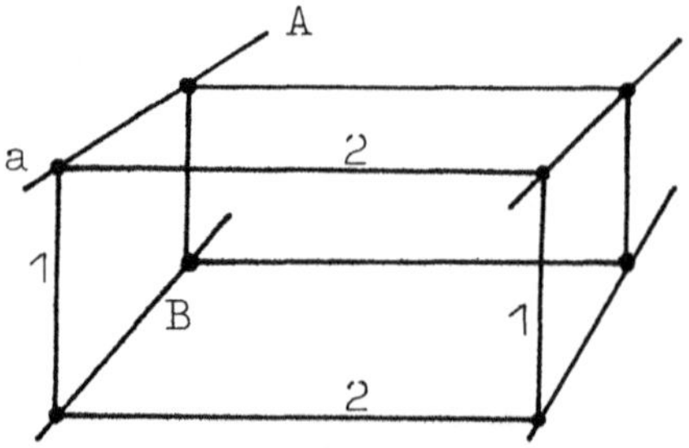


Fig. 5

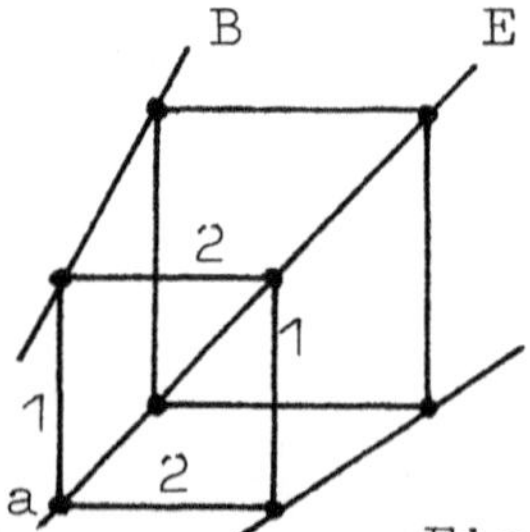


Fig. 6

$\underline{R}_i$ For any $B \in \mathfrak{R}$ the set
$\{[[[a]_1 \cap B]_2 \cap E]_1 \cap [a]_2 \mid a \in E\}$ is a chain of $\mathfrak{R}$.

For a 3-net or a (2,1)-net $\underline{R}_s$ is called the little or great REIDEMEISTER condition respectively.

From [19] we obtain the results (cf. also [3],[15],[32]):

(3.1) $\underline{R}_s$ holds for $(P, \mathfrak{G}_1 \cup \mathfrak{G}_2, \mathfrak{R})$ if and only if $\tilde{\mathfrak{R}}$ is a semigroup.

(3.2) $\underline{R}_i$ holds for $(P, \mathfrak{G}_1 \cup \mathfrak{G}_2, \mathfrak{R})$ if and only if $\tilde{\mathfrak{R}}^{-1} = \tilde{\mathfrak{R}}$.

For finite 2-chain nets the rectangle axiom $\underline{R}_s$ implies that $\tilde{\mathfrak{R}}$ is a group. In the infinite case this conclusion is not true. Let $E := \mathbb{Z}$, $\tau : \mathbb{Z} \longrightarrow \mathbb{Z}$, $x \longrightarrow x+1$ and $\Sigma = \{\tau^n \mid n \in \mathbb{N}\}$. Then $(E,\Sigma)$ is a permutation set where $\Sigma$ is a proper semigroup and the rectangle axiom $\underline{R}_s$ is valid in the derived geometric structures $\gamma(E,\Sigma)$.

From (2.4),(3.1) and (3.2) we obtain:

(3.3) The von STAUDT group $\Gamma_{12}(E)$ coincides with $\tilde{\mathfrak{R}}$ if and only if $\underline{R}_s$ and $\underline{R}_i$ are valid.

## § 4 THE VON STAUDT GROUPS OF A 3-NET

In the case of a 3-net $(P, \mathfrak{G}_1 \cup \mathfrak{G}_2 \cup \mathfrak{G}_3)$ we can consider the following four von STAUDT groups $\Gamma_{12}$, $\Gamma_{13}$, $\Gamma_{23}$ and $\Gamma_{123}$. Since by § 1 and § 2 to every loop L there belongs a 3-net we may understand by the von STAUDT groups $\Gamma_{12}(L)$, $\Gamma_{13}(L)$, $\Gamma_{23}(L)$ or $\Gamma_{123}(L)$ of the loop L the von STAUDT groups $\Gamma_{12}$, $\Gamma_{13}$, $\Gamma_{23}$ or $\Gamma_{123}$ of the associated 3-net $\gamma(L,L^+)$ (cf. (2.1)) respectively.

(4.1) For a loop $(E,+)$ and the associated 3-net $(P,\mathfrak{G}_1 \cup \mathfrak{G}_2 \cup \mathfrak{K})$ the following conditions are equivalent:

(1) $(E,+)$ is a group.

(2) In $(P,\mathfrak{G}_1 \cup \mathfrak{G}_2, \mathfrak{K})$ the little REIDEMEISTER condition $\underline{R}_s$ holds.

(3) $\Gamma_{12} = \tilde{\mathfrak{K}}$.

(4) $\Gamma_{12}$ is a regular permutation group.

Proof. (1) $\Leftrightarrow$ (2) was proved by REIDEMEISTER [32].
To show (1) $\Leftrightarrow$ (3): by definition we have $E^+ = \tilde{\mathfrak{K}}$. Therefore $\tilde{\mathfrak{K}}$ is a group and by (2.4) we have $\tilde{\mathfrak{K}} = \Gamma_{12}$ if and only if $(E,+)$ is a group.
(3) $\Leftrightarrow$ (4) because $\tilde{\mathfrak{K}}$ operates regularly on $E$ and $\tilde{\mathfrak{K}} \subseteq \Gamma_{12}$.

(4.2) For a group $(G,+)$ we have $\Gamma_{12}(G) \cong \Gamma_{13}(G)$.

Proof. This follows because the mapping $(x,y) \longrightarrow (x-y, -y)$ is an automorphism of the 3-net $\gamma(G,G^+)$.

A consequence of (4.2) is that the little {1,2}-REIDEMEISTER condition implies the little {1,3}- and {2,3}-REIDEMEISTER conditions.

Now we consider the von STAUDT group $\Gamma_{123}$ of a 3-net. $\Gamma_{123}$ contains the following projectivities:
for $i \in \{1,2\}$ let $i'$ denote the other index of $\{1,2\}$, hence $\{i,i'\} = \{1,2\}$. Then for $a,b \in E$ and $i,j \in \{1,2\}$ let $[a,i,j,b] := [E \xrightarrow{i} [a]_{i'} \xrightarrow{3} [b]_{j'} \xrightarrow{j} E]$. Each of these projectivities maps $a$ onto $b$ and we have the following rules
$[a,i,j,b]^{-1} = [b,j,i,a]$ and $[a,i,j,b] \cdot [b,j,k,c] = [a,i,k,c]$ for $i,j,k \in \{1,2\}$, $a,b,c \in E$.

From this we obtain:

(4.3) The two transitive permutation sets $\tilde{\mathfrak{R}} = \tilde{\mathfrak{S}}_3$ and $\mathfrak{S} := \{[a,2,1,b] \mid a,b \in E\}$ together generate the group $\Gamma_{123}$ and each $[a,2,1,b]$ interchanges $a$ and $b$. In the corresponding loop the projectivity $\sigma := [a,2,1,b]$ has the algebraic representation: for $c \in E$ let $c_a$ be the solution of $x + a = c$; then $\sigma(c)$ is the solution of $c_a + x = b$.

By specialisating (4.3) we have

(4.4) The von STAUDT group $\Gamma_{123}(G)$ of a group $(G,+)$ has the form

$\Gamma_{123}(G) = \{a_l \circ b_r \mid a,b \in G\} \cup \{a_l \circ \nu \circ b_r \mid a,b \in G\}$ where $a_l(x) = a + x$, $b_r(x) = x + b$ and $\nu(x) = -x$.

From the work of G. THOMSEN [34] we know that the 3-net of a commutative group $(G,+)$ is characterized by the following THOMSEN condition TH since this condition implies the REIDEMEISTER condition and commutativity of addition:

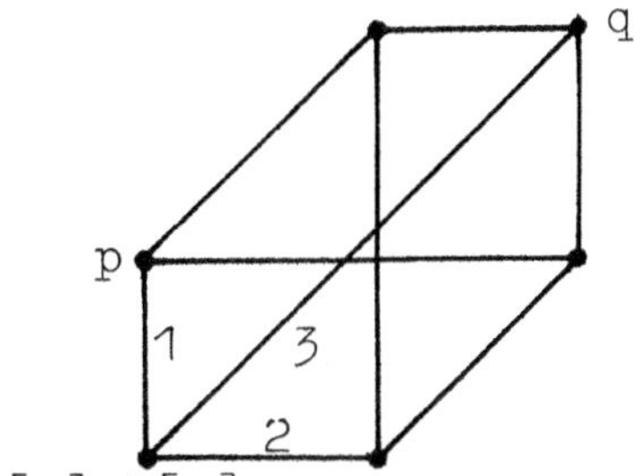


Fig. 7

TH For any two points $p,q \in P$,

$$[[q]_1 \cap [p]_2]_3 = [[[p]_3 \cap [q]_2]_1 \cap [[p]_1 \cap [q]_3]_2]_3$$

By (4.4), $\Gamma_{123}(G) = \{a_l \mid a \in G\} \cup \{a_l \circ \nu \mid a \in G\}$, and any map of $\Gamma_{123}$ interchanging two distinct elements $a,b$ is involutory. If $G$ does not contain any element of order 2, then $\sigma_{a,b} := [a,2,1,b]$ (with $\sigma_{a,b}(x) = (a+b) - x$) is the only involution interchanging the two distinct elements $a,b$.

Problem. Determine which loops are characterized by the property:
For any two distinct elements $a,b \in E$ there is exactly one involution $\sigma \in \Gamma_{123}$ with $\sigma(a) = b$.

## § 5 2-STRUCTURES AND (3.1)-NETS

From (1.1), (2.1) and (3.1) we can derive the following theorem for 2-structures (cf. [19]):

(5.1) For a 2-chain net $(P, \mathfrak{G}_1 \cup \mathfrak{G}_2, \mathfrak{R})$ the following statements are equivalent:

(1) $(P, \mathfrak{G}_1 \cup \mathfrak{G}_2, \mathfrak{R})$ is a 2-structure which fulfills the rectangle axiom $\underline{R}_s$.

(2) $(E, \tilde{\mathfrak{R}})$ is a sharply 2-transitive permutation group.

(3) $E$ can be provided with the structure of a near domain $(E,+,\cdot)$ such that $\tilde{\mathfrak{R}}$ is the affine group of the near domain.

Proof. By the mentioned theorems we have:
(3) $\Leftrightarrow$ (2) $\Rightarrow$ (1).
(1) implies that $(E, \tilde{\mathfrak{R}})$ is a sharply 2-transitive semigroup. Let $a,b$ be two distinct elements of $E$. For any $\alpha \in \tilde{\mathfrak{R}}$ let $\beta \in \tilde{\mathfrak{R}}$ be the permutation with $\beta(\alpha(a)) = a$, $\beta(\alpha(b)) = b$. By $\underline{R}_s$ $\beta \circ \alpha \in \tilde{\mathfrak{R}}$ and by the sharp 2-transitivity we have $\beta \circ \alpha = id$ and hence $\alpha^{-1} \in \tilde{\mathfrak{R}}$. Therefore $\tilde{\mathfrak{R}}$ is a group.

Next let $(P, \mathfrak{G}_1 \cup \mathfrak{G}_2 \cup \mathfrak{G}_3, \mathfrak{R})$ be a (3.1)-net where $0$ denotes the point of $P$ with $\mathfrak{R}(0) = \mathfrak{R}$ and let $E := [0]_3$ and $1 \in E \setminus \{0\}$. The permutation set $(E, \tilde{\mathfrak{L}})$ with $\mathfrak{L} := \mathfrak{G}_3 \cup \mathfrak{R}$ of the 2-chain net $(P, \mathfrak{G}_1 \cup \mathfrak{G}_2, \mathfrak{L})$ decomposes into the regular permutation set $(E, \tilde{\mathfrak{G}}_3)$ and the permutation set $(E, \tilde{\mathfrak{K}} \cup \{\tilde{E}\})$ which acts regularly on $E^* := E \setminus \{0\}$ with $\tilde{\mathfrak{G}}_3 \cap \tilde{\mathfrak{K}} = \emptyset$. Therefore by § 1 we can provide $E$ with an addition $+$ and a multiplication $\cdot$ making $(E,+)$ a loop and $(E,\cdot)$ a loop with $0$. For $a \in E\setminus\{0\}$ let $a^+ \in \tilde{\mathfrak{G}}_3$ with $a^+(0) = a$ and $a^\cdot \in \tilde{\mathfrak{K}} \cup \{id\}$ with $a^\cdot(1) = a$. Then $a + b := a^+(b)$ and $a \cdot b := a^\cdot(b)$.

Let $\Gamma_{12}$ denote the von STAUDT group of the 2-chain net $(P, \mathfrak{G}_1 \cup \mathfrak{G}_2, \mathfrak{L} := \mathfrak{G}_3 \cup \mathfrak{K})$ acting on $E$. We have the following theorems:

<u>(5.2)</u> $\Gamma_{12}$ acts 2-transitively on $E$.

Proof. Let $a, b \in E$ with $a \neq b$ and $c := (a^+)^{-1}(b)$. Then the map $a^+ c^\cdot \in \tilde{\mathfrak{G}}_3 \circ \tilde{\mathfrak{K}} \subset \Gamma_{12}$ maps $0$ onto $a$ and $1$ onto $b$.

<u>(5.3)</u> Let $(E, \Gamma_{12})$ be sharply 2-transitive and let $(E, \oplus, \odot)$ be the near domain which is associated to $(E, \Gamma_{12})$ by (1.1) and which has $0$ and $1$ as neutral elements. Then:

(1) $(P, \mathfrak{G}_1 \cup \mathfrak{G}_2, \mathfrak{G}_3 \cup \mathfrak{K})$ can be embedded in a 2-structure $(P, \mathfrak{G}_1 \cup \mathfrak{G}_2, \mathfrak{M})$ satisfying the rectangle axiom $\underline{R}_S$ with $\mathfrak{G}_3 \cup \mathfrak{K} \subset \mathfrak{M}$.

(2) $\Gamma_{12} = \tilde{\mathfrak{M}}$ is the affine group of the near domain $(E, \oplus, \odot)$ and consists of the mappings $[c,m] : E \to E$, $x \to c \oplus m \odot x$ with $c, m \in E$, $m \neq 0$.

(3) The multiplication $\odot$ of the near domain coincides with the multiplication $\cdot$ defined by $\tilde{\mathfrak{K}} \cup \{id\}$.

(4) The addition $\oplus$ of the near domain and the loop addition $+$ defined by $\tilde{\mathfrak{G}}_3$ coincide if and only if the following configuration theorem for the (3.1)-net $(P, \mathfrak{G}_1 \cup \mathfrak{G}_2 \cup \mathfrak{G}_3, \mathfrak{K})$ holds:

<u>A</u> For any $a \in E$ let
$b := [0]_1 \cap [a]_2$,
$c := [b]_3 \cap [0]_2$,
$-a := [c]_1 \cap E$,
$d := [-a]_1 \cap [a]_2$ and
$K \in \mathfrak{K}$ with $0, d \in K$;
then $[a]_1 \cap [-a]_2 \in K$.

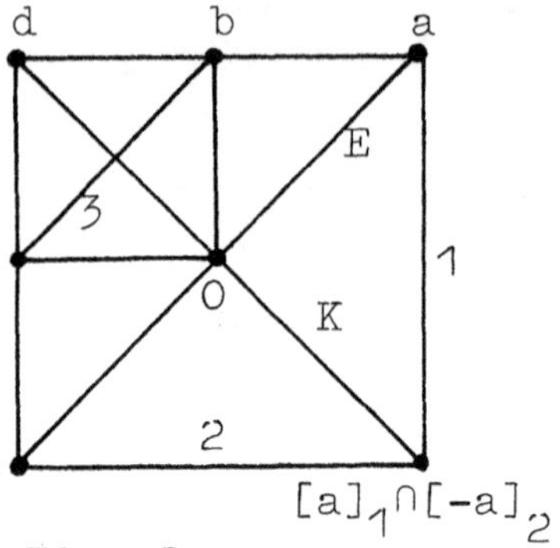

Fig. 8

(5) If $(P, \mathfrak{G}_1 \cup \mathfrak{G}_2 \cup \mathfrak{M})$ is an affine plane, then <u>A</u> is valid, the associated near domain is a planar nearfield and hence the little REIDEMEISTER condition $\underline{R}_s$ holds for $(P, \mathfrak{G}_1 \cup \mathfrak{G}_2 \cup \mathfrak{G}_3)$.

Proof. (1) Let $\mathfrak{M} := \{\{(x, \gamma(x)) \mid x \in E\} \mid \gamma \in \Gamma_{12}\}$. Since $(E, \Gamma_{12})$ is a sharply 2-transitive permutation group, the triple $(P, \mathfrak{G}_1 \cup \mathfrak{G}_2, \mathfrak{M})$ is a 2-structure which fulfills the rectangle axiom $\underline{R}_s$ by (5.1). Because $\widetilde{\mathfrak{G}}_3, \widetilde{\mathfrak{K}} \subset \Gamma_{12}$, we have $\mathfrak{G}_3, \mathfrak{K} \subset \mathfrak{M}$.
(3) Since both $\widetilde{\mathfrak{K}}$ and the stabilizer $\Gamma_{12}(0)$ of $0$ act regularly on $E \setminus \{0\}$ the multiplications coincide.
(4) For $a \in E$ define $b, c, -a, d$ and $K$ as in <u>A</u>. Then (for the loop addition) we have $a + (-a) = 0$ and $\widetilde{K}(-a) = a$. Now, let $\oplus = +$. Then $(-1) \cdot (-a) = a$ (by (1.4)) and therefore $(-1)^{\cdot} = \widetilde{K}$. This gives us $\widetilde{K}(a) = (-1) \cdot a = -a$, and hence $[a]_1 \cap [-a]_2 \in K$.

If <u>A</u> holds then to every $a \in E \setminus \{0\}$ there is a $K_a \in \mathfrak{K}$ with $\widetilde{K}_a(-a) = a$ and $\widetilde{K}_a(a) = -a$. Since $\Gamma_{12}$ is sharply 2-transitive, $\widetilde{K}_a$ is either the identity or the uniquely determined involution $(-1)^{\cdot}$ fixing the point $0$. In the first case $a^+ \in \widetilde{\mathfrak{G}}_3$ is an involution without fixed points, and so the near domain has characteristic 2 and

$a^+ = a^{\oplus}$ (cf. (1.2)). In the second case the characteristic of the near domain is different from 2 and $a^+ \widetilde{K}_a = a^+(-1)^{\cdot}$ interchanges the points 0 and a ; hence $a^+ \circ (-1)^{\cdot}$ is an involution and again by (1.2) we have $a^{\oplus} = a^+$.

Problem: For the 3-net $(P, \mathfrak{G}_1 \cup \mathfrak{G}_2 \cup \mathfrak{G}_3)$ does the little REIDEMEISTER condtion $\underline{R}_s$ imply $\underline{A}$ ? This problem is equivalent with the question as to whether there are near domains or non-planar nearfields E such that the affine group of E contains a subgroup different from $E^{\oplus} := \{a^{\oplus} | a \in E\}$ operating regularly on E.

From (5.1) and (2.4) we obtain

<u>(5.4)</u> For a 2-structure $(P, \mathfrak{G}_1 \cup \mathfrak{G}_2, \mathfrak{K})$ the rectangle axiom $\underline{R}_s$ is valid if and only if the von STAUDT group $\Gamma_{12}$ is sharply 2-transitive.

Remark: In (5.4) the condition $(P_2)$ for the von STAUDT group $\Gamma_{12}$ cannot be replaced by $(P_3)$ because there are free 2-structures where $(P_3)$ is fulfilled for $\Gamma_{12}$.

<u>(5.5)</u> Let $(P, \mathfrak{G}_1 \cup \mathfrak{G}_2 \cup \mathfrak{G}_3, \mathfrak{K})$ be a (3.1)-net where $\Gamma_{123}$ is sharply 2-transitive on E. Then $(P, \mathfrak{G}_1 \cup \mathfrak{G}_2 \cup \mathfrak{G}_3 \cup \mathfrak{K})$ is embeddable in a desarguesian affine plane $(P, \mathfrak{L})$ with point set P and $\mathfrak{G}_1 \cup \mathfrak{G}_2 \cup \mathfrak{G}_3 \cup \mathfrak{K} \subset \mathfrak{L}$.

Proof. By (5.2) we have $\Gamma_{12} = \Gamma_{123}$ and by (5.3) there is a 2-structure $(P, \mathfrak{G}_1 \cup \mathfrak{G}_2, \mathfrak{M})$ with $\underline{R}_s$ and $\mathfrak{G}_3 \cup \mathfrak{K} \subset \mathfrak{M}$. By (1.5) we have to show that the associated near domain $(E, \oplus, \cdot)$ is distributive. For this purpose we first prove that the double loop $(E, +, \cdot)$ is right distributive where $+$ is defined by $\widetilde{\mathfrak{G}}_3$, and $\cdot$ is the multiplication of the near domain $(E, \oplus, \cdot)$ and so defined by the group $\tilde{\mathfrak{K}}$.

Let $a, b, c \in E \setminus \{0\}$, $G := [[0]_1 \cap [a]_2]_3$, $H := [[0]_1 \cap [c]_2]_3$.

Then $a^+ = \tilde{G}$ and $c^+ = \tilde{H}$.

Case 1 $a + b \neq 0$. Let $B_1, B_2 \in \mathfrak{K}$ with $0, [b]_1 \cap G \in B_1$ and $0, [b]_2 \cap G \in B_2$ and

$\mu := [E \xrightarrow{2} [0]_1]$, $\nu := [[0]_1 \xrightarrow{3} B_2]$, $\rho := [B_2 \xrightarrow{2} E]$

$\sigma := [[0]_1 \xrightarrow{3} B_1]$, $\tau := [B_1 \xrightarrow{2} E]$, $\eta := [B_2 \xrightarrow{3} B_1]$ and

$\zeta := [B_1 \xrightarrow{1} E]$.

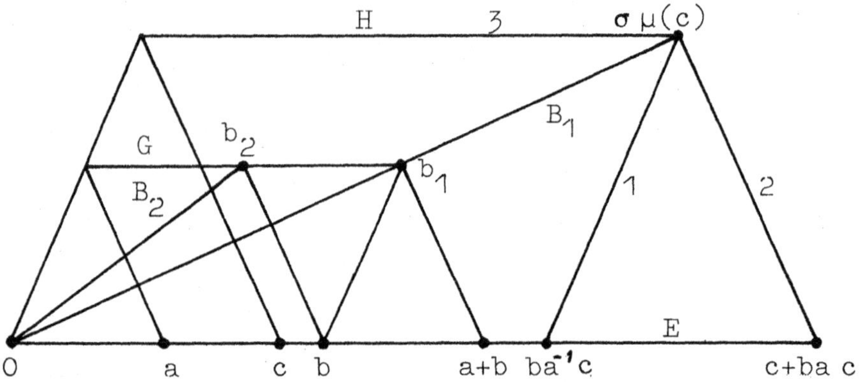


Fig. 9

Then we have $\eta\nu = \sigma$ and

$\mu(0) = \nu(0) = \rho(0) = \sigma(0) = \tau(0) = \eta(0) = \zeta(0) = 0$,

$\rho\nu\mu(a) = b = \zeta\sigma\mu(a)$, $\tau\sigma\mu(a) = a + b$ and $\zeta\eta\rho^{-1}(b) = b$;

hence $(ba^{-1})^{\bullet} = \rho\nu\mu = \zeta\sigma\mu$, $((a+b)a^{-1})^{\bullet} = \tau\sigma\mu$ and $\zeta\eta\rho^{-1} = \mathrm{id}$.

From this we obtain

$ba^{-1}c = \zeta\eta\rho^{-1}(ba^{-1}c) = \zeta\eta\rho^{-1}\rho\nu\mu(c) = \zeta\eta\nu\mu(c) = \zeta\sigma\mu(c)$,

and so $\sigma\mu(c) = \zeta^{-1}(ba^{-1}c)$, which implies

$\sigma\mu(c) = B_1 \cap H = [ba^{-1}c]_1 \cap H$ and therefore

$\tau\sigma\mu(c) = c + ba^{-1}c$. On the other hand

$(a + b)\, a^{-1}c = ((a+b)a^{-1})^{\bullet}(c) = \tau\sigma\mu(c)$, hence $(a+b)a^{-1}c = c + ba^{-1}c$ .

Case 2 $a+b=0$ . Here we put $B_1 := [0]_2$ and $B_2, \mu, \nu, \rho, \eta, \zeta$ as in case 1. (The perspectivity $\tau$ cannot be defined). Then as in case 1 $\sigma\mu(c) = [ba^{-1}c]_1 \cap H$ , and so $c + ba^{-1}c = 0 = (a+b)a^{-1}c$ .
Since $c$ is arbitrary and $(E^*, \cdot)$ is a group we obtain from $(a+b)a^{-1}c = c + ba^{-1}c$ the right distributive law.

Now we compare the addition $+$ and the addition $\oplus$ of the near domain $(E, \oplus, \cdot)$ . For any $a \in E^*$ the permutations $a^+$ and $a^{\oplus}$ have no fixed points and map $0$ onto $a$ ; hence $(a^{\oplus})^{-1} a^+ \in \tilde{\mathfrak{K}}$ and therefore there is an $m_a \in E$ with $m_a^{\cdot} = (a^{\oplus})^{-1} a^+$ .

For $a,b,c \in E$ with $a,c \neq 0$ we have $a+b = a \oplus m_a \cdot b$ and

$$(a \oplus m_a \cdot b) \cdot c = (a+b)c = ac + bc = ac \oplus m_{ac} \cdot bc .$$

For $b = m_a^{-1} \cdot (\ominus a)$ we obtain

$$0 = ac \oplus m_{ac} \cdot m_a^{-1} \cdot (\ominus a)c = ac \ominus m_{ac} \cdot m_a^{-1} \cdot ac$$

because of (1.4). This gives us $m_{ac} = m_a$ , i.e. $m := m_a$ is constant for all $a \in E^*$ . Hence

$$(a \oplus b)c = (a + m^{-1}b)c = ac + m^{-1}bc = ac \oplus bc .$$

As a corollary we obtain from (5.5) theorem (4.3) of [24]:

(5.6) For every 3-structure $(P, \mathfrak{G}, \mathfrak{K})$ the von STAUDT group $\Gamma_{123}$ is sharply 2-transitive if and only if $(P, \mathfrak{G} \cup \mathfrak{K})$ is a desarguesian affine plane.

## § 6 HYPERBOLA STRUCTURES AND MINKOWSKIAN PLANES

From (2.4) and (2.7) we obtain

(6.1) The von STAUDT group $\Gamma_{12}$ of a hyperbola structure is 3-transitive.

Let $(F,+,\cdot,\sigma)$ be a KT-field and $\Gamma$ the projectivity group of $F$, operating on $\overline{F} := F \cup \{\infty\}$ (cf. § 1). Then, by (2.7), $Hy(F) := \gamma(\overline{F},\Gamma)$ is a hyperbola structure. The von STAUDT group $\Gamma_{12}$ of $Hy(F)$ is isomorphic to $\Gamma$.

The following results have been previously shown [3], or [15] § 5.

(6.2) For a hyperbola structure $(P,\mathfrak{G},\mathfrak{K})$ the following are equivalent:

(1) $\Gamma_{12}$ is sharply 3-transitive.

(2) $\underline{R}_S$ holds for $(P,\mathfrak{G},\mathfrak{K})$.

(3) There exists a KT-field $(F,+,\cdot,\sigma)$ such that $Hy(F)$ and $(P,\mathfrak{G},\mathfrak{K})$ are isomorphic.

Remark: The condition (1) cannot be replaced by $(P_4)$ because there are free hyperbola structures and even free Minkowskian planes where $(P_4)$ is fulfilled for $\Gamma_{12}$.

From (6.2) we obtain:

(6.3) Let $(P,\mathfrak{G},\mathfrak{K})$ be a hyperbola structure where $\Gamma_{12}$ is sharply 3-transitive. Then $(P,\mathfrak{G},\mathfrak{K})$ is a Minkowskian plane if and only if for the associated KT-field $(F,+,\cdot,\sigma)$ the near domain $(F,+,\cdot)$ is a planar nearfield.

(6.4) For a hyperbola structure $(P,\mathfrak{G},\mathfrak{R})$ the following are equivalent:

(1) There is a commutative field $(F,+,\cdot)$ such that $(P,\mathfrak{G},\mathfrak{R})$ and $Hy(F,+,\cdot,\sigma)$ are isomorphic where $\sigma(x)=x^{-1}$ for all $x\in F^*$.

(2) $\Gamma_{12}$ has the property:

S Any element $\alpha\in\Gamma_{12}$ interchanging two distinct points is involutory.

Proof. By [15] Satz 17 we have (1) ⇒ (2). To show (2) ⇒ (1) : Since $\tilde{\mathfrak{R}}\subset\Gamma_{12}$, the condition S holds in particular for $\tilde{\mathfrak{R}}$. Hence by [20] there is a commutative field $(F,+,\cdot)$ with $\tilde{\mathfrak{R}}=PGL(2,F)$. With (2.4) we have $\Gamma_{12}=PGL(2,F)$ and therefore $(P,\mathfrak{G},\mathfrak{R})=Hy(F)$.

## § 7 VON STAUDT THEOREMS FOR CIRCLE GEOMETRIES

In addition to the hyperbola structures (in particular the Minkowskian planes) the Möbius and Laguerre planes are also circle geometries. Since for Möbius and Laguerre planes the underlying net is characterized by $I=\emptyset$ or $|I|=1$ the von STAUDT group $\Gamma_I$ is trivial. However it is possible to define other perspectivities in circle planes $(P,\mathfrak{G},\mathfrak{R})$ and therefore other von STAUDT groups. For this purpose we define the following concepts: besides the proper circles we need improper circles. By an improper circle we understand:
in the Möbius plane, any point $a\in P$, denoted by [a];
in the Laguerre plane, any union $[a]\cup[b]$ of not necessarily distinct generators [a], [b];
in the Minkowskian plane any union $[a]:=[a]_1\cup[a]_2$.

In the quadric model (cf. § 2 Example 4) of a circle

plane the improper circles are the non-empty plane sections, which are degenerate conics.

For $a,b \in P$ with $a \neq b$, $[a]_i \neq [b]_i$ we enlarge the <u>2-point-pencil</u> $\mathfrak{K}(a) \cap \mathfrak{K}(b)$ by improper circles:

$\mathfrak{K}(a,b) := \mathfrak{K}(a) \cap \mathfrak{K}(b)$ for $I = \emptyset$;

$\mathfrak{K}(a,b) := \mathfrak{K}(a) \cap \mathfrak{K}(b) \cup \{[a] \cup [b]\}$ for $I = \{1\}$;

$\mathfrak{K}(a,b) := \mathfrak{K}(a) \cap \mathfrak{K}(b) \cup \{[a]_1 \cup [b]_2, [a]_2 \cup [b]_1\}$ for $I = \{1,2\}$

We do the same for <u>touching pencils</u>: for $A \in \mathfrak{K}$, $a \in A$ let $\mathfrak{K}(a,A) := \{X \in \mathfrak{K} \mid X \cap A = \{a\}\} \cup \{A, [a]\}$.
The enlarged 2-point-pencils $\mathfrak{K}(a,b)$ have the property that to any point $x \in P \setminus \{a,b\}$ there is exactly one circle $X =: \mathfrak{K}(a,b,x)$ with $x \in X \in \mathfrak{K}(a,b)$. For a touching pencil $\mathfrak{K}(a,A)$ we can associate to any point $x \in P$ exactly one circle $X \in \mathfrak{K}(a,A)$ with $x \in X$ denoted by $\mathfrak{K}(a,A,x)$; for $x = a$ we have $\mathfrak{K}(a,A,a) = [a]$.

<u>Basic perspectivities</u>. In a circle geometry there are several possibilities to define basic perspectivities:
<u>Type 1</u>. Circle A and 2-point-pencil $\mathfrak{K}(a,b)$ with $a \in A$, $b \notin A$. The map

$$[A,a;b] : A \to \mathfrak{K}(a,b);\ x \to \begin{cases} \mathfrak{K}(a,b,x) & \text{for } x \neq a \\ \mathfrak{K}(a,A,b) & \text{for } x = a \end{cases}$$

is a bijection. The inverse map will be denoted by [b;a,A] (Fig. 10).

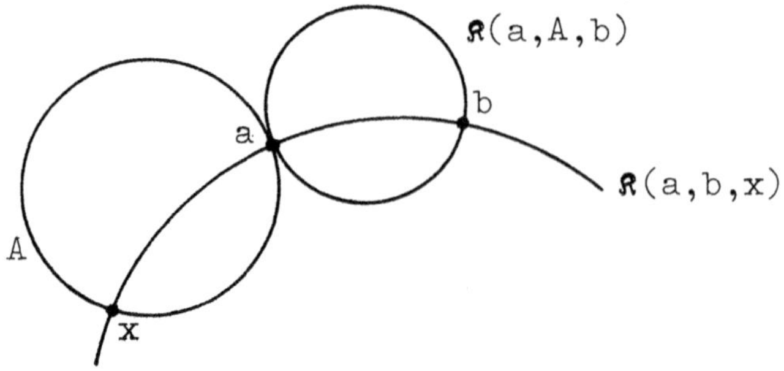


Fig. 10

Type 2. Circle $A$ and touching pencil $\mathfrak{K}(b,B)$ with $b \in A$ and $A \notin \mathfrak{K}(b,B)$. The map

$$[A;b,B] : A \longrightarrow \mathfrak{K}(b,B);\ x \longrightarrow \mathfrak{K}(b,B,x)$$

is bijective and the inverse is denoted by $[b,B;A]$ (Fig. 11).

Type 3. 2-point-pencil $\mathfrak{K}(a,c)$ and 2-point-pencil $\mathfrak{K}(c,b)$ with $[a]_i \neq [c]_i \neq [b]_i$. The map

$$[a,c,b] : \mathfrak{K}(a,c) \longrightarrow \mathfrak{K}(c,b);\ X \longrightarrow \mathfrak{K}(c,X,b)$$

is bijective and $[a,c,b]^{-1} = [b,c,a]$ (Fig. 12).

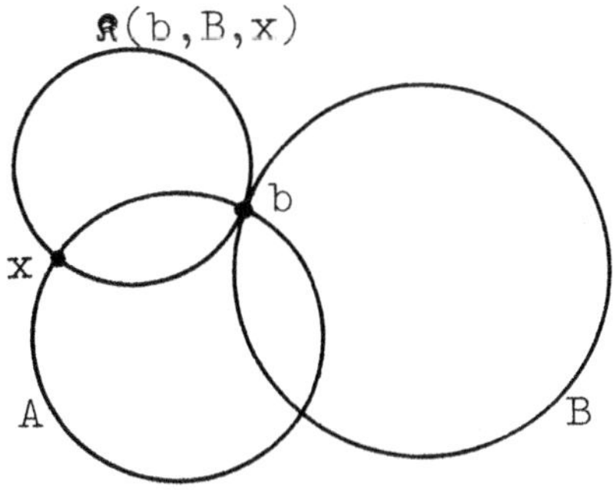


Fig. 11

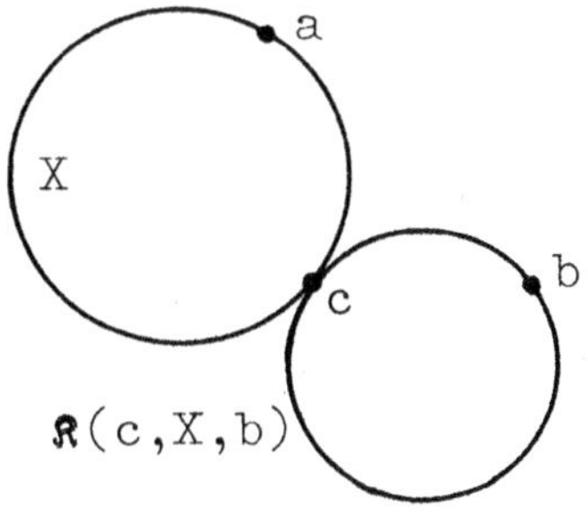


Fig. 12

Type 4. Circle $A$ onto itself with 2-point-pencil $\mathfrak{K}(b,c)$ with $b,c \notin A$. The map

$[A,b,c,A] : A \longrightarrow A$ ;

$$x \longrightarrow \begin{cases} \mathfrak{K}(b,c,x) \cap A \setminus \{x\} & \text{if } |\mathfrak{K}(b,c,x) \cap A| = 2 \\ x & \text{if } |\mathfrak{K}(b,c,x) \cap A| = 1 \end{cases}$$

is an involution (Fig. 13).

Type 5. Circle $A$ onto itself with touching pencil $\mathfrak{K}(b,B)$ with $b \notin A$. The map

$[A,b,B,A] : A \longrightarrow A$ ;

$$x \longrightarrow \begin{cases} \mathfrak{K}(b,B,x) \cap A \setminus \{x\} & \text{if } |\mathfrak{K}(b,B,x) \cap A| = 2 \\ x & \text{if } |\mathfrak{K}(b,B,x) \cap A| = 1 \end{cases}$$

is an involution (Fig. 14).

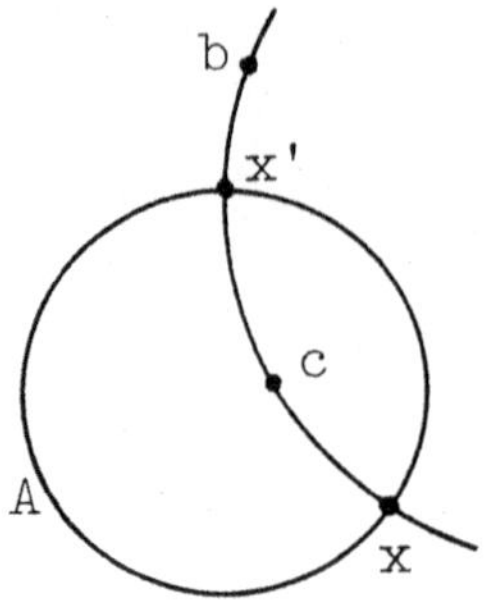


Fig. 13

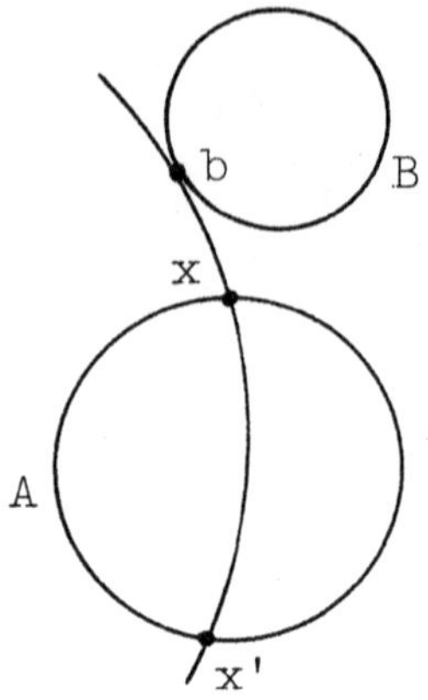


Fig. 14

With these basic perspectivities we can form the composition of perspectivities from one circle to another circle and projectivities. The following types were studied:

For $A, B \in \mathfrak{K}$, $a \in A \setminus B$, $b \in B \setminus A$ with $[a]_i \neq [b]_i$ the map $[A,a,b,B] := [a;b,B] \circ [A,a;b]$ is called a <u>proper perspectivity</u> with <u>centers</u> $\{a,b\}$ (Fig. 15). The set of all proper perspectivities will be denoted by $\Pi_p$.

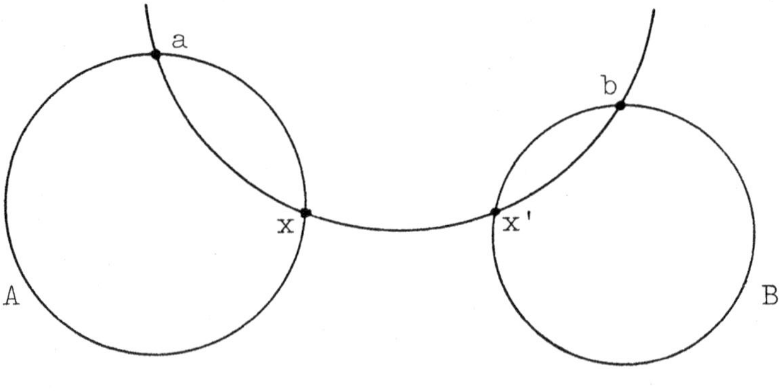


Fig. 15

Let the set of all proper perspectivities $\varphi := [A,a,b,B]$ with $\varphi(a) = b$ be denoted by $\Pi_o$. The elements of $\Pi_o$ are called <u>perspectivities of the stereographic type</u> (Fig. 16).

For $A,B \in \mathfrak{K}$ with $A \cap B \neq \emptyset$, $c \in A \cap B$ and $d \in P \setminus A \cup B$ with $[c]_i \neq [d]_i$ the bijection $[A,c,d,B] := [d;c,B] \circ [A,c;d]$ is called a <u>free perspectivity</u>. We denote the set of all free perspectivities by $\Pi_f$ (Fig. 17).

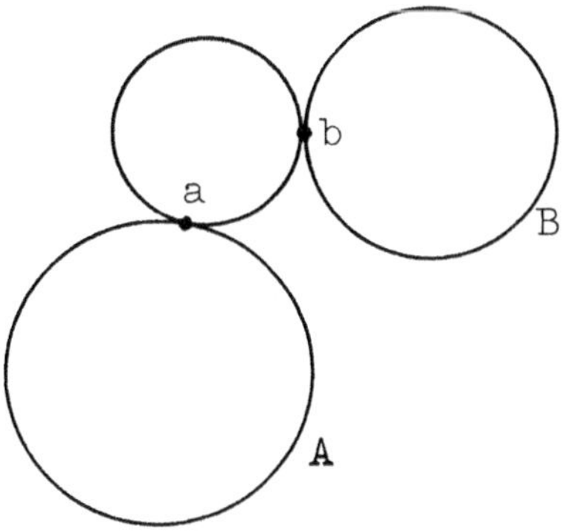


Fig. 16

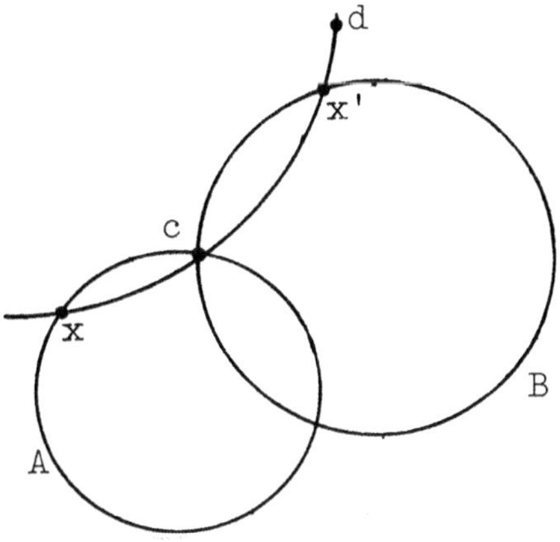


Fig. 17

For $A,B,C \in \mathfrak{K}$ with $A \cap B \cap C \neq \emptyset$, $c \in A \cap B \cap C$ and $A,B \notin \mathfrak{K}(c,C)$ the bijection $[A,c,C,B] := [c,C;B] \circ [A;c,C]$ is called an <u>affine perspectivity</u>. The set of all affine perspectivities is denoted by $\Pi_a$ (Fig. 18).

For $A,B \in \mathfrak{K}$ and $a \in A$, $b \in B$, $c \in P \setminus (A \cup B)$ with $[a]_i \neq [c]_i \neq [b]_i$ the bijection $[A,a,c,b,B] := [c;b,B] \circ [a,c,b] \circ [A,a;c]$ is called a <u>Herzer perspectivity</u>. The set of all Herzer perspectivities will be denoted by $\Pi_h$ (Fig. 19).

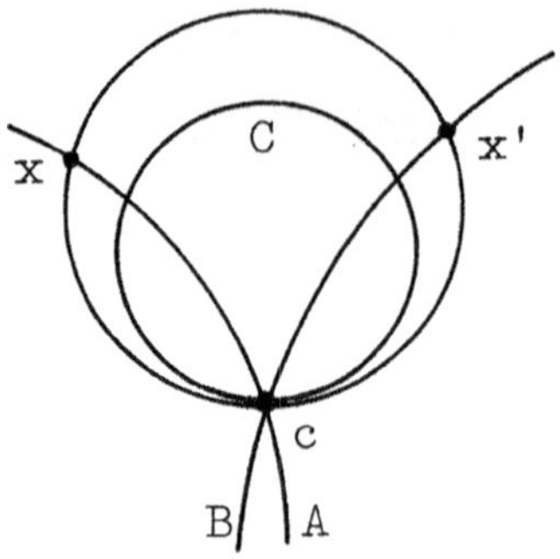


Fig. 18

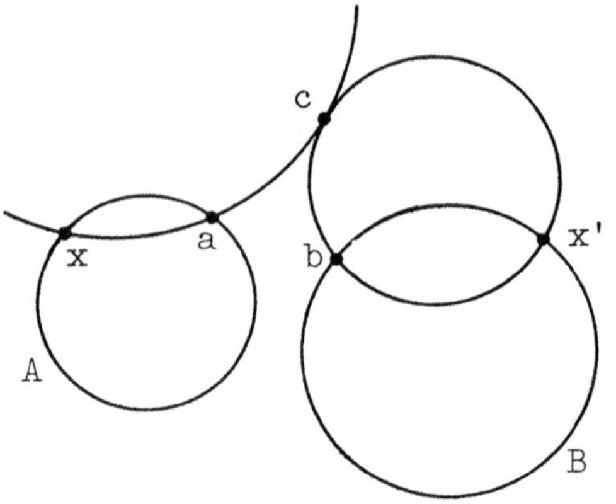


Fig. 19

A similiar type of perspectivities is the following: for $A, B \in \mathfrak{K}$ and $a \in A$, $b \in B \backslash A$, $c \in P \backslash A$ with $[b]_i \neq [a]_i \neq [c]_i$ the bijection

$$[A, a, c, a, b, B] := [a; b, B] \circ [c, a, b] \circ [A, a; c]$$

and its inverse bijection

$$[B, b, a, c, a, A] := [c; a, A] \circ [a, b, c] \circ [B, b; a]$$

are called <u>K-perspectivities</u>. Let $\Pi_k$ be the set of all K-perspectivities (Fig. 20).

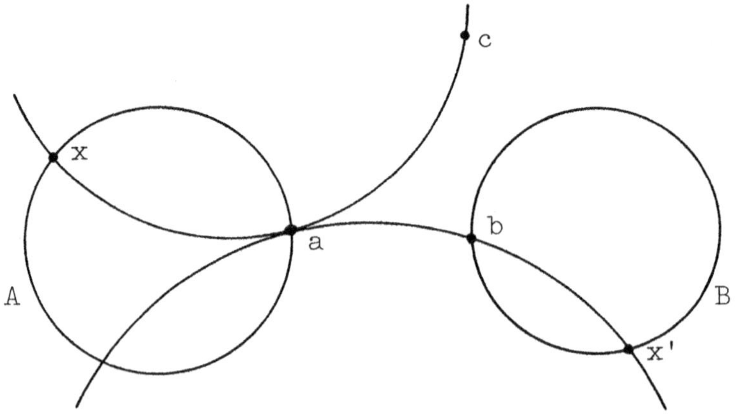


Fig. 20

(7.1) Any of the perspectivity sets $\Pi_p$, $\Pi_o$, $\Pi_f$, $\Pi_{fo}$, $\Pi_a$, $\Pi_h$ and $\Pi_k$ contains with every perspectivity also its inverse and we have $\Pi_{fo} \subset \Pi_f \subset \Pi_h$ and $\Pi_o \subset \Pi_p \subset \Pi_k$ because $[A,c,d,B]=[A,c,d,c,B]$, $[A,a,b,B]=[A,a,b,a,b,B]$.

Remark. If we replace in the case of a Laguerre or Minkowskian plane for two distinct points $a,b$ with $a \in [b]$ in the definition of a proper perspectivity the 2-point-pencil by the pencil of all improper circles through $a,b$ we obtain an i-perspectivity (cf. § 2).

The different types of perspectivities have interpretations in an affine or projective plane. Let us consider the derivation $D := (P,\mathfrak{G},\mathfrak{K})^a$ in a point $a \in P$, and let $\overline{D}$ denote the projective closure of the affine plane $D$ with $U$ as line of infinity. In the Laguerre and Minkowski cases ($I=\{1\}$ and $I=\{1,2\}$) let $u_i \in U$ denote the point determined by the generators $\mathfrak{G}_i^a$.

For the proper perspectivity $[A,a,b,B] \in \Pi_p$ the circle $A$ and the circles through $\{a,b\}$ appear as lines and $B$ as an oval $\overline{B}$ of $\overline{D}$ (Fig. 21) where $\overline{B}=B$ for $I=\emptyset$, $\overline{B}:=(B\setminus[a]) \cup \{u_1\}$ for $I=\{1\}$ or $\overline{B}=(B\setminus[a]) \cup \{u_1,u_2\}$ for $I=\{1,2\}$.

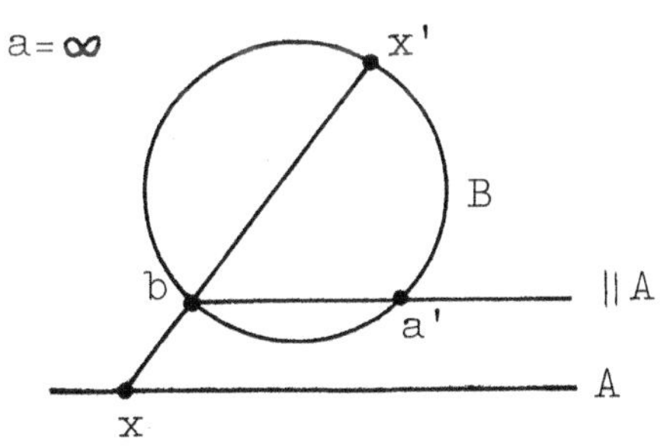


Fig. 21a

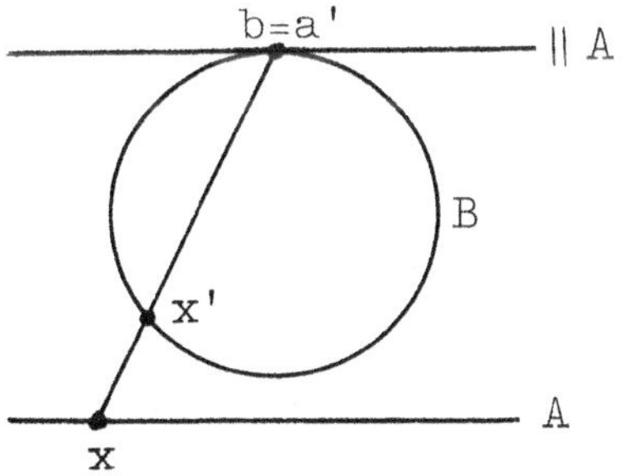


Fig. 21b

In the Möbius case, for $[A,a,b,B] \in \Pi_o$ we can restrict the perspectivity on the affine plane $D$ and obtain a stereographic projection of the line $A$ on the oval $B$. In the case of a free perspectivity $\varphi = [A,a,z,B] \in \Pi_f$ this mapping can be considered as a perspectivity in the projective plane $\overline{D}$ with center $z$ in $D$ because all circles appear as lines. For $\varphi \in \Pi_{fo}$ this mapping even induces an affine central perspectivity in $D$ because $A\setminus\{a\}$ and $B\setminus\{a\}$ are parallel.
In the same way an affine perspectivity $[A,a,C,B] \in \Pi_a$ appears as a parallel perspectivity of the affine plane $D$ with direction $C$.

For a HERZER perspectivity $[C,c,a,b,B] \in \Pi_h$ and a K-perspectivity $[A,a,c,a,b,B] \in \Pi_k$ the situation in the affine plane $D$ is given by the following figures (Fig. 22, Fig. 23).

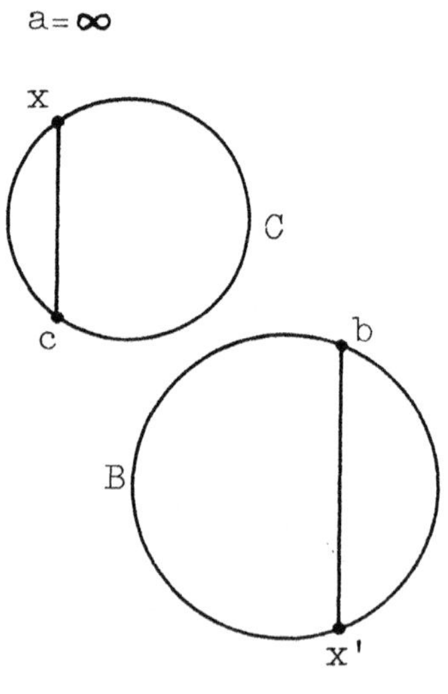


Fig. 22

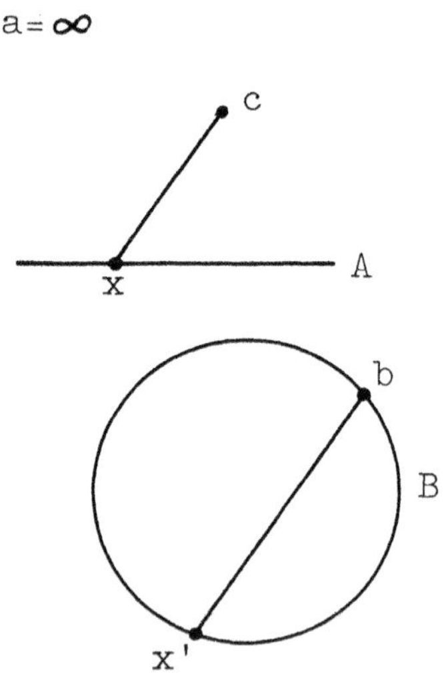


Fig. 23

The joining lines of c,x and b,x' are parallel in D.

For any subset $\Delta \subset \Pi_h \cup \Pi_k \cup \Pi_a$ a bijection $\gamma : A \longrightarrow B$ with $A, B \in \mathfrak{K}$ is called a <u>$\Delta$-projectivity</u> if there are $\pi_1, \pi_2, \ldots, \pi_n \in \Delta$ with $\gamma = \pi_1 \circ \ldots \circ \pi_n$ ; let $\overline{\Delta}$ be the set of all $\Delta$-projectivities. For $A \in \mathfrak{K}$ let $\Gamma_A(\Delta)$ be the semigroup of all $\Delta$-projectivities from $A$ onto itself. $\Gamma_A(\Delta)$ is a group if $\Delta = \Delta^{-1}$ .

To exclude trivial cases we assume that any circle $A \in \mathfrak{K}$ has at least 6 points. Then we obtain the following propositions:

<u>(7.2)</u> $\Pi_f \subset \overline{\Pi}_p$ , $\Pi_{fo} \subset \overline{\Pi}_{po}$

Proof. Let $\varphi := [A,c,d,B] = [d,c,B] \circ [A,c;d] \in \Pi_f$ or $\Pi_{fo}$ and $C := \mathfrak{K}(c,A,d)$ . For any $D \in \mathfrak{K}(d,C) \setminus \mathfrak{K}(c)$ we have $[D,d,c,B][A,c,d,D] = [A,c,d,B]$ with $[A,c,d,D] \in \Pi_{po}$ , $[D,d,c,B] \in \Pi_p$ and $[D,d,c,B] \in \Pi_{po}$ , if $A \cap B = \{c\}$ i.e. $\varphi \in \Pi_{fo}$ . Hence $\varphi \in \overline{\Pi}_p$ or $\varphi \in \overline{\Pi}_{po}$ .

<u>(7.3)</u> $\Pi_h \subset \overline{\Pi}_k$

Proof. Let $[A,a,c,b,B] \in \Pi_h$ . For any $C \in \mathfrak{K}(c) \setminus (\mathfrak{K}(a) \cup \mathfrak{K}(b)))$ we have $[A,a,c,b,B] = [C,c,a,c,b,B] \circ [A,a,c,C] \in \Pi_k \circ \Pi_p \subset \Pi_k \circ \overline{\Pi}_k \subset \overline{\Pi}_k$ by (7.1).

<u>(7.4)</u> Let $\Delta \subset \Pi_a \cup \Pi_h \cup \Pi_k$ , $A, B \in \mathfrak{K}$ . If there is a $\gamma \in \overline{\Delta}$ mapping $A$ onto $B$ with $\gamma^{-1} \in \overline{\Delta}$ then $\Gamma_A(\Delta) \cong \Gamma_B(\Delta)$ . In particular $\Gamma_A(\Delta)$ and $\Gamma_B(\Delta)$ are permutation groups which are isomorphic if $\Delta$ contains one of the sets $\Pi_p$, $\Pi_f$, $\Pi_a$ .

Remark. No example is known of a circle plane containing two circles $A, B \in \mathfrak{K}$ such that $\Gamma_A(\Pi_o)$ and $\Gamma_B(\Pi_o)$ are non-isomorphic.

(7.5) For $A \in \mathfrak{K}$ the permutation set $(A, \Gamma_A(\Pi_{fo}))$ is 3-transitive.

Proof. Let $a \in A$ and $\Gamma_a := \{\gamma \in \Gamma_A(\Pi_{fo}) \mid \gamma(a) = a\}$. It is sufficient to prove that $\Gamma_a$ acts 2-transitively on $A \setminus \{a\}$. Let $B \in \mathfrak{K}(a,A)$, $B \neq A$. In the derivation of $(P, \mathfrak{G}, \mathfrak{K})$ in the point $a$ the perspectivities $[X,a,y,Z] \in \Pi_{fo}$ appear as affine central perspectivities $[X\setminus\{a\} \xrightarrow{y} Z\setminus\{a\}]$. Hence, by (2.8), the subset $\Sigma_a := \{[B,a,y,A][A,a,x,B] \mid x,y \in P^a \setminus (A \cup B)\}$ of $\Gamma_a$ acts 2-transitively on $A\setminus\{a\}$.

In the same way we have as a consequence of (2.10):

(7.6) For $A \in \mathfrak{K}$ the permutation set $(A, \Gamma_A(\Pi_a))$ is 3-transitive.

From (7.4), (7.5) and (7.6) we obtain:

(7.7) Let $A, B \in \mathfrak{K}$ and $\Pi_{fo} \subset \Delta$ or $\Pi_a \subset \Delta$. Then the permutation set $(A, \Gamma_A(\Delta))$ is a 3-transitive permutation group and $\Gamma_A(\Delta) \cong \Gamma_B(\Delta)$.

(7.8) Let $(P, \mathfrak{G}, \mathfrak{K})$ be a miquelian circle plane and $X \in \mathfrak{K}$. Then $(X, \Gamma_X(\Delta))$ fulfills $(P_3)$.

Proof. Let $(P, \mathfrak{G}, \mathfrak{K})$ be represented according to the definition (cf. § 2) by the quadric $Q = P$ in the 3-dimensional projective space $(\overline{P}, \mathfrak{L})$ with $\overline{P}$ and $\mathfrak{L}$ as the sets of points and lines respectively. For a subset $M \subset \overline{P}$ let $\overline{M}$ denote the hull of $M$ in $(\overline{P}, \mathfrak{L})$. For a point $a \in Q = P$ let $T(a)$ denote the tangent plane of $Q$ in a. For $A \in \mathfrak{K}$, $a \in A$ let $\mathfrak{L}(A,a)$ and $\mathfrak{L}(a,a)$ be the pencil of lines passing through $a$ in the plane $\overline{A}$ and $T(a)$ respectively. For a line $L \in \mathfrak{L}$ let $\mathfrak{C}(L)$

denote the pencil of planes containing $L$. Any basic perspectivity $[A,a;b]$ of type 1 induces by means of the bijection $A \longrightarrow \mathfrak{L}(A,a)$; $x \longrightarrow \overline{a,x}$ for $x \nmid a$, $a \longrightarrow T(a) \cap \overline{A}$ a perspectivity $[A,a;b]': \mathfrak{L}(A,a) \longrightarrow \mathfrak{S}(\overline{a,b}); X \longrightarrow \overline{X \cup \{b\}}$.

In the same way any perspectivity $[A;b,B]$ of type 2 induces a perspectivity $[A;b,B]': \mathfrak{L}(A,b) \longrightarrow \mathfrak{S}(T(b) \cap \overline{B})$.

Let $[a,b,c]$ be a basic perspectivity of type 3. Then $[a,b,c]$ induces the perspectivity $[a,b,c]' = \beta\alpha$ with
$\alpha : \mathfrak{S}(\overline{a,c}) \longrightarrow \mathfrak{L}(c,c)$; $E \longrightarrow E \cap T(c)$ and
$\beta : \mathfrak{L}(c,c) \longrightarrow \mathfrak{S}(\overline{c,b})$; $X \longrightarrow \overline{X \cup \{b\}}$.
Since the cross ratio is invariant under the perspectivities $[A,a;b]'$, $[A;b,B]'$, $[a,b,c]'$ and since any projectivity of $\Delta$ is a product of the basic perspectivities of type 1,2,3 the permutation set $\Gamma_X(\Delta)$ fulfills $(P_3)$.

As a corollary of (7.7) and (7.8) we obtain

(7.9) Let $(P,\mathfrak{G},\mathfrak{K})$ be a miquelian circle plane and $X \in \mathfrak{K}$. Then $(X,\Gamma_X(\Delta))$ is sharply 3-transitive if $\Delta$ contains one of the sets $\Pi_{fo}$ and $\Pi_a$.

We have
$\Gamma_X(\Pi_{fo}) \leq \Gamma_X(\Pi_f) \leq \Gamma_X(\Pi_h)$, $\Gamma_X(\Pi_o) \leq \Gamma_X(\Pi_p) \leq \Gamma_X(\Pi_k)$ by (7.1)
$\Gamma_X(\Pi_{fo}) \leq \Gamma_X(\Pi_o)$, $\Gamma_X(\Pi_f) \leq \Gamma_X(\Pi_p)$ by (7.2) and
$\Gamma_X(\Pi_h) \leq \Gamma_X(\Pi_k)$ by (7.3) and any of these groups is 3-transitive. In the miquelian case all these groups are sharply 3-transitive, hence they coincide. From the proof of (7.8) we learn that in the miquelian case the group $\Gamma_X(\Pi_k)$ is isomorphic to the $PGL(2,K)$, where $K$ denotes the coordinatizing field of $(P,\mathfrak{G},\mathfrak{K})$.

From (7.5) one obtains immediately

(7.10) Let $(P,\mathfrak{G},\mathfrak{K})$ be a circle plane such that every derivation is desarguesian and $X \in \mathfrak{K}$. Then $\Gamma_X(\Pi_a) \leq \Gamma_X(\Pi_f)$.

Now we report on the following problem:
Let $\Gamma$ be one of the groups mentioned above. For which $n \in \mathbb{N}$ does the condition $(P_n)$ imply that the considered circle plane is miquelian?

M. FUNK proved in his thesis [12]

(7.11) For any free circle plane (Möbius, Laguerre, or Minkowskian plane) the groups $\Gamma_X(\Pi_p)$ and $\Gamma_X(\Pi_a)$ ful-fulfill $(P_6)$ and $(P_5)$ respectively.

(7.12) If for a circle plane $(P,\mathfrak{G},\mathfrak{K})$ and a circle $X \in \mathfrak{K}$ the group $\Gamma_X(\Pi_p)$ fulfills $(P_4)$ then $(P,\mathfrak{G},\mathfrak{K})$ is miquelian.

This theorem covers the corresponding theorems of H. FREUDENTHAL and K. STRAMBACH [13] and H.-J. KROLL [28] where $(P_3)$ was assumed for $\Gamma_X(\Pi_p \cup \Pi_a)$ and $\Gamma_X(\Pi_p)$ respectively.

A lemma in M. FUNK's proof of (7.12) is

(7.13) If $\Gamma_X(\Pi_f)$ fulfills $(P_5)$ then the derivation of the circle plane at any point is pappian.

For the affine group $\Gamma_X(\Pi_a)$ we have [29],[30]:

(7.14) Let $(P,\mathfrak{G},\mathfrak{K})$ be a circle plane and $X \in \mathfrak{K}$ such that $\Gamma_X(\Pi_a)$ fulfills $(P_3)$. Then the derivation at any point is pappian. Moreover, $(P,\mathfrak{G},\mathfrak{K})$ is miquelian in the case of a finite Möbius or Laguerre plane of odd order or a Minkowskian plane.

## § 8 THE VON STAUDT GROUP GENERATED BY BASIC PERSPECTIVITIES OF TYPE 4

For a distinct circle $A \in \mathfrak{K}$ and a distinct point $u \in P\backslash A$ let $v_i := A \cap [u]_i$, $i \in I$. We consider the permutation group $(A, \Gamma_A^u)$ generated by the set $\Delta_A^u := \{[A,u,x,A] \mid x \in P\backslash(A \cup [u]\}$ of basic perspectivities of type 4 and the permutation group $(A, \Gamma_A)$ generated by all groups $\Gamma_A^u$ with $u \in P\backslash A$. For $a,b \in A\backslash[u]$ with $a \neq b$ and $B := \mathfrak{K}(a,b,u)$ let

$\Gamma_{(a,b)} := \{\sigma \in \Gamma_A \mid \sigma(a) = a, \sigma(b) = b\}$,

$\Gamma^u_{(a,b)} := \{\sigma \in \Gamma^u_A \mid \sigma(a) = a, \sigma(b) = b\}$,

$\Delta^u_{A,a} := \{[A,u,x,A] \mid x \in \mathfrak{K}(a,A,u) \backslash\{a,u\}\}$,

$\Delta^u_{a,b} := \{[A,u,x,A] \mid x \in B\backslash\{a,b,u\}\}$.

<u>(8.1)</u> (a) $(A\backslash\{a,b\}, \Delta^u_{a,b} \circ \Delta^u_{a,b})$ is a transitive permutation set.

(b) $(A\backslash\{a,b\}, \Gamma^u_{(a,b)})$ is a transitive permutation group.

(c) $(A, \Gamma^u_A)$ and $(A, \Gamma_A)$ are 3-transitive permutation groups.

Proof. Since $\Delta^u_{a,b} \circ \Delta^u_{a,b} \subset \Gamma^u_{(a,b)} \subset \Gamma^u_A \subset \Gamma_A$ it is sufficient to prove (a). We consider the derivation in the point $u$, which is an affine plane. The circle $B = \mathfrak{K}(u,a,b)$ appears as a line. Since $|A| \geq 7$ there is a point $x \in A\backslash\{a,b,c,d\}$ such that none of the joining lines $\overline{c,x}$ and $\overline{d,x}$ is parallel to the line $B$.

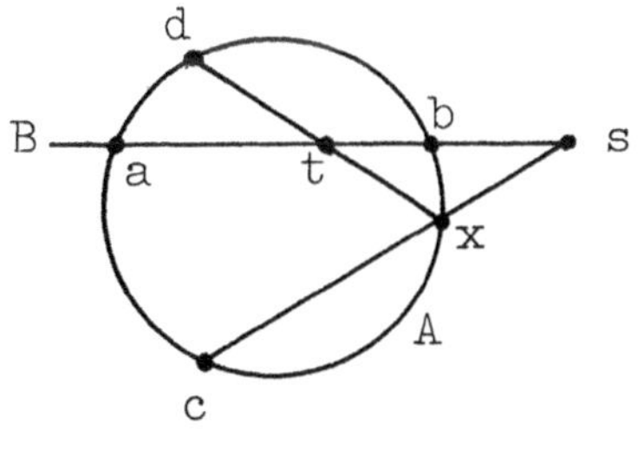

Fig. 24

Let $s := B \cap \overline{c,x}$ and $t := B \cap \overline{d,x}$. Then the permutation $[A,u,t,A] \circ [A,u,s,A] \in \Delta^u_{a,b} \circ \Delta^u_{a,b}$ maps c onto d.

(c) is obvious for Möbius planes and easy to show for Laguerre planes. In the case of a Minkowskian plane let (a,b,c) and (a',b',c') be two triples consisting of distinct points of A. If $b',c' \neq v_1,v_2$ then by (a) there exist $\sigma,\tau \in \Gamma^u_A$ with $\sigma(c) = c'$ and $\tau(c') = c'$, $\tau\sigma(b) = b'$ and $\rho \in \Delta^u_{b',c'} \circ \Delta^u_{b',c'}$ with $\rho(\tau\sigma(a)) = a'$. If $b' = v_1$ and $c' = v_2$ we choose a $\gamma \in \Gamma^u_A$ with $\gamma(v_1), \gamma(v_2) \neq v_1, v_2$. Then there exists an $\alpha \in \Gamma^u_A$ with $\alpha(a) = \gamma(a')$, $\alpha(b) = \gamma(v_1)$ and $\alpha(c) = \gamma(v_2)$. Hence the permutation $\beta := \gamma^{-1}\alpha$ maps the triple (a,b,c) onto the triple (a',b',c').

<u>(8.2)</u> $(A\backslash\{a\}, \Delta^u_{Aa} \circ \Delta^u_{Aa})$ is a transitive permutation set such that only the identity has fixed points.

Proof. Let $x,y \in A\backslash\{a\}$ and $T := \mathfrak{K}(a,A,u)$. Since $|A| \geq 6$ there is a point $z \in A\backslash\{a,x,y\}$ such that for $X := \mathfrak{K}(x,z,u)$ and $Y := \mathfrak{K}(y,z,u)$ we have $|X \cap T| = |Y \cap T| = 2$. For $r := (X \cap T)\backslash\{u\}$ and $s := (Y \cap T)\backslash\{u\}$ the permutation $[A,u,s,A] \circ [A,u,r,A] \in \Delta^u_{Aa} \circ \Delta^u_{Aa}$ maps x onto y. Obviously only the identity has fixed points in $A\backslash\{a\}$.

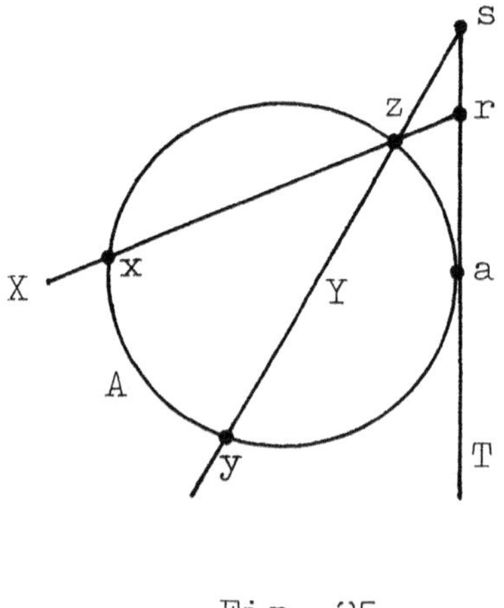


Fig. 25

From now on we assume the validity of $(P_3)$ for $(A,\Gamma^u_\Delta)$. Then we can prove the following theorems:

(8.3) Let $a,b \in A\setminus[u]$ with $a \neq b$, $B := \mathfrak{K}(a,b,u)$ and $\omega := [A,u,B,A]$ the (involutory) basic perspectivity of type 5 with direction $B$. Then:

a) there exist $\alpha,\beta,\gamma \in \Delta^u_{a,b}$ with $\alpha\beta\gamma = \omega$;

b) for any $\rho,\sigma,\tau \in \Delta^u_{a,b} \cup \{\omega\}$ we have $\rho\sigma\tau \in \Delta^u_{a,b} \cup \{\omega\}$;

c) $(A\setminus\{a,b\},\Gamma^u_{(a,b)})$ is a commutative regular permutation group.

Proof. First we prove the weaker statement:

b') for any $\alpha,\beta,\gamma \in \Delta^u_{a,b}$ we have $\alpha\beta\gamma \in \Delta^u_{a,b} \cup \{\omega\}$.

Let $\alpha = [A,u,r,A]$, $\beta = [A,u,s,A]$, $\gamma = [A,u,t,A]$ where $r,s,t$ are points of $B\setminus\{a,b,u\}$. For $x \in A\setminus\{a,b\}$, let $x' := \alpha\beta\gamma(x)$ and $X := \mathfrak{K}(u,x,x')$ if $x \neq x'$ and $X := \mathfrak{K}(u,A,x)$ if $x = x'$.

Case 1. There is an $x \in A\setminus\{a,b\}$ with $X \cap B \neq \{u\}$. For $q := (X \cap B)\setminus\{u\}$ we have $\delta := [A,u,q,A] \in \Delta^u_{a,b}$ and $\delta(x) = x$. Hence the permutation $\delta\alpha\beta\gamma$ fixes the three distinct points $a,b,x$ and by property $(P_3)$ $\delta\alpha\beta\gamma$ is the identity, i.e. $\alpha\beta\gamma = \delta \in \Delta^u_{a,b}$.

Case 2. For all $x \in A\setminus\{a,b\}$ we have $X \cap B = \{u\}$. Then the mapping $\alpha\beta\gamma$ coincides with the (involutory) basic perspectivity $\omega = [A,u,B,A]$ of type 5.

a) Let $x \in A\setminus\{a,b\}$, $X := \mathfrak{K}(u,B,x)$. $x' := (X \cap A)\setminus\{x\}$ if $X \cap A \neq \{x\}$ and $x' := x$ if $X \cap A = \{x\}$. Furthermore let $t \in B\setminus\{a,b,u\}$, $\gamma := [A,u,t,A]$ and $x_1 := \gamma(x)$. Since $|B| \geq 6$ there is an $s \in B\setminus\{a,b,u,t\}$ with $x' \notin \mathfrak{K}(u,s,x_1)$. Then $\beta := [A,u,s,A] \in \Delta^u_{a,b}$, $x_2 := \beta(x_1) \neq x,x'$, $r := B \cap \mathfrak{K}(u,x_2,x')\setminus\{u\} \neq \emptyset$ and

$\alpha := [A,u,r,A] \in \Lambda^u_{a,b}$. The permutation $\alpha\beta\gamma$ maps $x$ onto $x'$ and since $X \cap B = \{u\}$ we have $\alpha\beta\gamma = \omega$ by b')).

b). By a) there exist $\alpha,\beta,\gamma \in \Lambda^u_{a,b}$ with $\omega = \alpha\beta\gamma$. By b') it is enough to consider the cases:

1. $\rho = \omega\wedge\sigma, \tau \neq \omega$. We have $\omega\sigma\tau = \alpha\beta\gamma\sigma\tau$ and $\beta\gamma\sigma \in \Delta^u_{a,b} \cup \{\omega\}$ by a). If $\beta\gamma\sigma \in \Delta^u_{a,b}$ we are finished by a). Let $\beta\gamma\sigma = \omega$. Then $\omega\sigma\tau = \alpha\omega\tau = \alpha\alpha\beta\gamma\tau = \beta\gamma\tau \in \Delta^u_{a,b} \cup \{\omega\}$.
2. $\sigma = \omega\wedge\rho, \tau \neq \omega$. If $\rho\alpha\beta \in \Delta^u_{a,b}$ we are done by a). If $\rho\alpha\beta = \omega$ we have $\rho\sigma\tau = \rho\omega\tau = \rho\alpha\beta\gamma\tau = \omega\gamma\tau = \alpha\beta\gamma\gamma\tau = \alpha\beta\tau \in \Delta^u_{a,b} \cup \{\omega\}$.
3. $\rho = \tau = \omega\wedge\sigma \neq \omega$. Then $\omega\sigma\omega = \omega\sigma\alpha\beta\gamma$. If $\sigma\alpha\beta = \omega$ then $\omega\sigma\omega = \gamma$. If $\sigma\alpha\beta \in \Delta^u_{a,b}$ we have case 1.

Since $\Delta^u_{a,b}$ consists of involutions, by b) any two elements $\alpha\beta, \gamma\delta \in \Delta^u_{a,b} \circ \Delta^u_{a,b}$ commute: $(\alpha\beta\gamma)\delta = \gamma(\beta\alpha\delta) = \gamma\delta\alpha\beta$. Together with (8.1.b) and $(P_3)$, proposition c) is proved.

<u>(8.4)</u> a) The set $\Sigma^u_A := \{[A,u,C,A] \mid C \in \mathfrak{K}(u)\}$ of basic perspectivities of type 5 is contained in $\Gamma^u_A$.

b) For a Minkowskian plane $\alpha,\beta,\gamma \in \Sigma^u_A$ implies $\alpha\beta\gamma \in \Sigma^u_A$.

c) In a Laguerre plane $\alpha,\beta,\gamma \in \Sigma^u_A$ implies $\alpha\beta\gamma \in \Sigma^u_A \cup \{id\}$.

Proof. a) Case 1. There exists a $B \in \mathfrak{K}(u,C) \cap \mathfrak{K}$ with $|B \cap A| = 2$; let $\{a,b\} := B \cap A$. Then $a,b \notin [u]$ and $[A,u,C,A] = [A,u,B,A] \in (\Delta^u_{a,b})^3$ by (8.3.a)); hence $[A,u,C,A] \in \Gamma^u_A$.

Case 2. For all $X \in \mathfrak{K}(u,C) \cap \mathfrak{K}$ we have $|X \cap A| \leq 1$. Then for all Möbius and Laguerre planes $[A,u,C,A]$ is the identity. In a Minkowskian plane $[A,u,C,A]$ interchanges $v_1$ and $v_2$ ($\{v_1,v_2\} := A \cap [u]$) and fixes all other points of $A$. Let $x_1,x_2,x_3 \in A\backslash\{v_1,v_2\}$ be distinct, $X_1 := \mathfrak{K}(u,x_2,x_3)$ , $X_2 := \mathfrak{K}(u,x_3,x_1)$ , $X_3 := \mathfrak{K}(u,x_1,x_2)$ , $\xi_i := [A,u,X_i,A]$ and $T := \mathfrak{K}(u,C,x_1)$ . Then by case 1, all $\xi_i \in \Gamma_A^u$ and $\sigma := \xi_1\xi_2\xi_3 \in \Gamma_A^u$ fixes $x_1$ and interchanges $v_1$ and $v_2$. Since $|A| \geq 5$ there is a $y \in A\setminus\{v_1,v_2\}$ with $\sigma(y) \neq y$ because of $(P_3)$. For $Y := \mathfrak{K}(u,y,\sigma(y))$ and $\eta := [A,u,Y,A]$ we have $\eta \in \Gamma_A^u$ and the map $\eta\sigma \in \Gamma_A^u$ fixes the three distinct points $v_1,v_2,y$ ; hence $\eta = \sigma = \xi_1\xi_2\xi_3$ and therefore $\eta(x_1) = x_1$ , i.e. $T' := \mathfrak{K}(u,Y,x_1) \in \mathfrak{K}$ and $T' \cap A = \{x_1\}$ . Since $x_1 \neq v_1,v_2$ we have $T \in \mathfrak{K}(u,C) \cap \mathfrak{K}$ , hence $T \cap A = \{x_1\}$ . From $T,T' \in \mathfrak{K}(u)$ and $T \cap A = T' \cap A = \{x_1\}$ we obtain $T = T'$ ; hence $Y \in \mathfrak{K}(u,T') = \mathfrak{K}(u,T) = \mathfrak{K}(u,C)$ . $Y \in \mathfrak{K} \cap \mathfrak{K}(u,C)$ and $Y \cap A = \{y,\sigma(y)\}$ with $\sigma(y) \neq y$ contradicts the assumption of case 2 .

b) By a) $\sigma := \alpha\beta\gamma \in \Gamma_A^u$. Since in a Minkowskian plane any element of $\Sigma_A^u$ interchanges $v_1$ and $v_2$ we have $\sigma(v_1) = v_2$ and $\sigma(v_2) = v_1$ and by $(P_3)$ there is an $x \in A\setminus\{v_1,v_2\}$ with $\sigma(x) \neq x$. Let $X := \mathfrak{K}(u,x,\sigma(x)) \in \mathfrak{K}$ and $\tau := [A,u,X,A] \in \Sigma_A^u$. Then $\tau\sigma$ fixes the four distinct points $v_1,v_2,x,\sigma(x)$ ; hence $\tau\sigma = \mathrm{id}$ by $(P_3)$, i.e. $\sigma = \tau \in \Sigma_A^u$.

c) In a Laguerre plane any element of $\Sigma_A^u$ fixes $v = A \cap [u])$ , and so also $\sigma := \alpha\beta\gamma$ which is contained in $\Gamma_A^u$ by a). If $\sigma \neq \mathrm{id}$ , there is an $x \in A\setminus\{v\}$ with $\sigma(x) \neq x$. Then $X := \mathfrak{K}(u,x,\sigma(x)) \in \mathfrak{K}$ , $\tau := [A,u,X,A] \in \Sigma_A^u$ and $\tau\sigma$ fixes $v,x,\sigma(x)$ ; hence $\sigma = \tau \in \Sigma_A^u$.

<u>(8.5)</u> $(A,\Gamma_A^u)$ is a sharply 3-transitive permutation group and the associated KT-field $(F,+,\cdot,\sigma)$ (cf. § 1, (1.1),(1.2),(1.3)) is a commutative field (hence $\sigma(x) = x^{-1}$ for $x \in F\setminus\{0\}$) .

Proof. By $(P_3)$ and (8.1.c)) $(A, \Gamma_A^u)$ is sharply 3-transitive. For the construction of the associated near domain $(F,+,\cdot)$ one has to choose three distinct points $0, 1, \infty \in A$, put $F := A \setminus \{\infty\}$ and consider the group $G := \{\gamma \in \Gamma_A^u \mid \gamma(\infty) = \infty\}$ which acts sharply 2-transitively on $F$. The multiplication is determined by the group $\Gamma_{(0,\infty)}^u =: G_0$. By (8.3.c)) this group is commutative and hence by (1.6) $(F,+,\cdot)$ is a commutative field.

<u>(8.6)</u> a) $\alpha, \beta, \gamma \in \Delta_{Aa}^u$ implies $\alpha\beta\gamma \in \Delta_{Aa}^u \cup \{id\}$.

b) $(A \setminus \{a\}, \Delta_{Aa}^u \circ \Delta_{Aa}^u)$ is a commutative regular permutation group and $\Delta_{Aa}^u \circ \Delta_{Aa}^u$ is isomorphic to the additive group $(F,+)$.

Proof. b) Since $\Gamma_A^u$ is 3-transitive we may assume $a = \infty$. We know that any element $\alpha \in G := \{\gamma \in \Gamma_A^u \mid \gamma(\infty) = \infty\}$ has the representation $\alpha(x) = a + mx$ with $a, m \in F$ and $m \neq 0$, and that $\alpha$ is fixed point free in $F := A \setminus \{\infty\}$ if and only if $a \neq 0$ and $m = 1$ (notice that $(F,+,\cdot)$ is a commutative field). Since $\Delta_{Aa}^u \circ \Delta_{Aa}^u \subset G$ we have $\Delta_{Aa}^u \circ \Delta_{Aa}^u = \{p^+ : F \longrightarrow F;\ x \longrightarrow p + x \mid p \in F\}$ by (8.2), which proves b).

a) We have $\sigma := \alpha\beta\gamma \in G$ and by b) $\sigma^2 = \alpha(\beta\gamma) \circ (\alpha\beta)\gamma = \alpha\alpha\beta\beta\gamma\gamma = id$. Hence, if $\sigma(x) = a + mx$, then $m = -1$ for $\mathrm{char}(F) \neq 2$ and $m = 1$ for $\mathrm{char}(F) = 2$. By calculation we obtain now a).

<u>(8.7)</u> The set $\Delta_A^u \cup \Sigma_A^u$ (cf.(8.4)) contains all involutions of $\Gamma_A^u$ and fulfills the following three reflection theorem:

(*) $\alpha, \beta, \xi_i \in \Delta_A^u \cup \Sigma_A^u$, $i \in \{1,2,3\}$ with $\alpha \neq \beta$ and $\alpha\beta\xi_i$ involutory implies $\xi_1\xi_2\xi_3 \in \Delta_A^u \cup \Sigma_A^u \cup \{id\}$.

Proof. Let $\sigma$ be an involution of $\Gamma_A^u$, $b \in A\setminus[u]$ with $b \neq \sigma(b)$, $B := \mathfrak{K}(b,\sigma(b),u)$, $c \in A\setminus[u]$ with $c \neq b,\sigma(b)$ and $C := \mathfrak{K}(c,\sigma(c),u)$ for $c \neq \sigma(c)$ and $C := \mathfrak{K}(c,A,u)$ for $c = \sigma(c)$. If $r := B \cap C \setminus \{u\} \neq \emptyset$ then $\sigma = [A,u,r,A] \in \Delta_A^u$; if $B \cap C = \{u\}$ then $\sigma = [A,u,B,A] \in \Sigma_A^u$ by $(P_3)$. Since $\Gamma_A^u$ is equal to the projectivity group PGL(2,F) of the commutative field $(F,+,\cdot)$ and the set of all involutions of PGL(2,F) fulfills (*), theorem (8.7) is proved.

With theorems (8.3.b)), (8.4), (8.6.a)), (8.7) and the methods of [23] (cf. p. 43f and 50f) we can prove the following theorem:

<u>(8.8)</u> For the derivation $(P,\mathfrak{G},\mathfrak{K})^u$ in the point $u$ let $D := P^u$, $\mathfrak{D} := \bigcup_{i\in I} \mathfrak{G}_i^u \cup \mathfrak{K}^u$, $(\overline{D},\overline{\mathfrak{D}})$ the projective closure of the affine plane $(D,\mathfrak{D})$ and $U$ the line of infinity. Furthermore let $(F,+,\cdot)$ be the commutative field (cf. (8.5)) associated with $(A,\Gamma_A^u)$. Then the point set $\overline{D}$ can be provided with a line structure $\mathfrak{L}$ such that $(\overline{D},\mathfrak{L})$ is a pappian projective plane having $(F,+,\cdot)$ as coordinate field in which $\overline{A}$ (cf. § 7) is a conic and $X \in \mathfrak{L}$ or $X \in \overline{\mathfrak{D}}$ with $X \cap \overline{A} \neq \emptyset$ implies $X \in \overline{\mathfrak{D}}$ or $X \in \mathfrak{L}$ respectively.

For Möbius planes we have:
Let $w_1,w_2 \in U$ with $w_1 \neq w_2$, $W \in \mathfrak{L}$ with $w_1,w_2 \in W$. Then $\overline{A} = A$ is an ellipse in the pappian affine plane, which we obtain by omitting $W$ from the projective plane $(\overline{D},\mathfrak{L})$.

For Laguerre planes we have:
$U \in \overline{\mathfrak{D}} \cap \mathfrak{L}$ is a tangent of $\overline{A} = (A\setminus[u]) \cup \{u_1\}$ and $A\setminus[u]$ is a parabola in the affine plane obtained by omitting $U$ from $(\overline{D},\mathfrak{L})$.

For Minkowskian planes $U \in \mathfrak{D} \cap \mathfrak{L}$ is a secant of $\overline{A} = (A \setminus [u]) \cup \{u_1, u_2\}$ and $A \setminus [u]$ is a hyperbola in the affine plane obtained by omitting $U$ from $(\overline{D}, \mathfrak{L})$.

Problem: Find two non-isomorphic projective planes $(P, \mathfrak{G})$ and $(P, \mathfrak{L})$ with the same point set $P$ and a subset $A \subset P$ such that $(P, \mathfrak{G})$ is pappian, $A$ is a non-degenerate conic of $(P, \mathfrak{G})$, $X \in \mathfrak{G}$ with $X \cap A \neq \emptyset$ implies $X \in \mathfrak{L}$, and $X \in \mathfrak{L}$ with $X \cap A \neq \emptyset$ implies $X \in \mathfrak{G}$.

At the end of this paragraph we want to illustrate the perspectivities of type 4 and type 5 in the case of an ovoidal Möbius plane. Let $\mathcal{O}$ be an ovoid in a 3-dimensional projective space $(P, \mathfrak{L})$ and $(\mathcal{O}, \mathfrak{K})$ with $\mathfrak{K} = \{\mathcal{O} \cap E \mid E \text{ a plane}, |E \cap \mathcal{O}| \geq 2\}$ the Möbius plane determined by $\mathcal{O}$. For a subset $M \subset P$ let $\overline{M}$ denote the hull of $M$ in the projective space $(P, \mathfrak{L})$. For any point $a \in \mathcal{O}$ let $T(a)$ denote the tangent plane of $\mathcal{O}$ in $a$. Let $A \in \mathfrak{K}$, $b, c \in \mathcal{O} \setminus A$ with $b \neq c$, $s := \overline{b,c} \cap \overline{A}$ and $\gamma := [A, b, c, A]$. Then $A$ is an oval in the projective plane $\overline{A}$ and for $x \in A$ it holds

$$\{x, \gamma(x)\} = A \cap \{\overline{b,c,x}\} = A \cap \overline{s,x}$$

i.e. $\gamma$ coincides with the point reflection

$$\tilde{s} : A \longrightarrow A;\ x \longrightarrow \begin{cases} A \cap \overline{s,x} \setminus \{x\} & \text{for } |A \cap \overline{s,x}| = 2 \\ x & \text{for } |A \cap \overline{s,x}| = 1 \end{cases}.$$

Since $b, c$ are two distinct points of $\mathcal{O}$ we have $s \in \overline{A} \setminus T(b)$.

Also any perspectivity $[A, b, B, A]$ of type 5 and the point reflection $\tilde{t}$ at the point $t = \overline{A} \cap T(b) \cap \overline{B}$ coincide. In this way one obtains any point reflection $\tilde{s} : A \to A$ with $s \in \overline{A} \setminus A$, namely with $c = \overline{b,s} \cap \mathcal{O} \setminus \{b\}$ if $s \notin T(b)$ and with $B = \{\overline{b,s,d}\} \cap \mathcal{O}$ where $d \in \overline{A} \setminus T(b)$ if $s \in T(b)$. Hence for $u \in \mathcal{O} \setminus A$ the group $\Gamma_A^u$ coincides with the group generated by the point reflections $\tilde{s}$ with $s \in \overline{A} \setminus (T(u) \cup A)$. If $(P, \mathfrak{L})$ is pappian and $A$ a conic in $\overline{A}$, then $\Gamma_A = \Gamma_A^u$ is isomorphic to the PGL(2,K), where K denotes the coordinatizing field of $(P,\mathfrak{L})$.

Now let $(P,\mathfrak{L})$ be the projective closure of the 3-dimensional euclidean space $\mathbb{R}^3$ and C an oval in the (x,z)-plane which is symmetric with respect to the z-axis. Then the surface of revolution $\mathcal{O}$ obtained by revolving C about the z-axis is an ovoid which carries both conics and ovals which are not conics. If $A \in \mathfrak{K}$ is not a conic in $\overline{A}$, then the group generated by the point reflections is not sharply 3-transitive (cf.(8.8)). Hence this gives us an example of an ovoidal Möbius plane containing two circles $A,B \in \mathfrak{K}$ such that the groups $\Gamma_A$ and $\Gamma_B$ are non-isomorphic.

## § 9 BURAU GEOMETRIES

In § 7 and § 8 we studied perspectivities in plane circle geometries, in particular in Möbius planes. A generalization of Möbius planes to higher dimensions are the Burau geometries defined by H. KARZEL, G. KIST and H.-J. KROLL [21]. Von STAUDT introduced the concept of a chain to describe the circles in the compactified Gaußian plane, which can be considered as the projective line $\Pi(\mathbb{C}^2,\mathbb{C})$ over the field $\mathbb{C}$ of complex numbers. The circle determined by three distinct points a,b,c is the set of all points x such that the cross-ratio DV(a,b,c,x) is real.

The von Staudt concept of a chain was extended by C. JUEL [17] for the projective plane $\Pi(\mathbb{C}^3,\mathbb{C})$ and by W. BURAU [11] for projective spaces $\Pi(\mathbb{C}^{n+1},\mathbb{C})$ of arbitrary dimension n. BURAU stated and proved the fundamental theorems of the theory of chains.

The abstract axiomatic definition of a Burau geometry given by KARZEL, KIST, KROLL is the following:

Let $(P,\mathfrak{G})$ be a projective space with $\dim(P,\mathfrak{G}) \geqslant 2$ and $\mathfrak{K}$ a subset of the power set of $P$. The elements of $\mathfrak{K}$ are called <u>circles</u>. For $X \subset P$ the projective closure in $(P,\mathfrak{G})$ is denoted by $\overline{X}$. The triple $(P,\mathfrak{G},\mathfrak{K})$ is called a <u>Burau geometry</u> if the following axioms hold:

<u>B1</u> The projective closure $\overline{C}$ of any circle $C \in \mathfrak{K}$ is a line of the projective space $(P,\mathfrak{G}) : \overline{C} \in \mathfrak{G}$.

<u>B2</u> For any line $M \in \mathfrak{G}$ the pair $(M,\mathfrak{K}(M))$ with $\mathfrak{K}(M) := \{C \in \mathfrak{K} \mid C \subset M\}$ is a Möbius plane.

<u>B3</u> Let $a,b,c,d,e,f \in P$ be six distinct points such that $c = \overline{a,b} \cap \overline{d,e}$ and $f = \overline{a,e} \cap \overline{b,d}$. Then any line $X \in \mathfrak{G}$ with $f \in X$ and intersecting the circle $C \in \mathfrak{K}$ determined by $a,b,c$ intersects also the circle $D \in \mathfrak{K}$ through $c,d,e$.

Remark: Axiom <u>B3</u> tells us that a central perspectivity from a line $A$ into a line $B$ is circle preserving (cf. [21] (3.3)).

From the results of [21] one can obtain the following theorem:

<u>(9.1)</u> Let $(P,\mathfrak{G},\mathfrak{K})$ be a Burau geometry with $\dim(P,\mathfrak{G}) \geqslant 2$. Then:

a) in the projective space $(P,\mathfrak{G})$ the von STAUDT group $\Gamma$ consisting of all projectivities which map a distinct line $A \in \mathfrak{G}$ onto itself fulfills $(P_3)$;

b) for any line $A \in \mathfrak{G}$ the Möbius plane $(A,\mathfrak{K}(A))$ is miquelian.

BIBLIOGRAPHY

[1] ARTZY,R.: Eigenschaften von ebenen Viergeweben allgemeiner Lage. Math. Ann. 126 (1953) 336-342.

[2] ARTZY,R.: A Pascal theorem applied to Minkowski-geometry. J. Geometry 3 (1973) 93-105.

[3] BENZ,W.: Permutations and plane sections of a ruled quadric. In: Symposia Mathematica, Istituto Nazionale di Alta Matematica, Vol. V (1970) 325-339.

[4] BENZ,W.: Vorlesungen über Geometrie der Algebren. Berlin, Heidelberg, New York 1973.

[5] BENZ,W. and ELLIGER,S.: Über die Funktionalgleichung $f(1+x)+f(1+f(x))=1$. Aequationes Math. 1 (1968) 267-274.

[6] BLASCHKE,W.: Topologische Fragen der Differentialgeometrie. I. Thomsens Sechseckgewebe. Zueinander diagonale Netze. Math. Z. 28 (1928) 150-157.

[7] BLASCHKE,W. and BOL,G.: Geometrie der Gewebe. Berlin 1938.

[8] BOL,G.: Topologische Fragen der Differentialgeometrie.(65. Gewebe und Gruppen. Math. Ann. 114 (1937) 414-431.

[9] BONETTI,F. and MARCHI,M.: Proiettività di una retta in un S-spazio. Bolletino Unione Italiana. (to appear).

[10] BUEKENHOUT,F.: Ensembles quadriques des espaces projectifs. Math. Z. 110 (1969) 306-318.

[11] BURAU,W.: Mehrdimensionale projektive und höhere Geometrie. Berlin 1961.

[12] FUNK,M.: Regularität in Benz-Ebenen. Dissertation Univ. Erlangen-Nürnberg 1980.

[13] FREUDENTHAL,H. and STRAMBACH,K.: Schließungssätze und Projektivitäten in der Möbius- und Laguerre-Geometrie. Math. Z. 143 (1975) 213-234.

[14] HALDER,H. und HEISE,W.: Kombinatorik. München Wien 1976.

[15] HEISE,W. und KARZEL,H.: Symmetrische Minkowski-Ebenen. J. Geometry 3 (1973) 5-20.

[16] HEISE,W. und SÖRENSEN,K.: Freie Minkowski-Ebenenerweiterungen. J. Geometry 3 (1973) 1-4.

[17] JUEL,C.: Vorlesung über projektive Geometrie mit besonderer Berücksichtigung der v. Staudt'schen Imaginärtheorie. Berlin 1934.

[18] KARZEL,H.: Inzidenzgruppen. Vorlesungsausarbeitung von I. Pieper und K. Sörensen. Univ. Hamburg 1965.

[19] KARZEL,H.: Zusammenhänge zwischen Fastbereichen, scharf zweifach transitiven Permutationsgruppen und 2-Strukturen mit Rechtecksaxiom. Abh. Math. Sem. Univ. Hamburg 32 (1968) 191-206.

[20] KARZEL,H.: Symmetrische Permutationsmengen. Aequationes Math. 17 (1978) 83-90.

[21] KARZEL,H., KIST,G. und KROLL,H.-J.: Burau-Geometrien. Resultate der Mathematik 2 (1979) 88-104.

[22] KARZEL,H., SÖRENSEN,K. und Windelberg,D.: Einführung in die Geometrie. Göttingen 1973.

[23] KARZEL,H. und SÖRENSEN,K.: Die lokalen Sätze von Pappus und Pascal. Mitt. der Math. Gesellsch. Hamburg, Bd. X, Heft 1, (1971) 28-39.

[24] KARZEL,H. und KROLL,H.-J.: Gruppen von Projektivitäten in Zwei- und Hyperbelstrukturen. Lenz-Festband, Berlin, FU, Fachbereich Math. (1976) 125-134.

[25] KERBY,W.: On infinite sharply multiply transitive groups. Hamb. Math. Einzelschriften 6, Göttingen 1974.

[26] KERBY,W. und WEFELSCHEID,H.: Über eine scharf 3-fach transitiven Gruppen zugeordnete algebraische Struktur. Abh. Math. Sem. Univ. Hamburg 37 (1972) 225-235.

[27] KIST,G.: Zur Theorie der verallgemeinerten kinematischen Räume. Habilitationsschrift TU München, Mai 1980.

[28] KROLL,H.-J.: Die Gruppe der eigentlichen Projektivitäten in Benz-Ebenen. Geometriae Dedicata 6 (1977) 407-413.

[29] KROLL,H.-J.: Perspektivitäten in Benz-Ebenen. Beiträge zur Geometrischen Algebra (hrsg. von H.-J. ARNOLD, W. BENZ, H. WEFELSCHEID). Basel, Stuttgart (1977) 203-207.

[30] KROLL,H.-J.: Möbius planes with sharply 3-transitive groups of projectivities. To appear in the proceedings of the symposium on foundations of geometry, 1978 Silivri. Istanbul 1980.

[31] REIDEMEISTER,K.: Topologische Fragen der Differentialgeometrie. V. Gewebe und Gruppen. Math. Z. 29 (1929) 427-435.

[32] REIDEMEISTER,K.: Vorlesungen über Grundlagen der Geometrie. Berlin 1930.

[33] THOMSEN,G.: Un teorema topologico sulle schiere di curve e una caratterizzazione geometrica delle superficie isotermo-asintotiche. Boll. Un. Mat. Ital. 6 (1927) 80-85.

[34] THOMSEN,G.: Topologische Fragen der Differentialgeometrie. XII. Schnittpunktsätze in ebenen Geometrien. Abh. Math. Sem. Univ. Hamburg 7 (1930) 99-106.

[35] WEFELSCHEID,H.: Die Automorphismengruppe der Hyperbelstrukturen. Beiträge zur geometrischen Algebra: proceedings d. Symposiums über Geometrische Algebra vom 29.3.-3.4.1976 in Duisburg hrsg. von H.-J. ARNOLD, W. BENZ, H. WEFELSCHEID, Basel, Stuttgart (1977) 337-343.

[36] WEFELSCHEID,H.: Zur Charakterisierung einer Klasse von Hyperbelstrukturen. J. Geometry 9 (1977) 127-133.

CROSS-RATIOS IN PROJECTIVE AND AFFINE PLANES

J. C. Ferrar

The Ohio State University

In this chapter, the concepts of cross-ratio and harmonic position in Moufang planes are discussed, generalizing classical concepts for desarguesian planes. Beginning with algebraic, rather than geometrical formulations of these concepts, we study the group of harmonicity preserving permutations of the points on a line in a Moufang plane. The main result is a generalization of von Staudt's theorem relating this group to the group of projectivities of the line. A substantial portion of this exposition is devoted to the resolution of algebraic difficulties which arise when the classical results are reinterpreted in a plane coordinatized by a properly alternative, rather than associative, division ring.

## 0. INTRODUCTION

The classical definition of the cross ratio [9, p. 117] based on interplay between homogeneous and inhomogeneous coordinates is not available in a proper (non-desarguesian) plane lacking homogeneous coordinates. We take as definition the strictly algebraic condition

$$(A,B:C,D) = \ulcorner((a-d)^{-1}(b-d))((b-c)^{-1}(a-c))] \qquad (1)$$

where $A = (a,0)$, $B = (b,0)$, $C = (c,0)$, $D = (d,0)$ and $\ulcorner x]$ denotes the conjugacy class of $x$ in the coordinate ring. With suitable provision for the cases when one of A, B, C, or D is the ideal point on the line $y = 0$, this definition provides a basis for deriving in the Moufang plane many properties of the classical cross ratio (§§2, 3). The cross ratio provides an



*P. Plaumann and K. Strambach (eds.), Geometry - von Staudt's Point of View, 101–125.*

algebraic formulation of the concept of harmonic position for point quadruples (§§4, 5) which leads to a generalization of von Staudt's theorem on the group of harmonicity preserving transformations of the projective line (Theorem 9).

Throughout the chapter we are led in our algebraic considerations by the view that the natural structure for the coordinate ring of a projective line with harmonicity is that of a Jordan ring. We thus view inversion as the operation of central importance in the proof of von Staudt's theorem and in the general development of classical harmonic point theory.

Throughout the chapter, Moufang planes will be understood in the inclusive sense of projective planes in which the little Desargues theorem holds. We thus allow desarguesian planes in our discussion unless otherwise specified. We assume always that the coordinate ring has characteristic other than 2.

For the most part this chapter is an exposition of the work of Havel and Schleiermacher with some new proofs and a slightly different point of view. Several generalizations of the results of this chapter have been published by other authors. In particular, Braun [1] has introduced a concept of cross-ratio in a Jordan algebra which reduces to that of §2 in the case at hand. Havel [5] has generalized somewhat the class of projective planes in which a version of our definition of harmonicity can be defined and in which a version of von Staudt's theorem can be proved.

I wish to express my thanks to John R. Faulkner, who made numerous valuable suggestions regarding the organization of this material from the Jordan point of view.

## 1. ALGEBRAIC PRELIMINARIES

Throughout the paper $\underline{D}$ will denote an alternative division ring of characteristic $\neq 2$, hence by the Bruck-Kleinfeld theorem either an associative division ring or a Cayley division algebra over its center $k$. In this section we collect some basic definitions and results under the assumption that $\underline{D}$ is a Cayley algebra. Recall that $\underline{D}$ admits a quadratic form $n(x)$ which is <u>multiplicative</u> is the sense that $n(ab) = n(a)n(b)$ for all $a, b \in \underline{D}$ and for which the associated bilinear form $n(x,y) = 1/2(n(x+y) - n(x) - n(y))$ is nondegenerate. Indeed, Cayley algebras can be characterized as 8-dimensional algebras over $k$ with nondegenerate quadratic form which is multiplicative [10]. If $1$ is the multiplicative identity of $\underline{D}$, we set $\underline{D}_0 = \{a \in \underline{D} \mid n(a,1) = 0\} = k^{\perp}$. Then

$\underline{D} = k \oplus \underline{D}_o$ and the mapping $-: \alpha + a_o \to \alpha - a_o$, $\alpha \in k$, $a_o \in \underline{D}_o$ is a k-linear involution, i.e. $\overline{ab} = \bar{b}\,\bar{a}$ for $a,b \in \underline{D}$. This mapping is called the <u>canonical involution</u> of $\underline{D}$. The norm form $n(x)$ of $\underline{D}$ can be computed via $n(x) = x\bar{x}$. In particular, since $\underline{D}$ is a division ring, $n(a) \neq 0$ for $a \neq 0$ in $\underline{D}$. The <u>trace</u> linear form on $\underline{D}$ is defined by $t(x) = 1/2(x + \bar{x}) = \overline{n(x,1)}$. The trace form is symmetric $(t(xy) = t(yx))$ and associative $(t(x(yz))) = t(((xy)z))$.

The Cayley algebra $\underline{D}$ is alternative, hence satisfies the identity $(xy)x = x(yx)$ and the <u>Moufang identities</u>:

$$x(y(xz)) = (xyx)z$$

$$((yx)z)x = y(xzx)$$

$$(xy)(zx) = x(yz)x .$$

Moreover, it is known that the subalgebra generated by 1, a and b is associative for any $a,b \in \underline{D}$. It follows from the Moufang identities and the fact that for any $a$, $a^{-1} = n(a)^{-1}\bar{a} = n(a)^{-1}(2t(a) - a)$ that the Moufang identities remain valid when one $x$ is replaced by its inverse, e.g. $x^{-1}(y(xz)) = (x^{-1}yx)z$. As a useful consequence we have

$$t((ab)(b^{-1}c)) = t((b^{-1}c)(ab)) = t(b^{-1}(ca)b) \tag{2}$$

$$= t(b(b^{-1}(ca))) = t(ca) = t(ac)$$

A second application of the associativity of the subalgebra generated by $a$ and $b$ is

$$(x(yx))x^{-1} = xy \tag{3}$$

which shows that $xy$ is <u>conjugate</u> to $yx$ in $\underline{D}$. We will make consistent use of a further conjugacy result for Cayley algebras.

<u>Lemma 1</u>. If $t(a) = t(a')$ and $n(a) = n(a')$ then $a$ is conjugate to $a'$.

<u>Proof</u>: Set $a = \alpha + a_o$, $a' = \alpha' + a_o'$, $\alpha,\alpha' \in k$, $a_o, a_o' \in \underline{D}_o$. $\alpha = t(a) = t(a') = \alpha'$ so $a\bar{a} = n(a) = n(a') = a'\bar{a}'$ implies $a_o^2 = (a_o')^2$. One immediately calculates that $(a_o + a_o')^{-1}a(a_o + a_o') = a'$ if $a_o' \neq -a_o$, i.e. if $x \neq \bar{x}$. If $x = \bar{x}$ the result follows from

$$yxy^{-1} = \bar{x} \quad \text{if} \quad n(x,y) = n(1,y) = 0 . \tag{4}$$

(4) follows in turn since $n(1,y) = y + \bar{y} = 0$ implies $\bar{y} = -y$ and $n(x,y) = x\bar{y} + y\bar{x} = -xy + y\bar{x} = 0$.

(4) implies as well

<u>Lemma 2</u>. Let $\underline{E}_0 \subseteq \underline{D}$ be of k-dimension 2, $1 \in \underline{E}_0$, and $\underline{E}_1 = \underline{E}_0^{\perp}$. Then $\underline{D} = \underline{E}_0 \oplus \underline{E}_1$ and if $x \in \underline{E}_0$ satisfies $n(1,x) = 0$, $\sigma: y \to x^{-1}yx$ has the property $\sigma|_{\underline{E}_0} = \mathrm{Id}$, $\sigma|_{\underline{E}_1} = -\mathrm{Id}$.

<u>Proof</u>: Since $n|_{\underline{E}_0}$ is nondegenerate because $n(x) \neq 0$ for all $x \neq 0$, $\underline{D} = \underline{E}_0 \oplus \underline{E}_1$ follows from standard linear algebra. Since $\underline{E}_0$ is spanned by 1 and $x$, both commuting with $x$, $\sigma|_{\underline{E}_0} = \mathrm{Id}$. By (4) $\sigma|_{\underline{E}_1} = -\mathrm{Id}$.

We will denote by $[x]$ the conjugacy class $\{y^{-1}xy \mid y \in \underline{D}\}$. Lemma 1 and the multiplicativity of $n(x)$ imply that conjugacy is an equivalence relation, hence if $y \in [x]$, $[y] = [x]$.

We will denote by $\hat{\underline{D}}$ the set $\underline{D} \cup \infty$ which will coordinatize the projective line in later sections.

## 2. DEFINITION AND PROPERTIES OF THE CROSS-RATIO

If $a, b, c, d$ are distinct elements of $\underline{D}$ we define the <u>cross-ratio</u> $(a,b:c,d)$ by

$$(a,b:c,d) = [((a-d)^{-1}(b-d))((b-c)^{-1}(a-c))]. \tag{5}$$

In the context of Jordan algebras it will be useful to be able to write (5) in terms of addition, subtraction and inversion.

<u>Lemma 3</u>. $(a,b:c,d) = [((a-b)^{-1} - (a-d)^{-1})((a-b)^{-1} - (a-c)^{-1})^{-1}]$.

<u>Proof</u>: By Lemma 1, the multiplicativity of $n(x)$, and the associativity of $t(x)$, $((a-d)^{-1}(b-d))((b-c)^{-1}(a-c))$ is

conjugate to $u = (((a-d)^{-1}(b-d))(b-c)^{-1})(a-c)$. Thus $(a-d)^{-1}(b-d) = (u(a-c)^{-1})(b-c)$ so $((a-d)^{-1}(b-d))(a-b)^{-1} = ((u(a-c)^{-1})(b-c))(a-b)^{-1}$. The left side of this equation, viewed as $((a-d)^{-1}((a-d)-(a-b)))(a-b)^{-1}$ is $(a-b)^{-1}-(a-d)^{-1}$. The right side of the equation is of form $((ux^{-1})y)z^{-1}$ where $x^{-1}$, $y$, and $z^{-1}$ all lie in a quaternion subalgebra of $\underline{D}$ if $\underline{D}$ is Cayley (if $\underline{D}$ is associative no problem arises here). Writing $\underline{D} = \underline{E} \oplus \underline{E}v$ for $v \in \underline{E}^{\perp}$ and computing explicitly one can show that $((ux^{-1})y)z^{-1} = u((x^{-1}y)z^{-1})$ since $x^{-1}yz^{-1} = z^{-1}yx^{-1}$. Thus, arguing as above for $((a-c)^{-1}(b-c))(a-b)^{-1}$ we have the right side equal to $u((a-b)^{-1}-(a-c)^{-1})$. The lemma is thus proved.

We extend the cross-ratio definition to $a,b,c,d \in \hat{\underline{D}}$ as follows:

$$\begin{aligned}(\infty,b:c,d) &= [(b-d)(b-c)^{-1}]\\(a,\infty:c,d) &= [(a-d)^{-1}(a-c)]\\(a,b:\infty,d) &= [(a-d)^{-1}(b-d)]\\(a,b:c,\infty) &= [(b-c)^{-1}(a-c)]\end{aligned}\tag{6}$$

(as a rule of thumb, use (5) and drop every factor involving $\infty$).

Several classical results about cross-ratios remain valid in this setting.

<u>Lemma 4</u>. If $a$, $b$, $c$, $d$ are distinct in $\underline{D}$ and $[x]^{-1}$ is defined to be $[x^{-1}]$ and $1-[x] = [1-x]$, then

(i) $(a,b:c,d) = (b,a:c,d)^{-1}$

(ii) $1-(a,b:c,d) = (a,c:b,d)$

<u>Proof</u>: (i) If $a,b,c,d \in \underline{D}$, $(b,a:c,d)^{-1} = [(((b-d)^{-1}(a-d))((a-c)^{-1}(b-c)))^{-1}] = [((b-c)^{-1}(a-c))((a-d)^{-1}(b-d))] = [((a-d)^{-1}(b-d))((b-c)^{-1}(a-c))]$ since $xy$ is conjugate to $yx$.

If $d=\infty$, we have $[((a-c)^{-1}(b-c))^{-1}] =$

$[(b-c)^{-1}(a-c)]$ which is the desired result. $c=\infty$ is treated similarly, while $a=\infty$ and $b=\infty$ can be handled by direct computation using the conjugacy of $xy$ and $yx$.

(ii) If $a, b, c, d \in \underline{D}$, $1-(a,b:c,d) = [1-((a-b)^{-1}-(a-d)^{-1})\cdot((a-b)^{-1}-(a-c)^{-1})^{-1}] = [((a-c)^{-1}-(a-d)^{-1})((a-c)^{-1}-(a-b)^{-1})^{-1}] = (a,c:b,d)$. If $d=\infty$, $1-(a,b:c,d) = [1-(b-c)^{-1}(a-c)] = [(b-c)^{-1}(b-a)] = (a,c:b,d)$. Similar arguments handle the cases $a=\infty$, $b=\infty$, and $c=\infty$.

As a consequence of Lemma 4, there are at least six values attained by the cross-ratio when applied to four points, permuted in all possible manners. If $(a,b:c,d) = w$, one also can attain $w^{-1}$, $1-w$, $1-w^{-1}$, $(1-w)^{-1}$, and $(1-w^{-1})^{-1}$. That at most six values are possible follows from the invariance of the cross-ratio under the four permutations I, (12)(34), (13)(24), (14)(23). This invariance follows from

<u>Theorem 1</u>. $(a,b:c,d) = (b,a:d,c) = (c,d:a,b) = (d,c:b,a)$.

<u>Proof</u>: $(a,b:c,d) = [((a-d)^{-1}(b-d))((b-c)^{-1}(a-c))] = [((b-c)^{-1}(a-c))((a-d)^{-1}(b-d))] = (b,a:d,c)$ since $[xy] = [yx]$. Now $(c,d:a,b) = 1-(c,a:d,b) = 1-(a,c:b,d) = 1-(1-(a,b:c,d)) = (a,b:c,d)$ by Lemma 4, (ii) and the first equality. The final equality follows from the first two.

That the invariance group is exactly of order 4 will follow if we can find $a, b, c, d$ with $(a,b:c,d) = w$ where the six expressions represent distinct elements of $\underline{D}$. For sufficiently large $\underline{D}$ this follows from

<u>Theorem 2</u>. Let $r \in \underline{D}$, $r \neq 0$, $r \neq 1$. If $a,b,c \in \hat{\underline{D}}$ are distinct, then there is $d \in \hat{\underline{D}}$ such that $(a,b:c,d) = [r]$. If $r$ is in the center $k$ of $\underline{D}$ then $d$ is unique.

<u>Proof</u>: Suppose first that $a,b,c \in \underline{D}$. For any $s \in [r]$ the equation $((a-d)^{-1}(b-d))u = s$ has the unique solution $d = (as-bu)(s-u)^{-1}$ if $u \neq s$. In particular, if $s \neq (b-c)^{-1}(a-c)$ this yields a solution to $(a,b:c,d) = [r]$ when $u = (b-c)^{-1}(a-c)$. If $s = (b-c)^{-1}(a-c)$, then $d=\infty$ is the unique solution. If $r \in k$, so $[r] = \{r\}$, the solution $d$ is unique.

If $s \in [r]$ and $c = \infty$, we need only solve $(a-d)^{-1}(b-d) = s$ for $d$. This equation again has a unique solution in $\underline{D}$. The remaining cases $(b = \infty, a = \infty)$ reduce to the case $c = \infty$ by Theorem 1.

## 3. ALGEBRAIC TRANSFORMATIONS PRESERVING CROSS-RATIOS

In later sections, the transformations $t_a: x \to x+a$; $r_u: x \to x_u$ for $u \neq 0$; $i: x \to x^{-1}$, $0 \leftrightarrow \infty$; and $\gamma: x \to \bar{x}$ (if $\underline{D}$ is Cayley) will have special geometric significance. Their importance is due to

Theorem 3. If $\sigma = t_a, r_u$, $i$ or $\gamma$ then $(a,b:c,d) = (\sigma(a),\sigma(b):\sigma(c),\sigma(d))$ for all $a,b,c,d \in \hat{\underline{D}}$.

Proof: The result for $t_a$ is immediate from the definition (5) of the cross-ratio. For $\gamma$ we note that (4) implies $[x] = [\bar{x}]$ so

$$
\begin{aligned}
(a,b:c,d) &= [\overline{((a-b)^{-1} - (a-d)^{-1})((a-b)^{-1} - (a-c)^{-1})^{-1}}] \\
&= [((\bar{a}-\bar{b})^{-1} - (\bar{a}-\bar{c})^{-1})^{-1}((\bar{a}-\bar{b})^{-1} - (\bar{a}-\bar{d})^{-1})] \\
&= [((\bar{a}-\bar{b})^{-1} - (\bar{a}-\bar{d})^{-1})((\bar{a}-\bar{b})^{-1} - (\bar{a}-\bar{c})^{-1})^{-1}] \\
&\qquad \text{since } [xy] = [yx] \\
&= (\bar{a},\bar{b}:\bar{c},\bar{d}) \quad \text{if } a,b,c,d \in \underline{D}.
\end{aligned}
$$

If one of $a$, $b$, $c$, $d$ is $\infty$ we assume without loss of generality that $a = \infty$ by Theorem 1. Then $(a,b:c,d) = [\overline{(b-d)(b-c)^{-1}}] = [(\bar{b}-\bar{d})(\bar{b}-\bar{c})^{-1}] = (\bar{a},\bar{b}:\bar{c},\bar{d})$ where $\bar{\infty}$ is defined to be $\infty$.

For $r_u$ with $a,b,c,d \in \underline{D}$ we have $(au,bu:cu,du) =$

$$
\begin{aligned}
&= [((au-bu)^{-1} - (au-du)^{-1})((au-bu)^{-1} - (au-cu)^{-1})^{-1}] \\
&= [(u^{-1}((a-b)^{-1} - (a-d)^{-1}))(((a-b)^{-1} - (a-c)^{-1})^{-1}u)] \\
&= (a,b:c,d).
\end{aligned}
$$

If $a = \infty$, $(au,bu:cu,du) = [((b-d)u)(u^{-1}(b-c)^{-1}] = (a,b:c,d)$ by Lemma 1, the multiplicativity of $n$ and the associativity of $t$.

We now turn to $i$, assuming first that $a,b,c,d \in \underline{D}\backslash\{0\}$.

We will apply Lemma 1 to $(a^{-1},b^{-1}:c^{-1},d^{-1}) = [s]$ where $s = ((a^{-1}-d^{-1})^{-1}(b^{-1}-d^{-1}))((b^{-1}-c^{-1})^{-1}(a^{-1}-c^{-1})$. Writing $s = ((d^{-1}(d-a)a^{-1})^{-1}(d^{-1}(d-b)b^{-1}))((c^{-1}(c-b)b^{-1})^{-1}(c^{-1}(c-a)a^{-1})) = ((a(d-a)^{-1}d)(d^{-1}(d-b)b^{-1}))((b(c-b)^{-1}c)(c^{-1}(c-a)a^{-1}))$ we see easily that $n(s) = n(r)$ for $[r] = (a,b:c,d)$, $((a-d)^{-1}(b-d))((b-c)^{-1}(a-c)) = r$. By the same argument as in the proof of Lemma 3, since $d$ and $b$ (resp. $a$ and $c$) lie in a quaternion subalgebra of $\underline{D}$ if $\underline{D}$ is Cayley, $s = (a((d-a)^{-1}((d-b)b^{-1})))(((b(c-b)^{-1})(c-a))a^{-1}))$. By associativity and symmetry of $t(x)$ we have

$$\begin{aligned}
t(s) &= t(((d-a)^{-1}((d-b)b^{-1}))((b(c-b)^{-1})(c-a)))\\
&= t(((a-d)^{-1}((b-c)b^{-1}))((b(c-b)^{-1})(c-a)))\\
&\quad - t(((a-d)^{-1}((a-c)b^{-1}))((b(c-b)^{-1})(c-a)))\\
&\quad + t(((a-d)^{-1}((a-d)b^{-1}))((b(c-b)^{-1}(c-a)))\\
&= -t((a-d)^{-1}(c-a)) \qquad \text{(by (2))}\\
&\quad - t(((c-a)((a-d)^{-1}((a-c)b^{-1})))(b(c-b)^{-1}))\\
&\qquad \text{(by symmetry and associativity)}\\
&\quad + t((c-b)^{-1}(c-a))\\
&= -t((a-d)^{-1}(c-a)) - t((((c-a)(a-d)^{-1})(a-c)) \cdot\\
&\quad (c-b)^{-1}) + t((c-b)^{-1}(c-a)) \quad \text{(applying the}\\
&\qquad \text{Moufang identities and (2))}\\
&= -t(((a-d)^{-1}(b-c))((b-c)^{-1}(c-a)))\\
&\quad + t(((a-d)^{-1}(a-c))((b-c)^{-1}(c-a)))\\
&\quad - t(((a-d)^{-1}(a-d))((b-c)^{-1}(c-a)))\\
&= t(((a-d)^{-1}(d-b))((b-c)^{-1}(c-a)) = t(r).
\end{aligned}$$

Since $t(r) = t(s)$ and $n(r) = n(s)$ we have $(a^{-1},b^{-1}:c^{-1},d^{-1}) = (a,b:c,d)$. If $a = \infty$, $b,c,d \in \underline{D}\backslash\{0\}$ we compare $[(b-d)(b-c)^{-1}]$ and $(0,b^{-1}:c^{-1},d^{-1}) = [(d(b^{-1}-d^{-1}))((b^{-1}-c^{-1})^{-1}c^{-1})]$. Since $(d(b^{-1}-d^{-1}))((b^{-1}-c^{-1})^{-1}c^{-1}) = (d(d^{-1}(d-b)b^{-1}))(b(c-b)^{-1}c)c^{-1})$, application of Lemma 1 yields the desired result. The cases $b = \infty$, $c = \infty$, $d = \infty$ follow by Theorem 1. There remain the cases $a = 0$, $b = \infty$; $a = 0$, $c = \infty$; and $a = 0$, $d = \infty$. These can be verified by direct computation.

## 4. HARMONICITY IN $\hat{\underline{D}}$

Following the results in classical desarguesian geometry, we define a quadruple a, b, c, d of elements of $\hat{\underline{D}}$ to be harmonic if $(a,b:c,d) = [-1]$. For convenience we let $h(a,b,c,d)$ represent the statement: a,b,c,d are harmonic. From §2 we deduce

Theorem 4. (i) If a, b, c are distinct elements of $\hat{\underline{D}}$, then there is a unique fourth element d such that $h(a,b,c,d)$.

(ii) The relation $h(a,b,c,d)$ is invariant under the action of the dihedral group generated by (12) and (13)(24).

Proof: (i) follows from Theorem 2 with $r = -1 \in k$. By Theorem 1 h is invariant under (13)(24). Invariance under (12) follow from Lemma 4, (i) and $(-1)^{-1} = -1$.

We conclude further

Theorem 5. $h(a,b,c,d)$ implies $h(\sigma(a),\sigma(b),\sigma(c),\sigma(d))$ if $\sigma = t_a, r_a$, i, or $\gamma$.

Proof: Immediate from Theorem 3 and the definition of $h(a,b,c,d)$.

For "affine" elements $a,b,c,d \in \underline{D}$, Lemma 3 provides an alternate definition of $h(a,b,c,d)$, namely $h(a,b,c,d)$ if and only if $2(a-b)^{-1} = (a-c)^{-1} + (a-d)^{-1}$. If $d = \infty$ the definition (6) leads to $h(a,b,c,d)$ if and only if $a-c = c-b$. With these definitions we can view harmonicity as a Jordan ring phenomenon, rather than an alternative ring phenomenon since the alternative produce in $\underline{D}$ does not enter in the definition. It is also easy to see from this formulation that the harmonic conjugates c and d are never equal if char. $\underline{D} \neq 2$ while in characteristic 2 they are always equal.

By straightforward calculation one can verify from (5) and (6)

Lemma 5. (i) $h(0,a,\infty,a/2)$.

(ii) $h(a,b,\infty,(a+b)/2)$.

(iii) $h(a,-a,\infty,0)$.

(iv) $h(1,-1,a,a^{-1})$.

## 5. HARMONIC POINTS IN THE PROJECTIVE PLANE

The projective planes $\pi$ we study are those coordinatized by alternative division rings $\underline{D}$. The points of the plane are the pairs $(a,b)$, $a,b \in \underline{D}$; singletons $(a)$, $a \in \underline{D}$; and one exceptional point $(\infty)$. The lines are $[m,r] = \{(a,b) \mid b = ma + r\} \cup \{(m)\}$; $[c] = \{(a,b) \mid a = c\} \cup \{(\infty)\}$; and $[\infty] = \{(a) \mid a \in D\} \cup \{(\infty)\}$, the line at infinity. We will denote throughout by $L$ the line $[0]$ with equation $y = 0$. Since $L = \{(a,0) \mid a \in \underline{D}\} \cup \{(0)\}$, with $(0)$ the ideal point of $L$, we can identify the points of $L$ with elements of $\hat{\underline{D}}$ via $(a,0) \leftrightarrow a$, $(0) \leftrightarrow \infty$. We shall do this throughout, bearing in mind that some confusion may arise from the use of the symbols $(\infty)$ and $\infty$ to represent two different points. If $P$ and $Q$ and distinct points of $\pi$ we will denote by $P \wedge Q$ the unique line lying on both $P$ and $Q$. Since we have assumed $\underline{D}$ to be alternative, $\pi$ is a Moufang plane [9]. In this chapter we will use two important consequences of that fact. First, if A, B, and C are distinct points of $L$ and $D$ is constructed from A, B, C, $P_1$ and $P_2$ via the configuration (7), the point D

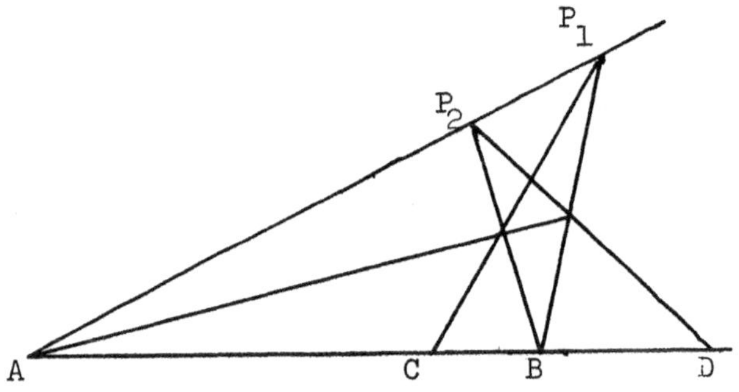


(7)

is uniquely determined by A, B, and C (independent of the choice of $P_1$ and $P_2$). Second, if $\underline{1}$ is a line and $P$ and $Q$ are points not on $\underline{1}$, then there is an elation fixing all points of $\underline{1}$ and mapping $P$ to $Q$.

Addition and multiplication are defined geometrically by the configurations (8) and (9) respectively: In (8), $Q = (b,b)$, so $P = [1,-a] \cap [0,b] = (a + b,b)$. This construction can be carried out with the point $(1,1)$ replaced by any point $(1,q)$ and a simple calculation shows that the resulting sum is independent of the choice of $q$.

In (9) we have $P = [1,-1] \cap [b] = (b,b - 1)$, $Q = [1,-a] \cap [(b-1)b^{-1},0] = (ba,b(a-1))$. In contrast to the

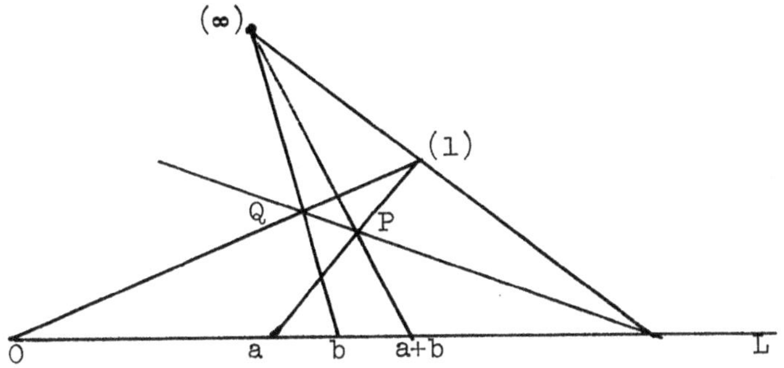


(8)

situation for addition, carrying out this construction with $(1,1)$ replaced by $(1,q)$ we find $P = [q^{-1},-q^{-1}] \cap [b] = (b,q^{-1}(b-1))$. Then $Q = [q^{-1},-q^{-1}a] \cap [q^{-1}(1-q^{-1}),0] = ((bq)(q^{-1}a),--)$. The new choice of $E$ results in a change of product from $ab$ to $a*b = (aq)(q^{-1}b)$. While the choice of $E$ affects the alternative produce in $\underline{D}$ it does not affect the Jordan product $a \cdot b = 1/2(ab + ba)$ since this can be obtained by linearizing $a*a = (aq)(q^{-1}a) = a^2$ (since $a$ and $q$ generate an associative subalgebra).

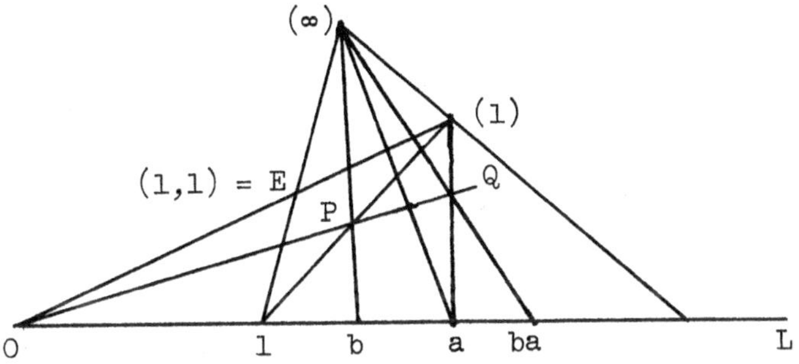


(9)

We call points $A$, $B$, $C$, $D$ on $L$ in <u>harmonic position</u> if they can be embedded in a configuration (7) (a complete quadrilateral). $C$ and $D$ are <u>harmonic conjugates</u> relative to $A$ and $B$. The important observation allowing us to handle some geometric questions by algebraic methods is

<u>Theorem 6</u>. The distinct points $a$, $b$, $c$, $d$ (possibly $\infty$) are in harmonic position if and only if $h(a,b,c,d)$.

<u>Proof</u>: Suppose first that $a,b,c,d \in L$ are in harmonic position. Without loss of generality we may assume $P_1 = (\infty)$ and $P_2 = (a,1)$. Then $B \wedge P_2 = [(a-b)^{-1},-(a-b)^{-1}b]$ while

$C \wedge P = [c]$ so $E = (c,(a-b)^{-1}(c-b))$. Now $P_2 \wedge D$ has equation $y = (d-a)^{-1}(d-x)$ and $A \wedge E$ has equation $y = (((b-a)^{-1}(b-c))(a-c)^{-1})(a-x)$. Since these lines are concurrent with $B \wedge P_1 = (b)$ we have the relation $(d-a)^{-1}(d-b) = (((b-a)^{-1}(b-c))(a-c)^{-1})(a-b)$ so $((b-a)^{-1}(b-c))(a-c)^{-1} = ((d-a)^{-1}(d-b))(a-b)^{-1}$. Writing $b-c = (b-a)+(a-c)$ and $d-b = (d-a)+(a-b)$ we obtain $(a-c)^{-1} + (b-a)^{-1} = (a-b)^{-1} + (d-a)^{-1}$ so $(a,b:c,d) = -1$ by Lemma 3.

If $d = \infty$, we utilize the same computations with the exception $P_2 \wedge D$ has equation $y = 1$ so $(((b-a)^{-1}(b-c))(a-c)^{-1})(a-b) = 1$. Simple manipulation yields $(b-c)(a-c)^{-1} = -1$ as desired.

If $b = \infty$, $E = (c,1)$, $F = ((c-a)^{-1})$, so $F \wedge D$ has equation $y = (c-a)^{-1}(x-d)$ and contains $(a,1)$ so again $(a,b:c,d) = -1$.

If $c = \infty$, $F = (b,(d-a)^{-1}(d-b))$ as above. $A \wedge F$ thus has equation $y = (((d-a)^{-1}(d-b))(a-b)^{-1})(a-x)$ while $P_2 \wedge B$ has equation $y = (b-a)^{-1}(b-x)$ so $((d-a)^{-1}(d-b))(a-b)^{-1} = (b-a)^{-1}$. Again we have $h(a,b:c,d)$.

Finally, for $a = \infty$, we take $P_2 = (1)$ and find $F = (b,b-d)$, $E = (c,c-b)$ so $b-d = c-b$ and the proof is complete, the converse following from Theorem 2.

Corollary. If a, b, c, d are in harmonic position, so is any quadruple consisting of the same four points permuted by an element of the dihedral group generated by (12) and (13)(24).

Proof: a, b, c, d are in harmonic position if and only if $h(a,b,c,d)$ by the theorem and we know from Theorem 4 that $h(a,b,c,d)$ is invariant under the action of the group in question.

It is not difficult to prove this corollary by purely geometric means [4], hence reducing the algebraic computations necessary for the proof of Theorem 6. We have chosen a method of proof for Theorem 6 which emphasizes the role of the cross-ratio in harmonic point theory.

## 6. ALGEBRAIC DESCRIPTION OF THE PROJECTIVITIES OF L

A perspectivity in $\pi$ is a bijection from a line $L_1$ to a line $L_2$ obtained by projecting through a point $P$, $P \notin L_1$, $P \notin L_2$ (see (10)). We denote such a perspectivity by

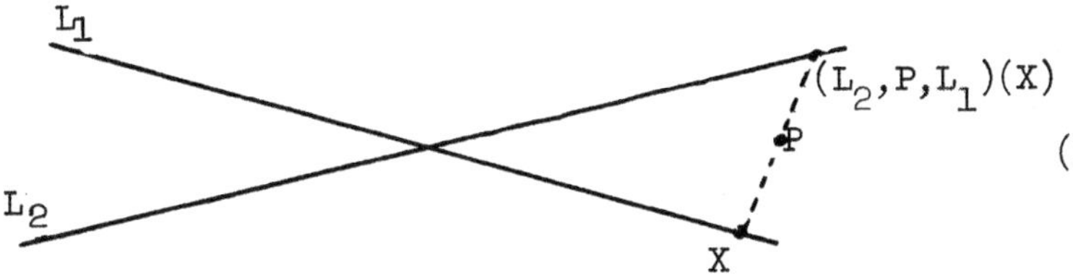

(10)

$\sigma(L_2,P,L_1)$. A mapping induced on L by a product $\prod_{i=0}^{n-1} \sigma(L_{i+1},P_i,L_i)$ of perspectivities with $L_o = L_n = L$ is a projectivity of L. In this section we will obtain an algebraic description of the group of projectivities of L. In §11 we will give a somewhat more geometric description. An important property of Moufang planes [9; p. 67] states that every projectivity of L is induced by a collineation $\mu$ of $\pi$ which is a product of central collineations (collineations fixing all points on a line and all lines on a point).

With the notation of Theorem 3 we have

Lemma 6. The transformations $t_a$, $r_a$ $(a \neq 0)$ and i are projectivities of L.

Proof: For $t_a$ we note from (8) that

$$t_a = \sigma(L,(\infty),a \wedge (1))\sigma(a \wedge (1),\infty,0 \wedge 1)\sigma(0 \wedge 1,(\infty),L).$$

From (9) we have $r_a = \sigma(L,(\infty),a \wedge 1)\sigma(a \wedge 1,0,(1) \wedge 1)\sigma((1) \wedge 1,(\infty),L)$. For i we consider the configuration (11) which, as with (8) and

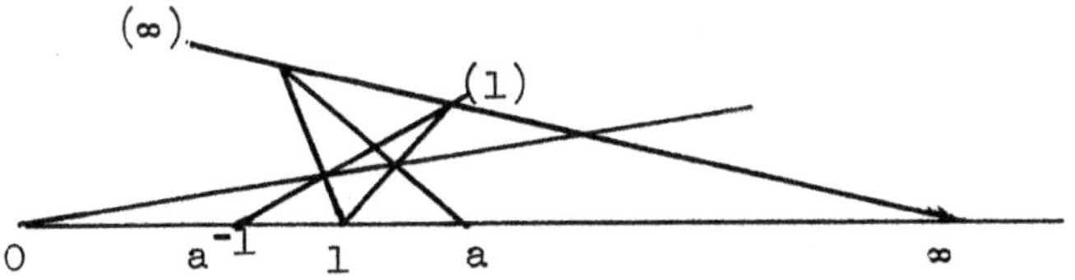

(11)

(9) can easily be justified by computing coordinates. (11) yields $i = \sigma(L,(1),1 \wedge (\infty))\sigma(1 \wedge (\infty),0,1 \wedge (1))\sigma(1 \wedge (1),(\infty),L)$.

We denote by $\underline{P}_o$ the group of transformations of L generated by $\{t_a\} \cup \{r_a\} \cup \{i\}$. Note that for $a \neq 0$, $l_a: x \to ax$ is in $\underline{P}_o$ since $ax = (x^{-1}a^{-1})^{-1}$ implies $l_a = ir_{a^{-1}}i$.

Lemma 7. $\underline{P}_o$ is transitive on ordered triples of distinct points of $L$.

Proof: It suffices to show for distinct $a, b, c \in L$ that there is $\sigma \in \underline{P}_o$ such that $\sigma(a) = 0$, $\sigma(b) = 1$, and $\sigma(c) = \infty$. If $c = \infty$, $\sigma = r_{(b-a)^{-1}}t_{-a}$ will suffice. If $c = 0$, we first apply $i$, then the case $c = \infty$. If $c \neq 0, \infty$, $t_{-c}$ reduces considerations to the case $c = 0$.

We denote by $\underline{P}$ the group of all projectivities of $L$ and assume that $\pi$ is infinite (this is no restriction in the case of a proper Moufang plane since there exist no Cayley division algebras over finite fields).

Theorem 7. $\underline{P} = \underline{P}_o$.

Proof: Lemma 6 implies $\underline{P}_o \subseteq \underline{P}$. Suppose $\tau = \prod_{i=0}^{n-1} \sigma(L_{i+1}, P_i, L_i)$ with $L = L_o = L_n$. Since $\pi$ is infinite there is $P \in \pi$ such that $P \neq P_i$ and $P \notin L_i$ for all $i$. Then $\tau = \prod_{i=0}^{n-1} \sigma(L,P,L_{i+1})\sigma(L_{i+1},P_i,L_i)\sigma(L_i,P,L)$. It suffices to show that $\sigma(L,P,L_{i+1})\sigma(L_{i+1},P_i,L_i)\sigma(L_i,P,L) \in \underline{P}_o$. We thus consider a general element $\sigma(L,P,L'')\sigma(L'',Q,L')\sigma(L',P,L)$.

Case (i). $Q \notin L$. Then $\sigma(L'',Q,L') = \sigma(L'',Q,L)\sigma(L,Q,L')$ and it suffices to consider transformations of type $\tau = \sigma(L,P,L'')\sigma(L'',Q,L)$. Such a transformation is pictured in (12).

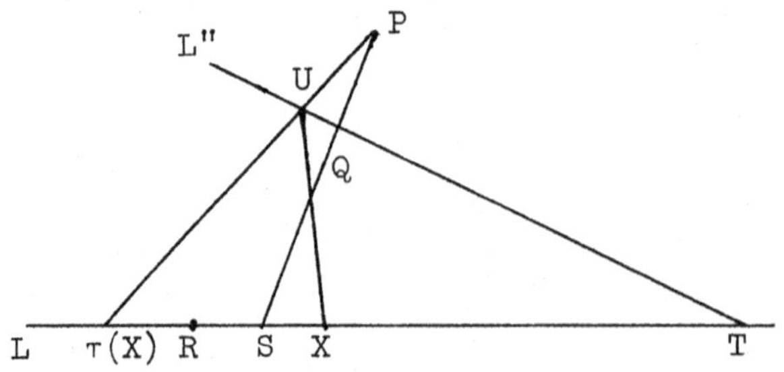


(12)

By Lemma 7 there is $\sigma \in \underset{\multimap}{P}$ such that $\sigma(R) = 0$, $\sigma(S) = 1$, and $\sigma(T) = \infty$. Let $\tilde{\sigma}$ be a collineation of $\pi$ inducing $\sigma$ on L. Conjugating $\tau$ by $\sigma$ we obtain a new mapping defined by the configuration (12) where $R = 0$, $S = 1$ and $T = \infty$. This mapping will not be altered if the entire configuration is acted upon by an elation with axis L mapping P to $(\infty)$, so we may assume as well that $P = (\infty)$. Thus $Q = (1,q)$. If $L' = [0,s]$, $U = [0,s] \cap [q(1-a)^{-1}, -q(1-a)^{-1}a] = (q^{-1}s + a(1-q^{-1}s), s)$ if $X = (a,0)$. Thus the mapping described by (12) is $t_{q^{-1}s}\, r_{(1-q^{-1})s}$ which lies in $\underset{\multimap}{P}$. $\tau$ is thus in $\underset{\multimap}{P}$ as well.

Case (ii). $Q \in L$. We must analyze $\tau = \sigma(L,P,L'')\sigma(L'',Q,L')\sigma(L',P,L)$ which is described in terms of the configuration (13). As in case (i) we can without loss of generality assume $Q = 0$, $S = 1$, $T = \infty$, and $P = (\infty)$. $L' = [0,s]$, $L'' = [q,-q]$. If $X = (a,0)$, $E = L' \cap (X \wedge (\infty)) = (a,s)$. $F = L'' \cap (0 \wedge E) = ((q - ra^{-1})^{-1}q, --)$ so $\tau(X) = ((q - sa^{-1})^{-1}q, 0)$ so the mapping in question is $r_q i t_{q^1 - s^i} \in \underset{\multimap}{P}$ so we are finished.

<u>Corollary</u>. If $\sigma: L \to L$ is a projectivity then $\sigma$ preserves harmonic position.

<u>Proof</u>: By Theorem 5, the elements of $\underset{\multimap}{P}$ preserve $h(a,b,c,d)$. Since $h(a,b,c,d)$ implies a, b, c, d are in harmonic position and since $\sigma \in \underline{P}$ implies $\sigma \in \underset{\multimap}{P}$, the result follows.

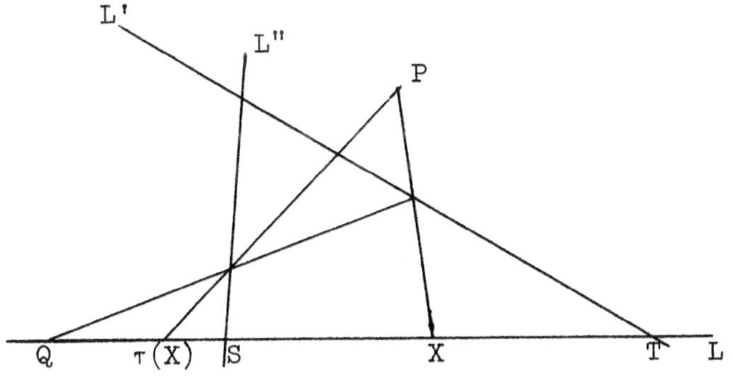


(13)

Of course, this corollary is geometrically obvious once one knows that every projectivity of L is induced by a collineation of $\pi$ since collineations preserve the configuration (7).

## 7. THE JORDAN STRUCTURE

A nonassociative ring is a Jordan ring if it satisfies the identities

$$a \cdot b = b \cdot a \tag{14}$$

$$(a^{\cdot 2} \cdot b) \cdot a = a^{\cdot 2} \cdot (b \cdot a) .$$

The simplest example of a nontrivial Jordan ring is the ring $A^+$ where $A$, with product $ab$, is an associative ring and the Jordan multiplication is defined by $a \cdot b = 1/2(ab + ba)$ (char. $\neq 2$). It is not difficult to check directly that the identities (14) are consequences of the associativity of $A$. Precisely the same procedure makes of an alternative ring $\underline{D}$ a Jordan ring $\underline{D}^+$ since the identities involve only two elements of $\underline{D}$. In a Jordan ring with identity element 1, $b$ is a Jordan inverse of $a$ if $b \cdot a^{\cdot 2} = a$ and $b \cdot a = 1$. It is a simple matter to verify that the alternative inverse $a^{-1}$ is also a Jordan inverse.

In the classical literature [7] Jordan algebras are studied from one of two equivalent points of view (char. $\neq 2, 3$), either in terms of the binary structure controlled by (14) or in terms of the Jordan triple product $\{abc\} = (a \cdot b) \cdot c + (b \cdot c) \cdot a - (a \cdot c) \cdot b$. In the algebras $\underline{D}^+$, $\underline{D}$ associative or alternative, the triple product reduces to $\{abc\} = 1/2(abc + cba)$. More recently, Springer [13] has introduced a class of algebraic objects called J-structures, with algebraic structure based on the existence of an inversion operator satisfying certain axioms. This class of structures is equivalent to the class of Jordan algebras. The connection between the two classes is based on the Hua identity

$$(x^{-1} + (x^{-1} - y)^{-1})^{-1} = x - \{xyx\} \tag{15}$$

which is easily verified for $\underline{D}^+$ because of the underlying associativity in $\underline{D}$.

An automorphism of a Jordan ring $J$ is a bijective mapping $\sigma: J \to J$ such that for all $a, b \in J$:

$$\text{(i)} \quad \sigma(a + b) = \sigma(a) + \sigma(b) \tag{16}$$

$$\text{(ii)} \quad \sigma(a \cdot b) = \sigma(a) \cdot \sigma(b) .$$

In the classical proofs of various versions of van Staudt's theorem, one encounters similar conditions on mappings of an alternative or associative ring, namely bijections $\sigma$ satisfying (16) (i) and

$$\sigma(ab + ba) = \sigma(a)\sigma(b) + \sigma(b)\sigma(a). \tag{17}$$

Since (16) (i) implies $\sigma(2a) = 2\sigma(a)$, whence $\sigma(a/2) = \sigma(a)/2$, it is clear that the mappings satisfying (17) are just the Jordan automorphisms of $\underline{D}^+$.

We assume now that $J$ is a Jordan division ring, i.e. every $a \neq 0$ is invertible in $J$.

Lemma 8. If $\sigma: J \to J$ is a bijective mapping satisfying (16) (i) and $\sigma(1) = 1$, then the following are equivalent:

(i) $\sigma$ is an automorphism of $J$

(ii) $\sigma(x^{\cdot 2}) = \sigma(x)^{\cdot 2}$ for all $x \in J$

(iii) $\sigma(\{xyx\}) = \{\sigma(x)\sigma(y)\sigma(x)\}$ for all $x,y,z \in J$

(iv) $\sigma(x^{-1}) = \sigma(x)^{-1}$ for all $x \in J$.

Proof: Assume (i). Then $\sigma(xx^{-1}) = \sigma(x) \cdot \sigma(x^{-1}) = \sigma(1) = 1$ and $\sigma(x) = \sigma((x^{\cdot 2}) \cdot x^{-1}) = \sigma(x^{\cdot 2}) \cdot \sigma(x^{-1}) = \sigma(x)^{\cdot 2} \cdot \sigma(x^{-1})$. Thus $\sigma(x^{-1}) = \sigma(x)^{-1}$ so (iv) holds. Now assume (iv). Applying $\sigma$ to the left side of (15) yields $(\sigma(x)^{-1}+(\sigma(y)^{-1}-\sigma(x))^{-1})^{-1} = \sigma(x) - \{\sigma(x)\sigma(y)\sigma(x)\}$. Applied to the right side of (15), $\sigma$ yields $\sigma(x) - \sigma(\{xyx\})$, whence (iii) holds. Now if (iii) holds, choose $y = 1$ so $\{xyx\} = x^{\cdot 2}$ and since $\sigma(1) = 1$, (ii) must hold. Finally, if (ii) holds $(x+y)^{\cdot 2} - x^{\cdot 2} - y^{\cdot 2} = xy + yx$ so $\sigma$ satisfies (17) and by the above remarks, $\sigma$ is an automorphism, hence (i) holds.

Note that an automorphism of a Jordan division ring always satisfies $\sigma(1) = 1$ since by (16) (ii) with $x = y = 1$ we have $\sigma(1) = \sigma(1) \cdot \sigma(1)$ and since $\sigma(1) \neq 0$ we can multiply both sides of the equation by $\sigma(1)^{-1}$. Note also that the result of Lemma 7, and the proof, remain valid in much more generality than we have presented. Clearly it suffices to drop the assumption that domain $\sigma$ = range $\sigma$, that $\sigma$ is $1-1$, and that $\sigma$ is onto.

In the context of harmonicity we have

Theorem 8. If $\sigma$ is an automorphism of the Jordan ring $\underline{D}^+$ and $h(a,b,c,d)$, then $h(\sigma(a),\sigma(b),\sigma(c),\sigma(d))$.

Proof: If $a,b,c,d \in \underline{D}$ we have $h(a,b,c,d) \Leftrightarrow 2(a-b)^{-1} = (a-c)^{-1} + (a-d)^{-1} \Leftrightarrow h(\sigma(a),\sigma(b),\sigma(c),\sigma(d))$ by Lemma 8. If

$d = \infty$, $h(a,b,c,d)$ implies $[(b-c)^{-1}(a-c)] = [-1]$ so $b-c = c-a$. Since we define $\sigma(\infty) = \infty$, and $\sigma(b)-\sigma(c) = \sigma(b-c) = \sigma(c-a) = \sigma(c) - \sigma(a)$ we have $h(\sigma(a),\sigma(b),\sigma(c),\sigma(d))$. The remaining cases follow from Theorem 4, (ii).

Buekenhout [2] has generalized Theorem 8 to allow $\sigma$ a homomorphism (place) from $\hat{\underline{D}}^+$ to $\hat{\underline{D}}^+$. To do so he introduces a somewhat more inclusive definition of harmonic position, admitting quadruples $a$, $b$, $c$, $d$ not all distinct. Basic assumptions then include: If three of $a$, $b$, $c$, $d$ are equal, then $h(a,b,c,d)$; and if two of $a$, $b$, $c$, $d$ are equal and $h(a,b,c,d)$ then three of $a$, $b$, $c$, $d$ are equal. The major difficulty in proving this generalization lies in the proof that if $a$, $b$, $c$, $d$ are distinct and $h(a,b,c,d)$ then equality of two of the $\sigma(a)$, $\sigma(b)$, $\sigma(c)$, $\sigma(d)$ implies equality of three.

We shall later use

Lemma 9. ([12]) Let $\underline{D}$ be a Cayley division algebra over $k$ with norm form $n(x)$. Then $\sigma: \underline{D} \to \underline{D}$, bijective, is an automorphism of $\underline{D}^+$ if and only if $\sigma(1) = 1$ and $n(\sigma(a)) = \sigma'(n(a))$ for all $a \in \underline{D}$, $\sigma'$ an automorphism of $k$.

Proof: Suppose that $\sigma$ is an automorphism of $\underline{D}^+$. By Lemma 8 $\sigma(xyx) = \sigma(x)\sigma(y)\sigma(x)$ and by the comments preceding Lemma 8 $\sigma(xy + yx) = \sigma(x)\sigma(y) + \sigma(y)\sigma(x)$. One checks directly using associativity of the subalgebra generated by $x$ and $y$ that $(xy - yx)^2 = ((xyx)y + y(xyx)) - ((xy)(yx) + (yx)(xy))$. It follows that $(\sigma(xy - yx))^2 = (\sigma(x)\sigma(y) - \sigma(y)\sigma(x))^2$, so $x \in k$ if and only if $\sigma(x) \in k$. $\sigma' = \sigma|_k$ is thus an automorphism of the field $k$. For $a \in k$, $\sigma(ax) = \sigma((ax + xa)/2) = \sigma(a)\cdot\sigma(x) = \sigma(a)\sigma(x)$. Now for $x \in \underline{D}$ we have $x^2 = 2t(x)x - n(x)$ so one has both $\sigma(x)^2 = 2t(\sigma(x))\sigma(x) - n(\sigma(x))$ and $\sigma(x^2) = \sigma(x)^2 = 2\sigma(t(x))\sigma(x) - \sigma(n(x))$. If $x \notin k$, $1$ and $x$ are $k$-linearly independent so $\sigma'(n(x)) = n(\sigma(x))$. If $x \in k$, $n(x) = x\bar{x} = x^2$ and $\sigma'(n(x)) = \sigma(x^2) = \sigma(x)^2 = n(\sigma(x))$ so the condition stated in the lemma holds. Moreover, since $t(x) = 1/2(n(x+1) - n(x) - n(1)) = n(x,1)$ it follows that $t(\sigma(x)) = \sigma'(t(x))$ since $\sigma(1) = 1$.

Conversely, if $n(\sigma(x)) = \sigma'(n(x))$ and $\sigma(1) = 1$, then $t(\sigma(x)) = \sigma'(t(x))$, whence $\sigma(x)^2 = \sigma(x^2)$ by the above computation so $\sigma$ is an automorphism by Lemma 8.

## 8. VON STAUDT'S THEOREM IN THE MOUFANG PLANE

As preparation for the proof of von Staudt's theorem we derive the converse of Theorem 8 for a Moufang plane $\pi$ with coordinate ring $\underline{D}$. L denotes as before the line [0].

Lemma 10. ([3]) Let $\sigma: L \to L$ be a bijection satisfying

(i) $h(a,b,c,d)$ implies $h(\sigma(a),\sigma(b),\sigma(c),\sigma(d))$ and

(ii) $\sigma(0) = 0$, $\sigma(1) = 1$, $\sigma(\infty) = \infty$.

Then $\sigma|_{\underline{D}}$ is an automorphism of $\underline{D}^+$.

Proof: By Lemma 5, $h(a,b,\infty,(a+b)/2)$. (i) thus implies, in conjunction with $\sigma(\infty) = \infty$, that $h(\sigma(a),\sigma(b),\infty,\sigma((a+b)/2))$. Since also $h(\sigma(a),\sigma(b),\infty,(\sigma(a)+\sigma(b))/2)$ and the fourth harmonic point is unique (Theorem 4) we have $\sigma((a+b)/2) = (\sigma(a)+\sigma(b))/2$. From $h(0,a,\infty,a/2)$ we similarly deduce that $\sigma(a/2) = \sigma(a)/2$. It follows that (16) (i) holds. From $h(a,-a,\infty,0)$ follows $\sigma(-a) = -\sigma(a)$ so in particular $\sigma(-1) = -1$. Now from $h(1,-1,a,a^{-1})$ follows $\sigma(a^{-1}) = \sigma(a)^{-1}$ so $\sigma|_{\underline{D}}$ is an automorphism of $\underline{D}^+$ by Lemma 8.

Combining the Lemmas 6, 7 and 10 we obtain

Theorem 9. (von Staudt's theorem) Let $\sigma: L \to L$ be a bijection satisfying: $h(a,b,c,d)$ implies $h(\sigma(a),\sigma(b),\sigma(c),\sigma(d))$. Then $\sigma = \tau\alpha$ where $\tau$ is a projectivity of L and $\alpha$ is an automorphism of $\underline{D}^+$ extended so that $\alpha(\infty) = \infty$.

Proof: Since $\sigma$ is bijective, $\sigma(0)$, $\sigma(1)$, $\sigma(\infty)$ are distinct points of L. By Lemmas 6 and 7 there is a projectivity $\tau: L \to L$ such that $\tau^{-1}\sigma(0) = 0$, $\tau^{-1}\sigma(1) = 1$, $\tau^{-1}\sigma(\infty) = \infty$. Since $\tau^{-1}$ preserves harmonic position by the corollary to Theorem 7, $\tau^{-1}\sigma$ satisfies the hypotheses of Lemma 10 so $\alpha = \tau^{-1}\sigma$ is an automorphism of $\underline{D}^+$.

Theorem 9 has been generalized [2] to weaken the hypothesis that $\sigma$ is bijective. With the assumption that $\sigma$ has at least three distinct image points, one concludes that $\sigma = \sigma_1\sigma_2\sigma_3$ where $\sigma_1$ and $\sigma_3$ are projectivities and $\sigma_2$ is a place from $\underline{D}^+$ to itself.

Theorem 9, as stated, is somewhat less than satisfactory since the conclusion is neither purely algebraic nor purely geometric. Since the group of projectivities of L is generated by algebraic transformations (Theorem 7) we can reformulate the theorem in the purely algebraic form

Theorem 9'. The group of bijective mappings of $\hat{\underline{D}}$ to itself satisfying: $(a,b:c,d) = -1$ implies $(\sigma(a),\sigma(b):\sigma(c),\sigma(d)) = -1$ is generated by $\{t_a\} \cup \{r_a\} \cup \{i\} \cup \mathrm{Aut}\ \underline{D}^+$ (Aut $\underline{D}^+$ the group of all automorphisms of $\underline{D}^+$).

In §10 we will consider a geometric analog of Theorem 9.

## 9. JORDAN RINGS (CONTINUED)

While the statement of Theorem 9' is purely algebraic, it is not Jordan algebraic since $r_a$ appears to be an alternative, not Jordan, entity. To formulate von Staudt's theorem in Jordan terms we need to introduce the concept of isotopy in Jordan rings. If $J$ is a Jordan ring and $u \in J$ is invertible, the u-isotope $J^{(u)}$ of $J$ shares its additive group structure with $J$ but has product $x \cdot_u y = \{xuy\}$. It is well known [7] that $J^{(u)}$ is again a Jordan ring (with identity $u^{-1}$). An isotopy of $J$ is an isomorphism from $J$ to some u-isotope. In particular, an isotopy maps 1 to $u^{-1}$. Since for $J = A^+$, $A$ associative, $x \cdot_a y = \{xay\} = 1/2((xa)y + (ya)x)$ it follows easily that $r_{a^{-1}}: x \to xa^{-1}$ is an isotopy of $J$. A similar result holds for $J = \underline{D}^+$, $D$ alternative. It follows that every isotopy of $\underline{D}^+$ is of form $\sigma = r_a\tau$ for some $\tau \in \mathrm{Aut}\ \underline{D}^+$. If we denote by $\Gamma(J)$ the group of all isotopies of the Jordan ring $J$ ($\Gamma(J)$ is the structure group of $J$) we have shown that $\Gamma(\underline{D}^+)$ is generated by $\{r_a\} \cup \mathrm{Aut}\ \underline{D}^+$. We can thus reformulate Theorem 9 once more as

Theorem 9". The group of bijective mappings of $\hat{\underline{D}}$ to itself preserving harmonicity is generated by $\{t_a\} \cup \{i\} \cup \Gamma(\underline{D}^+)$.

The linear analog of the group generated by $\{t_a\} \cup \{i\} \cup \Gamma(J)$ has been introduced and studied by Koecher [8] in another context. It has been applied by Braun [1] in her study of cross-ratios in Jordan algebras.

## 10. A GEOMETRIC VERSION OF VON STAUDT'S THEOREM

In this section we restrict out attention to proper Moufang planes, so $\underline{D}$ is assumed to be a Cayley algebra over its center k.

Recall that a <u>duality</u> of $\pi$ is a bijective, incidence preserving correspondence between the set of points of $\pi$ and the set of lines of $\pi$, i.e. a collineation from $\pi$ to the dual plane $\pi^*$. If $\delta$ is a duality of $\pi$ such that $\delta(L) \notin L$, $\delta$ induces a bijective mapping $\delta_o$ of $L$ by $\delta_o(P) = \delta(P) \cap L$ for $P \in L$. We denote by $\underline{I}$ the group of transformations of $L$ generated by all $\delta_o$, $\delta$ a duality of $\pi$, and all $\sigma_o = \sigma|_L$ where $\sigma$ is a collineation of $\pi$ leaving $L$ invariant. Using the existence of all elations in $\pi$ with axis $L$, one can easily show that each element of $\underline{I}$ induced by a collineation $\sigma$ is in fact induced by a collineation fixing $(\infty)$. Similarly, the mapping induced by a duality $\delta$ is induced by a duality with $\delta(L) = (\infty)$, $\delta((\infty)) = L$. We denote by $C_\infty$ the group of collineations fixing $(\infty)$ and leaving $L$ invariant, by $\Delta_\infty$ the set of dualities exchanging $L$ and $(\infty)$, and by $\underline{I}_o$ the subgroup of $\underline{I}$ induced by $C_\infty$.

We show first that $\Delta_\infty$ is not empty. The mapping $\delta$ defined by $\delta((a,b)) = [-\bar{a},\bar{b}]$, $\delta((m)) = [\bar{m}]$, $\delta((\infty)) = \infty \wedge (\infty)$ is easily checked to be a duality exchanging $[0]$ and $\infty$. The mapping $\sigma\colon \pi \to \pi$ defined by $\sigma((a,b)) = (b,a)$, $\sigma((m)) = (m^{-1})$ if $m \neq 0$, $\sigma(\infty) = (\infty)$, $\sigma((\infty)) = \infty$ is a collineation of $\pi$. If $\sigma^*$ denotes the collineation of $\pi^*$ induced by $\sigma$, $\delta' = \sigma^*\delta\sigma$ is a duality of $\pi$ lying in $\Delta_\infty$.

<u>Lemma 11</u>. $\gamma\colon a \longrightarrow \bar{a}$, $\gamma(\infty) = \infty$ is in $\underline{I} \setminus \underline{I}_o$.

<u>Proof</u>: The duality $\delta'$ above maps $(a,0)$ to $[\bar{a}]$, hence $\delta'_o(a) = [\bar{a}] \cap [0,0] = \bar{a}$. Similarly $\delta'_o(\infty) = \infty$.

Suppose $\gamma \in \underline{I}_o$ so there is $\sigma \in C_\infty$ with $\sigma_o = \gamma$. Since $\gamma$ fixes $0$, $1$, and $\infty$, $\sigma((1,1)) = (1,q)$ for some $q \neq 0$. By remarks in §5 we see that $\sigma$ induces an isomorphism of $\underline{D}$ with the ring structure on $(\underline{D},+)$ with product $a*b = (aq)(q^{-1}b)$, hence $\overline{ab} = (\bar{a}q)(q^{-1}\bar{b})$. If $q \in k$ this implies that $\underline{D}$ is a commutative algebra, a contradiction. If $q \notin k$, there is $p \in \underline{D}$ with $pq \neq qp$. Then $qp = \bar{\bar{q}}\,\bar{\bar{p}} = \bar{\bar{p}}\,\bar{\bar{q}} = (\bar{\bar{p}}q)(q^{-1}\bar{\bar{q}}) = pq$ which again is a contradiction.

<u>Lemma 12</u>. $\underline{I} = \underline{I}_o \cup \gamma\underline{I}_o$.

<u>Proof</u>: By definition $\underline{I}_o \subseteq I$. Suppose that $\delta \in \Delta_\infty$ and let $\delta'$ be the duality defined above. $\delta' \in \Delta_\infty$ implies $(\delta')^{-1} \in \Delta_\infty$. It follows that $(\delta')^{-1}\delta \in C_\infty$ and a simple

computation shows that $((\delta')^{-1}\delta)_o = ((\delta')^{-1})_o\delta_o = (\delta'_o)^{-1}\delta_o$. Thus $\delta_o = \delta'_o\sigma = \gamma\sigma$ for $\sigma \in \underline{I}_o$.

Since I is by definition generated by $\underline{I}_o \cup \gamma\underline{I}_o$, it suffices to conclude that $\underline{I}_o \cup \gamma\underline{I}_o$ is a group. Since for $\delta_1, \delta_2 \in \Delta_\infty$, $(\delta_1)_o(\delta_2)_o = (\delta_1\delta_2)_o = \tau_o$ for $\tau \in C_\infty$ and $\sigma\delta_1 \in \Delta_\infty$ if $\sigma \in C_\infty$ the result follows from the conclusion of the first part of the proof.

Corollary 1. $\underline{I}_o$ is of index 2 in $\underline{I}$.

Proof: By Lemma 12, $[\underline{I}:\underline{I}_o] \leq 2$. By Lemma 11, $\underline{I} \neq \underline{I}_o$.

Corollary 2. If $\sigma \in \underline{I}$ then $h(a,b,c,d)$ implies $h(\sigma(a),\sigma(b),\sigma(c),\sigma(d))$.

Proof: For $\sigma \in \underline{I}_o$ this is clear since any collineation inducing $\sigma$ will preserve the configuration (7). Theorem 5 implies that $\gamma$ preserves the harmonic relation, hence $\gamma\sigma$ does so for all $\sigma \in \underline{I}_o$.

Our aim in this section is to prove the converse to Corollary 2.

Theorem 10. Suppose $\sigma: L \to L$ is a bijection preserving harmonic position. Then $\sigma \in \underline{I}$.

Proof: Lemmas 6 and 7 imply that there is a projectivity (hence element of $\underline{I}_o$) $\tau$ such that $\tau\sigma$ fixes the points 0, 1, and $\infty$ and preserves harmonic position (Corollary to Theorem 7). Lemma 10 then implies $\tau\sigma$ is an automorphism of $\underline{D}^+$. The proof will be complete when we show that every automorphism of $\underline{D}^+$ is in $\underline{I}$. This we do in the following sequence of lemmas.

Lemma 13. Let $O(n)$ be the orthogonal group of the norm form $n(x)$ of $\underline{D}$, $O^+(n)$ the subgroup of transformations of determinant 1. If $\sigma$ is a bijection of L such that $\sigma|_{\underline{D}} \in O^+(n)$ then $\sigma$ is a projectivity of L.

Proof: $O^+(n)$ is generated by mappings $\sigma_S: S \oplus S^\perp \to S \oplus S^\perp$ with $\sigma_S(x) = x$ for $x \in S$, $\sigma_S(x) = -x$ for $x \in S^\perp$ as S runs over all two dimensional subspaces of $\underline{D}$. For given S and $a \in S$, $Sa^{-1}$ is a two dimensional subspace of $\underline{D}$ containing 1. Moreover $S^\perp a^{-1} = (Sa^{-1})^\perp$ since

$t(x)$ is associative. It follows from Lemma 2 that $\sigma_S = r_{a}1_{b^{-1}}r_b r_{a^{-1}}$ for $b \in Sa^{-1}$, $n(b,1) = 0$. It then follows from Lemma 6 and the succeeding remark that $\sigma_S$ is a projectivity.

Corollary. $O(n) \subseteq \underline{I}$.

Proof: $O(n) = O^+(n) \cup \gamma O^+(n)$. Since $\gamma \subset \underline{I}$ and $O^+(n) \subseteq I$ we are finished.

Lemma 14. If $\sigma$ is an automorphism of $\underline{D}^+$, then $\sigma \in \underline{I}$.

Proof: Define on $\underline{D}$ a new product $a * b = \sigma(\sigma^{-1}(a)\sigma^{-1}(b))$ and denote this ring by $\underline{D}^*$. By Lemma 9 $n(a*b) = \sigma(n(\sigma^{-1}(a))n(\sigma^{-1}(b))) = \sigma(\sigma^{-1}(n(a))\sigma^{-1}(n(b))) = n(ab)$ so by the opening remarks of §1, $\underline{D}^*$ is a Cayley algebra over $k$ with norm form $n(x)$. It is not difficult to prove [6] that there is an isomorphism $\tau: \underline{D} \to \underline{D}^*$ which is necessarily in $O(n)$. By construction, $\sigma^{-1}: \underline{D}^* \to \underline{D}$ is an isomorphism of alternative rings, so $\sigma^{-1}\tau = \alpha$ is an automorphism of $\underline{D}$, hence there is a collineation $\hat{\alpha}$ of $\pi$ $(\hat{\alpha}((a,b)) = (\alpha(a),\alpha(b))$; $\hat{\alpha}((\infty)) = (\infty)$; $\hat{\alpha}((m)) = (\alpha(m)))$ such that $\hat{\alpha}$ induces $\sigma^{-1}\tau$ on $L$. It follows that $\sigma^{-1}\tau \in \underline{I}$ and since $\tau \in \underline{I}$ by the corollary to Theorem 10, $\sigma \in \underline{I}$.

We thus complete the proof of Theorem 10, which we view as a geometric formulation of von Staudt's theorem (Theorem 9).

## 11. A FINAL LOOK AT CROSS-RATIOS

We began this chapter by introducing an algebraic concept, the cross-ratio, which allowed algebraic formulation of the concept of harmonic position. This, in turn, facilitated the description of the group of harmonicity preserving mappings of the projective line. We conclude by describing the group $\underline{R}$ of bijections of $L$ preserving cross-ratios.

In §§3 and 4 we have showed that the group $\underline{P}$ of projectivities of $L$ is contained in $\underline{R}$. By Theorem 3, $\gamma: a \to \bar{a}$ is also in $\underline{R}$. Since clearly any $\sigma$ preserving the cross-ratio also preserves harmonicity, Theorem 10 implies $\sigma \in \underline{I} = \underline{I}_o \cup \gamma \underline{I}_o$. To determine $\underline{R}$, it thus suffices to prove

Theorem 11. [11] $\sigma$ is a collineation of $\pi$ leaving L invariant and satisfying: $(\sigma_o(a),\sigma_o(b):\sigma_o(c),\sigma_o(d)) = (a,b:c,d)$ for all $a,b,c,d \in L$, if and only if $\sigma_o \in \underline{P}$.

Proof: By Lemmas 6 and 7, we may assume $\sigma_o$ fixes 0, 1, and $\infty$. Since $\sigma_o$ preserves harmonic position, $\sigma_o$ is an automorphism of $\underline{D}^+$ by Lemma 10. By Lemma 9, $n(\sigma_o(a)) = \sigma_o'(n(a))$ for all $a \in \underline{D}$. Now $[\sigma_o(a)] = (\infty,0:1,a) = [a]$ so $n(\sigma_o(a)) = n(a)$ since $\sigma_o(a)$ is a conjugate of $a$. It follows that $\sigma|_k = Id$ so $\sigma_o \in O(n)$. If $\sigma_o \notin O^+(n)$, Lemma 13 implies that $O(n) = O^+(n) \cup \sigma O^+(n) \subseteq \underline{I}_o$. Since $\gamma \in O(n)$, $\gamma \notin \underline{I}_o$ by Lemma 11, this leads to a contradiction. Thus $\sigma_o \in O^+(n) \subseteq \underline{P}$ as desired.

The converse is proved in the remarks preceding the statement of the theorem.

An interesting consequence of Theorem 11, paralleling the classical result for desarguesian planes is

Corollary. $\underline{P}$ acts transitively on the set of quadruples with fixed cross-ratio $[\lambda]$, $\lambda \in \underline{D}$.

Proof: By Theorem 11, if $(a,b:c,d) = [\lambda]$ and $\sigma \in \underline{P}$, $(\sigma(a),\sigma(b):\sigma(c),d(d)) = [\lambda]$. Conversely, if $(a,b:c,d) = [\lambda] = (a',b':c',d')$ we may assume by Lemma 7 and Theorem 11 that $a = a' = 0$, $b = b' = 1$, $c = c' = \infty$, so $(a,b:c,d) = [\lambda] = [(-d)^{-1}(1-d)] = [1-d^{-1}] = [1-(d')^{-1}] = (a',b',c',d')$. It follows that $d'$ is conjugate to $d$ in $\underline{D}$, i.e. $d' = u^{-1}du$ where we may assume $u \notin k$. By Lemma 2 and the multiplicativity of the norm form we see that $\sigma: x \to u^{-1}xu$ is in $O^+(n)$, hence in $\underline{P}$ by Lemma 13.

## REFERENCES

1. Braun, H.: Doppelverhältnisse in Jordan-Algebren, 1968, Abh. Math. Sem. Hamburg, 32, pp. 25-51.

2. Buekenhout, F.: Une généralisation du théorème de Von Staudt-Hua, 1965, Acad. Roy. Belg. Bull. Cl. Sci. (51), pp. 446-457.

3. Havel, V.: Eine Bemerkung zum Staudtschen Satz in der Moufang-Ebene, 1957, Czech. Math. J. 7(82), pp. 314-317.

4. Havel, V.: Harmonical quadruplets in Moufang planes, 1955, Czech. Math. J. 5(80), pp. 76-82.

5. Havel, V.: One generalization of the fourth harmonic point, 1968, Czech. Math. J. 18(93), pp. 294-300.

6. Jacobson, N.: Composition algebras and their automorphisms, 1958, Rend. Circ. Mat. Palermo 7, pp. 55-80.

7. Jacobson, N.: Structure and representations of Jordan algebras, 1968, Amer. Math. Soc. Colloq. Publ. 34.

8. Koecher, M.: Uber eine Gruppe von rationalen Abbildungen, 1967, Invent. Math. 3, pp. 136-171.

9. Pickert, G.: Projective Ebene, 1955, Springer-Verlag, Berlin.

10. Schafer, R.: An introduction to nonassociative algebras, 1966, Academic Press, New York.

11. Schleiermacher, A.: Doppelverhältnisse auf einer Geraden in einer Moufang-Ebene, 1965, Indag. Math. 27, pp. 482-496.

12. Smiley, M.: Von Staudt projectivities of Moufang planes, 1959, Coll. on alg. and top. foundations of geom., Utrecht.

13. Springer, T.: Jordan algebras and algebraic groups, 1973, Ergib. Math., vol. 75, Springer-Verlag, New York.

# CROSS RATIOS AND A UNIFYING TREATMENT OF VON STAUDT'S NOTION OF REELLER ZUG

Walter Benz, Hans-Joachim Samaga,
Helmut Schaeffer

Hamburg

On the basis of a complex projective geometry K.G.Chr. von Staudt defined the notion of reeller Zug or what was called later on von Staudt'sche Kette. Those chains can be represented for instance in the completed complex plane as circles or lines extended by the point at infinity.
Given four distinct points A,B,C,D: They are on a common chain iff their cross ratio is an element of $\mathbb{R}$ (Neutraler Wurf). This definition of von Staudt carries over to other plane geometries, namely to those of Laguerre and Minkowski: Replacing the projective line over $\mathbb{C}$ by those over $\mathbb{D}$ (ring of duals) or $\mathbb{A}$ (ring of double numbers) one again has von Staudt's notion of reeller Zug thus representing Laguerre-, Minkowski-circles respectively. The following report deals with a unifying theory of von Staudt's basic notion of reeller Zug collecting various results which were obtained by different authors. However, restrictions are inevitable according to the purpose of this paper. Especially we had to exclude theorems characterizing different classes of circle geometries axiomatically. We also had to exclude the finite case as well as topological considerations. There are many results in those areas: a report in book form would be more appropriate to cover all these investigations.



*P. Plaumann and K. Strambach (eds.), Geometry - von Staudt's Point of View, 127–150.*

## § 1 The Geometry $A_K$ .

Let K be a commutative field and A a commutative and associative algebra over K with 1 such that $\dim_K A \geq 2$ . By identifying $k \in K$ and $k \cdot 1 \in A$ we can consider K as a subset of A. We call $\frac{x}{y}$ a quotient over A if $x,y \in A$ and if A is the ideal generated by x,y in A. Put $\frac{x}{y} = \frac{x'}{y'}$ if $xy' = x'y$ and define U to be the group of units of A.

<u>Lemma:</u> $\frac{x}{y} = \frac{x'}{y'}$ iff there exists $u \in U$ such that $x'=ux$ and $y' = uy$ .

<u>Proof:</u> Consider $\frac{x}{y} = \frac{x'}{y'}$ . Then there exist $a,b \in A$ such that $1=ax+by$ since A is the ideal generated by x,y.Putting $u:=ax'+by'$ we get $x' = ux$, $y' = uy$. Since also A is the ideal generated by x',y' it follows $u \in U$ .

This Lemma implies that the equality relation on the set of quotients over A must be an equivalence relation. In the sequel we do not distinguish between a quotient and the equivalence class generated by that. By identifying $a \in A$ and $\frac{a}{1}$ we can consider A as a subset of the set $\bar{A}$ of all quotients over A. Given $u \in U$ and $a \in A$ then obviously

$$\frac{au^{-1}}{1} = \frac{a}{u} .$$

This "old quotients" $au^{-1} \in A$ can be considered as new ones.
The elements of $\bar{A}$ are called points of the geometry $A_K$ . Given $a,b,c,d \in A$ such that $ad-bc \in U$ then

$$\frac{x}{y} \rightarrow \frac{ax+by}{cx+dy}$$

turns out to be a permutation of $\bar{A}$ . By $\Gamma = PGL(2,A)$ we denote the group of all those permutations.
$\Gamma$ is isomorphic to the group $GL(2,A)$ modulo its center. Define $\pi$ to be the set of all quotients which can be written in form $\frac{x}{y}$ such that $x,y \in K$. Given $\gamma \in \Gamma$ the set of points

$$\pi^\gamma := \{ p^\gamma | p \in \pi \}$$

is called a chain of the geometry $A_K$.

Assume for a moment $\dim_K A = n < \infty$. Denote by L the quotient field of $K[x_1,\ldots,x_n]$. Then the group of birational transformations of the n-dimensional space over K is given by the group $\mathrm{Aut}_K L$ of all automorphisms of L fixing K elementwise. Such an automorphism is determined as soon as one knows the images

$$\frac{f_i(x_1,\ldots,x_n)}{g_i(x_1,\ldots,x_n)} =: x_i' \text{ of } x_i,\ i=1,\ldots,n \quad .$$

Given $(\xi_1,\ldots,\xi_n)$, $\xi_i \in K$, such that

$$\prod_{i=1}^{n} g_i(\xi_1,\ldots,\xi_n) \neq 0$$

then one can associate to $(\xi_1,\ldots,\xi_n)$ an image point

$(\xi_1',\ldots,\xi_n')$ with respect to an automorphism in $\mathrm{Aut}_K L$.

Let G be a subgroup of $\mathrm{Aut}_K L$ such that not all $g \in G$ are line preserving. The geometry of G in the sense of Kleins Erlangen Programme is then called a Cremonian geometry. Since $\Gamma$ is induced by a subgroup of $\mathrm{Aut}_K L$ the geometries $A_K$ are Cremonian geometries:
Let $e_1,\ldots,e_n$ be a basis of A over K.

Then
$$\frac{x}{1} \to \frac{ax+b}{cx+d}$$

leads to an automorphism in $\mathrm{Aut}_K L$ replacing x by

$$\sum_{i=1}^{n} x_i e_i \quad .$$

For example:
$K = \mathbb{Q}$, $A = \mathbb{Q}(\sqrt[3]{2})$, $e_1=1$, $e_2=\sqrt[3]{2}$, $e_3=\sqrt[3]{4}$,
$\frac{x}{y} \to \frac{y}{x}$ we get the automorphism given by

$$x_1 \to x_1' = \frac{x_1^2 - 2x_2x_3}{\rho}$$

$$x_2 \to x_2' = \frac{2x_3^2 - x_1x_2}{\rho}$$

$$x_3 \to x_3' = \frac{x_2^2 - x_1x_3}{\rho} \quad ,$$

where $\rho$ is equal to

$$x_1^3 - 6x_1 x_2 x_3 + 2x_2^3 + 4x_3^3 .$$

The chains of $A_K$ turn out to be algebraic curves as far as their part in A is concerned.

We say that $\{P,Q,R,\ldots\} \subset \bar{A}$ is cocircular in case there exists a chain containing $\{P,Q,R,\ldots\}$ as a subset. The points $P \neq Q$ are called parallel, $P \parallel Q$, iff $\{P,Q\}$ is not cocircular. We also define $P \parallel P$ for all $P \in \bar{A}$. The points $\frac{x}{y}, \frac{x'}{y'}$ are called parallel iff $\begin{vmatrix} x x' \\ y y' \end{vmatrix} \notin U$. The parallel relation on $\bar{A}$ is reflexive and symmetrical. It is transitive iff A is a local ring. If $\gamma \in \Gamma$ and $P \parallel Q$ then $P^\gamma \parallel Q^\gamma$. $\Gamma$ operates transitively on the set of points and furthermore sharply transitively on the set of ordered triples of pairwise non-parallel points P,Q,R. Through any three pairwise non-parallel points there is exactly one chain.
Consider four pairwise non-parallel points P,Q,R,S and let $\gamma \in \Gamma$ be a mapping such that

$$P^\gamma = \frac{1}{0}, \quad Q^\gamma = 0, \quad R^\gamma = 1 .$$

Then $S^\gamma$ is called the cross ratio $\begin{bmatrix} P & Q \\ S & R \end{bmatrix}$ of the ordered quadruple P,Q,R,S. By putting $P = \frac{p_1}{p_2}, \ldots,$ we get

$$(*) \qquad \begin{bmatrix} P & Q \\ S & R \end{bmatrix} = \frac{\begin{vmatrix} p_1 r_1 \\ p_2 r_2 \end{vmatrix}}{\begin{vmatrix} p_1 s_1 \\ p_2 s_2 \end{vmatrix}} \cdot \frac{\begin{vmatrix} q_1 s_1 \\ q_2 s_2 \end{vmatrix}}{\begin{vmatrix} q_1 r_1 \\ q_2 r_2 \end{vmatrix}} ,$$

which turns out to be an element of U since P,Q,R,S are pairwise non-parallel.
Observing that distinct points P,Q on a chain c must be non-parallel we get von Staudt's characterization of chains:
Four pairwise non-parallel points are cocircular iff their cross ratio is in K.
We like to mention that the expression (*) is also well-defined in case $P \nparallel S$ and $Q \nparallel R$.
The vector space A over K defines an affine geometry $\Delta$ over K. The set of points of $\Delta$ can be identified with A and is thus a subset of $\bar{A}$. We sometimes call a point $P \in \bar{A} \setminus A$ an improper point and a point $Q \in A$ an affine point of $A_K$. Obviously, $\bar{A} \setminus A = \{X \in \bar{A} \mid X \parallel \frac{1}{0}\}$.

The geometry $A_K$ is called of type

a) Möbius
b) Laguerre
c) Minkowski

respectively iff

a) $P \parallel Q$ iff $P = Q$
b) $\parallel$ is an equivalence relation. To every chain c and to every point $P \in \bar{A}$ there exists $Q \in \bar{A}$ such that $c \ni Q \parallel P$.
c) There exist two equivalence relations $\parallel_+$, $\parallel_-$ on $\bar{A}$ such that

$c_1$) $P \parallel Q$ iff $P \parallel_+ Q$ or $P \parallel_- Q$

$c_2$) $P = Q$ iff $P \parallel_+ Q$ and $P \parallel_- Q$

$c_3$) To every chain c and to every point P there exist $P_+, P_- \in c$ such that $P_+ \parallel_+ P$ and $P_- \parallel_- P$ .

The following can be proved: Given $A_K$, then

a) iff A is a field
b) iff A is local such that $A = K \oplus (A \setminus U)$
c) iff $A = K \oplus K$ .

The set $A \setminus U$ of points is called the singular cone C of $A_K$. It is C the union of all maximal ideals of A and thus a union of subspaces of the vector space A. Given a line g of $\Delta$ . If the line $h \ni 0$ parallel to g is in C, $h \subset C$, we call g singular and regular otherwise. Put $W = \frac{1}{0}$ and consider a chain $c \ni W$. Then $c \setminus \{W\}$ is a regular line of $\Delta$ . If on the other hand g is a regular line then $g \cup \{W\}$ is a chain passing through W.
Define B to be the set of all ordered triples (c,P,d), where c,d are chains and where P is a point of $c \cap d$. A contact relation of $A_K$, is by definition a subset of B. By assuming $|K|>2$ the following can be proved: There is exactly one contact relation $B_o$ such that the following conditions hold true (writing aPb instead of $(a,P,b) \in B_o$).

1) aPb and $a \neq b$ imply $|a \cap b| = 1$
2) $P \in a$ implies aPa
3) aPb $\Rightarrow$ bPa
4) aPb and bPc imply aPc

5) If t is a chain and $P \nparallel Q$ are points such that $P \in t \not\ni Q$ then there is exactly one chain $t' \ni P,Q$ with $tPt'$
6) $\forall_{\gamma\in\Gamma}$ $aPb \Rightarrow a^\gamma P^\gamma b^\gamma$ .

This basic contact relation $B_o$ can be described explicitly as follows:
Given $(a,P,b) \in B$ and given $\gamma \in \Gamma$ such that $P^\gamma = W := \frac{1}{0}$ . Then aPb iff the lines $a^\gamma \setminus \{W\}$, $b^\gamma \setminus \{W\}$ are parallel in $\Delta$ .

This contact relation $B_o$ is also useful in case $|K|=2$. By means of cross ratios the relation $B_o$ can be characterized as follows:
Consider chains a,b through a point P. Then the following conditions are equivalent:

1) a P b
2) For distinct points $Q,Q' \in a \setminus \{P\}$ and for distinct points $R,R' \in b \setminus \{P\}$

$$\begin{bmatrix} P & Q \\ R & Q' \end{bmatrix} \equiv \begin{bmatrix} P & Q \\ R' & Q' \end{bmatrix} \quad (\text{mod } K^+)$$

holds true.

3) There exist distinct points $Q,Q' \in a \setminus \{P\}$ and distinct points $R,R' \in b \setminus \{P\}$ such that

$$\begin{bmatrix} P & Q \\ R & Q' \end{bmatrix} \equiv \begin{bmatrix} P & Q \\ R' & Q' \end{bmatrix} \quad (\text{mod } K^+) .$$

There exist geometries $A_K$ such that $B_o$ is not invariant under all automorphisms of $A_K$. However, $B_o$ is invariant if the following condition holds true.

(R) To every $a \in A$ there exist distinct $\alpha,\beta \in K$ that $a+\alpha$ and $a+\beta$ are in U.

If the number of maximal ideals of A is finite, say n, and if $|K| \geq n + 2$, then (R) is a consequence. If $A_K$ is of Möbius-, Laguerre- or Minkowski type then $n \leq 2$.
$B_o$ is called sharp iff aPb and $a \neq b \Leftrightarrow a \cap b=\{P\}$.
It can be proved that $B_o$ is sharp iff $[A:K] = 2$ or $K^* = U$.
If A is the ring of polynomials of one variable over K then obviously $K^* = U$.
Assume $\dim_K A < \infty$ .
Consider the affine part $c_{aff}$ of a chain c and

define the dimension of c, dim c, to be the dimension of the linear closure of $c_{aff}$. If $K(\gamma), \gamma \in A$, denotes the smallest algebra in A containing $K \cup \{\gamma\}$ then all numbers $\dim_K K(\gamma)$ occur as dimensions of chains. If A is quadratic then all chains are at most 2-dimensional. The material presented in § 1 can be found more explicitly in [ 5 ],[ 6 ] .

§ 2 The real case in 3 dimensions.

Consider $K = \mathbb{R}$ and A of form $\mathbb{C}$, $\mathbb{D}$, $\mathbb{A}$ respectively. Then $A_K$ is the classical plane geometry of Möbius, Laguerre, Minkowski respectively. For arbitrary K and $\dim_K A=2$ we get the miquelian Möbius-,Laguerre-, Minkowski planes. We now like to deal with $K=\mathbb{R}$ and $\dim_{\mathbb{R}} A=3$. There are exactly five algebras A of this kind, namely

$$\mathbb{C} \oplus \mathbb{R}, \mathbb{D} \oplus \mathbb{R}, \mathbb{A} \oplus \mathbb{R}, \mathbb{R}[x]/\langle x^3\rangle,$$

$$\mathbb{R}[x,y]/\langle x^2, xy, y^2\rangle.$$

The set of affine points is in all cases the real space $\mathbb{R}^3$ and any chain carries at most three improper points. Therefore a chain can be characterized by its affine points. We are now describing the chains in all cases as far as their affine part is concerned, omitting the regular lines. For more details see [ 22 ] .

1) $A=\mathbb{C} \oplus \mathbb{R}$. The singular cone C is the union of the $x_1x_2$-plane and the $x_3$-axis. Chains are suitable hyperbolas and cubic ellipses.
2) $\mathbb{D} \oplus \mathbb{R}$. It is C the union of the $x_1x_2$-plane and the $x_2x_3$-plane. Chains are suitable parabolas,hyperbolas and cubic hyperbolic parabolas.
3) $\mathbb{A} \oplus \mathbb{R}$. The union of the coordinate planes equals C. The chains are suitable hyperbolas and cubic hyperbolas.

4) $\mathbb{R}[x]/_{<x^3>}$ . It is C the $x_2x_3$-plane. Chains are suitable parabolas and cubic parabolas.

5) $\mathbb{R}[x,y]/_{<x^2,xy,y^2>}$ . Again C is the $x_2x_3$-plane. The chains are suitable parabolas.

Chains which are not lines but plane curves of the $\mathbb{R}^3$, must be conic sections. The remaining chains, the cubic ones, are representable as intersections $(Q_1 \cap Q_2) \setminus g$, where $Q_i$ are quadrics having exactly one line g in common.

a) A cubic hyperbolic parabola is the intersection of a parabolic and a hyperbolic cylinder or a hyperbolic paraboloid:

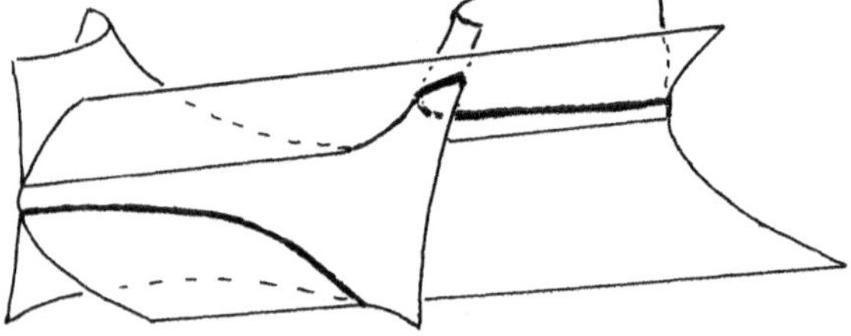

b) A cubic hyperbola is the intersection of two hyperbolic cylinders.

c) A cubic ellipse is the intersection of an elliptic cylinder and a quadratic cone or a hyperbolic paraboloid:

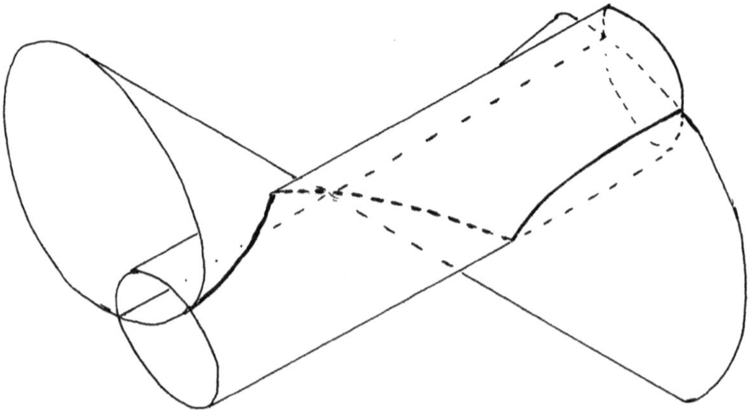

d) A cubic parabola is the intersection of a parabolic cylinder and a cone or a hyperbolic paraboloid:

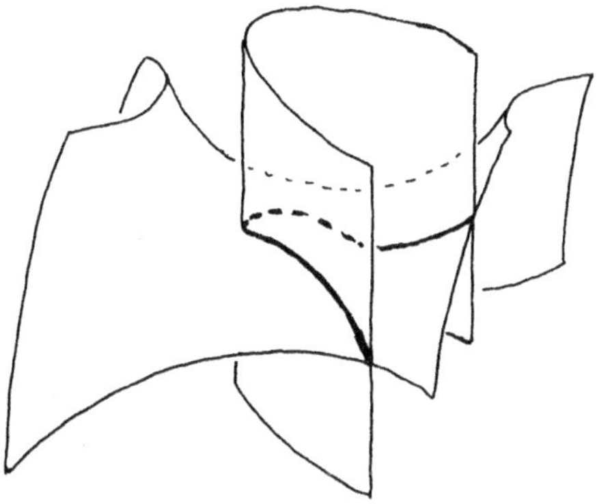

All types of cubic cone sections (s. Fladt-Baur [9]) occur as chains in the $\mathbb{R}^3$. Lines and quadratic parabolas must be completed by one improper point, quadratic hyperbolas by two.

Given a set $\mathbb{P}$ of points and a set $\mathbb{K} \subset 2^{\mathbb{P}}$ of chains. Define $P \parallel Q$ for distinct $P, Q \in \mathbb{P}$ iff there is no chain joining $P, Q$. Put also $P \parallel P$ for all $P \in \mathbb{P}$. Given chains $a, b$ passing through $P$ such that $a \cap b = \{P\}$. Define $aPb$ ("a contacts b in P") iff there exist distinct points $A_1, A_2 \in a \setminus \{P\}$ and distinct points $B_1, B_2 \in b \setminus \{P\}$ and a point $F \neq P$ such that each of the quadruples $PFA_iB_i$, $i=1,2$, is cocircular. Put also $aPa$ for $P \in a$.

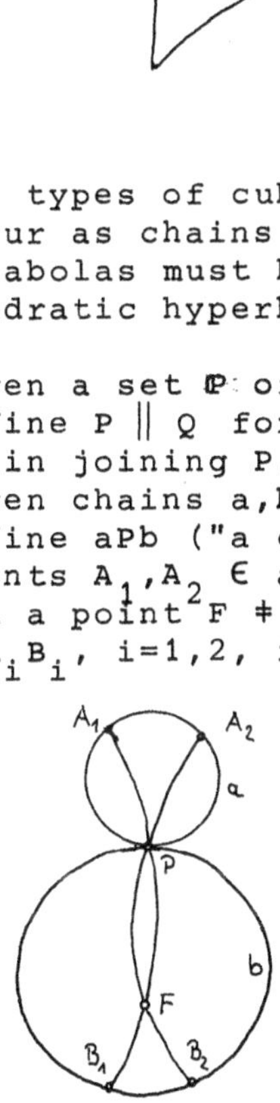


Assume the following axioms:

(1) Through three pairwise non-parallel points there is exactly one chain

(2) $aPb$, $bPc$ imply $aPc$. Given a chain $k$ and points $P \neq Q$ such that $P \in k \not\ni Q$. Then there exists exactly one chain $l \ni P, Q$ with $kPl$.

(3) $|k| \geq 3$ for all $k \in \mathbb{K}$. There exist four pairwise non-parallel points not cocircular.

Given such a chain structure $(\mathbb{P}, \mathbb{K})$. A vector space $V_K$ over a field $K$ is called a derivation in $W \in \mathbb{P}$ if the following conditions hold true:

(i) There is a bijection $\varphi$ between $\{P \in \mathbb{P} \mid P \not\parallel W\}$ and $V$.

(ii) $k \setminus \{W\}$ is a line under that bijection for all $k \in \mathbb{K}$ with $k \ni W$.

(iii) $(k \setminus \{W\})^{\varphi} \parallel (1 \setminus \{W\})^{\varphi}$ implies kWl.

Question: Determine all chain structures having the $\mathbb{R}^3$ as a derivation.

§ 3 von Staudt's Theorem

Let $A \neq 0$ be an arbitrary commutative and associative ring with 1, U the group of units of A, and V a 2-dimensional module over A. The set of points $\bar{A}$ of the projective line over A is defined to be the set of all (free) 1-dimensional direct summands of V. The groups GL(2,A), ΓL(2,A), Γ:= PGL(2,A) and PΓL(2,A) are defined as in the classical case, where A is a field. PΓL(2,A) consists of mappings $\gamma: \bar{A} \to \bar{A}$, which are induced by semilinear bijections of V over A. In order to relate these definitions to those given in § 1 we state the following lemma:

(3.1) If $\{z_1, z_2\}$ is a basis of V over A and

$p = p_1 z_1 + p_2 z_2 \in V$, $p_i \in A$, then the following

conditions are equivalent :

(i) p is unimodular, i.e. there exists a linear mapping $\lambda: V \to A$ such that $\lambda(p) = 1$

(ii) $\frac{p_1}{p_2}$ is a quotient (in the sense of § 1)

(iii) The submodule $P = A \cdot p$ generated by p is free and a direct summand in V.

When $P, Q \in \bar{A}$ are points of the projective line we have $P \nparallel Q \Leftrightarrow P+Q=V$ (or equivalently $P \oplus Q = V$).
The cross-ratio of any quadruple of pairwise non-parallel points is then uniquely obtained by choosing a basis $\{z_1, z_2\}$ of V such that $P = Az_1$, $Q = Az_2$,

$R = A(z_1+z_2)$, $S = A(rz_1+z_2)$, $r \in A$, and $\begin{bmatrix} P & Q \\ S & R \end{bmatrix} = r$ .

Let now two projective lines over rings A,A' be given, 1+1 a unit in both rings. Any semilinear bijection $\gamma: V \to V'$ induces a bijection $\bar{\gamma}: \bar{A} \to \bar{A}'$ by $\bar{\gamma}(P) := \gamma(P)$.

The set of all such $\overline{\gamma}$ is denoted by $\Lambda(A,A')$.
In case A=A', V=V' we have $\Lambda(A,A') = P\Gamma L(2,A)$.
A quadruple of points (P,Q,R,S) of the projective line is called harmonic

iff $\begin{bmatrix} PQ \\ RS \end{bmatrix} = -1$. The classical von Staudt Theorem

says that, in case A,A' are fields, a bijection $\lambda : \overline{A} \to \overline{A}'$ belongs to $\Lambda(A,A')$ iff it is harmonic in the following sense:

$\lambda$ and $\lambda^{-1}$ map harmonic quadruples onto harmonic quadruples.

Several authors have generalized the von Staudt Theorem to rings under various additional assumptions
(see [5] [26] [14] [15] [16] [21])

We mention the following result, the proof of which can be found in [16].

(3.2) Theorem. Let $\overline{A},\overline{A}'$ be the sets of points of projective lines over rings A,A' (1+1 a unit in both rings) and let $\Lambda(A,A')$ be defined as above.

Each of the following conditions (I),(II) on A implies that a bijection $\gamma : \overline{A} \to \overline{A}'$ is harmonic if and only if $\gamma \in \Lambda(A,A')$:

(I) If $P_o, P_1, \ldots, P_5 \in \overline{A}$ are any points, there exists $Q \in \overline{A}$ with $Q \not\parallel P_i$ $(i=0,\ldots,5)$

(II) A is a semilocal ring (i.e. the number of maximal ideals of A is finite) and $|A/I| > 3$ for every maximal ideal I of A.

Clearly, A being a finite dimensional algebra over K, $|K| > 3$, condition (II) holds for A.
We will see in § 4 that in general conditions (I) and (II) in the theorem cannot be dropped.
However, it might be possible to weaken both conditions.
In [21] is proved another generalization of the von Staudt Theorem:

A quadruple of distinct points $(P_1,P_2,P_3,P_4)$ is called "harmonic" iff there exist unimodular elements $z_i \in V$, $i=1,\ldots,4$ and a,b $\in$ A such that $P_i = A z_i$

$Z_3 = aZ_1+bZ_2$, $Z_4 = aZ_1-bZ_2$ .

If furthermore bijections $\varphi : \overline{A} \to \overline{A'}$ such that $\varphi$ and $\varphi^{-1}$ map quadruples of harmonic points into harmonic points are called "harmonic", then the von Staudt Theorem is valid without assumptions (I) or (II).

## § 4 The Automorphism Group of $A_K$

An isomorphism of $A_K$ onto $A'_K$ is by definition a bijection $\varphi: \overline{A} \to \overline{A}'$ such that $\varphi$ and $\varphi^{-1}$ map chains into chains.

For algebras A over K we will consider conditions (III), (IV) (in addition to (I), (II) of § 3 ):

(III) There exists $s \in A \setminus K$ such that $s, s-1 \in U$ (equivalently: There are four pairwise non-parallel points in $A_K$ which are not cocircular)

(IV) A is semilocal and $|K| > 3$.

It is (III) and (II) a consequence of (IV). Furthermore (I) implies (III): If $t \in A \setminus K$ is arbitrary we can find by (I) $s \in A$ with $s \nparallel 0,1,t,t+1$. Because of $s \notin K$ or $s-t \notin K$ and $s, s-1, s-t, (s-t)-1 \in U$ the conclusion follows.

(4.1) Let $A_K, A'_{K'}$ be geometries such that (III) holds for A over K, char $K \neq 2$. If $\varphi: \overline{A} \to \overline{A}'$ is an isomorphism of $A_K$ onto $A'_{K'}$, fixing $\frac{1}{0}$, 0, 1 respectively, then $\varphi(-1) = -1$.

Proof: Let $s \in A$ be as in (III). We get $\varphi(-1)=-1$ by looking to the image points of $0,1,-1,s,-s,s-1,s-2$ (see fig.5) under $\varphi$ .

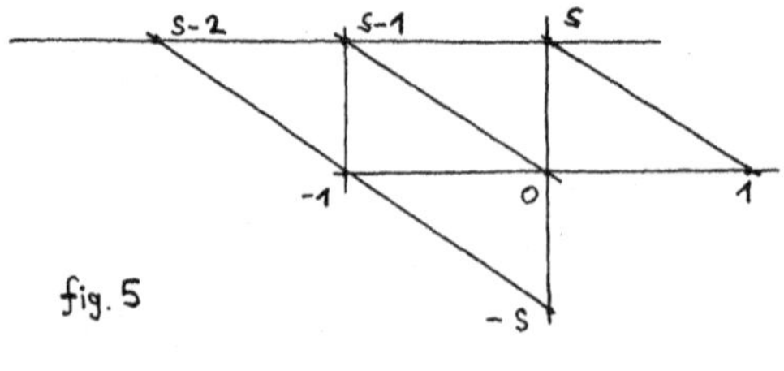


fig. 5

It is a consequence of (4.1) that isomorphisms of $A_K$ onto $A'_{K'}$ are hamonic mappings when (III) as well as char $K \neq 2$ is assumed.
$A_K$ and $A'_{K'}$ being geometries such that char $K \neq 2$ and (I) or (IV) hold for A, we can apply (3.2) and get a "Fundamental Theorem" for circle geometry:

(4.2) $A_K$ and $A'_{K'}$ are isomorphic iff the rings $A, A'$ are isomorphic with respect to an isomorphism $\alpha: A \to A'$ such that $K' = \alpha(K)$.
Let $P\Gamma L_K(2,A)$ be the group of all $\lambda \in P\Gamma L(2,A)$ with the following property: The automorphism associated to $\lambda$ maps K onto K. The same arguments proving (4.2) lead to

(4.3) It is $P\Gamma L_K(2,A)$ the group of automorphism of $A_K$, whenever char $K \neq 2$ and (I) or (IV) hold.

It has been proved that for geometries of Möbius type Theorems (4.2) and (4.3) hold for arbitrary A over K with the only assumption that $|K| \geq 3$.
(See [ 3 ] [ 17 ] [ 27 ]) This is also true for plane Laguerre- and Minkowski-geometries.

Let us finally consider an example. If $A := \mathbb{R}[x]$ the automorphism group of $A_{\mathbb{R}}$ contains $P\Gamma L_{\mathbb{R}}(2,A)$ as a proper subgroup. This can be seen as follows (compare [ 2 ]):

Define $\varphi : \bar{A} \to \bar{A}$ by $\varphi(0) = 0$, $\varphi(1) = 1$, $\varphi(\frac{1}{0}) = \frac{1}{0}$ and for $f(x), g(x) \in \mathbb{R}[x] \setminus \{0\}$ with $\frac{f(x)}{g(x)} \in \bar{A}$,

$$\varphi(\frac{f(x)}{g(x)}) := \frac{f(x)}{g(x)} \quad \text{iff } \deg(f(x)) \equiv \deg(g(x)) \bmod 2,$$

$$\varphi(\frac{f(x)}{g(x)}) = \frac{r.f(x)}{g(x)}$$ otherwise, here $r \neq 0,1$ is a fixed real number. We have $\varphi \notin P\Gamma L_{\mathbb{R}}(2,A)$, but $\varphi$ preserves all cross-ratios and therefore $\varphi$ is an automorphism of $A_{\mathbb{R}}$ and a harmonic mapping as well.

## § 5 EMBEDDING OF $A_K$

H is known that the plane miquelian geometries $A_{\mathbb{R}}$ can be represented as 2-dimensional varieties $\mathcal{O}$ of the 3-dimensional projective space over $\mathbb{R}$, the chains correspond to intersections of $\mathcal{O}$ with suitable planes of the projective space. For details see [ 5 ] .
At first this was generalized by Hotje [ 11 ],[ 12 ].
An algebra A over K, $|K| > 2$ is called quadratic (kinematische Algebra), if there exist $k_x, l_x \in K$ such that $x^2+k_x x+l_x = 0$ for all $x \in L$.

The elements $k_x$ and $l_x$ are defined uniquely if $x \in A \setminus K$. Otherwise in case of $x \in K$ we determine $k_x := -2x$.

$-: A \to A$ defined by $\bar{x} := -x-k_x$ is an involution and $N : A \to K$; $N(x) := x\bar{x}$ is a quadratic form.
All algebras A of the 2-dimensional geometries $A_K$ as well as $A := K[x_1, \ldots, x_n] / \langle x_1^2, \ldots, x_n^2, x_1x_2, x_1x_3, \ldots, x_{n-1}\, x_n \rangle$ are examples for quadratic algebras. We need some denotations: Let $\Pi$ be the projective space $(A\times K\times K)^* / K^*$ .

Let Q be the quadratic form $A\times K\times K \to K$ defined by $Q(a,s,t) := N(a)-st$; $\mathrm{Rad}_Q$ the radical belonging to Q, $\mathrm{Rad}_Q\Pi := (\mathrm{Rad}_Q)^* / K^*$ and $\mathcal{O} := \{[a,s,t] \in \Pi \mid Q(a,s,t)=0\}$

Finally let $\varepsilon$ be the projective plane generated by $[0,1,0]$, $[1,1,1]$, $[0,0,1] \in \Pi$.

The following remarks can be proved:

1. The mapping $\Delta: \bar{A} \to \Pi$ defined by $(\frac{a}{b})^\Delta := [a\bar{b}, N(a), N(b)]$ is injective and it is $\bar{A}^\Delta = \mathcal{O} \setminus \mathrm{Rad}_Q\Pi$, $\bar{K}^\Delta = \mathcal{O} \cap \varepsilon$
2. There is a monomorphism $\Lambda : \Gamma \to PGL(n+2,K)$; n denotes the dimension of A.
3. For all $\gamma \in \Gamma$ we have $(\bar{A}^\Delta)^{\Lambda(\gamma)} = \bar{A}^\Delta$, $\mathcal{O}^{\Lambda(\gamma)} = \mathcal{O}$

4. Two different points $P, Q \in \bar{A}$ are parallel iff $P^{\Delta}$, $Q^{\Delta}$ are lying on a line of $\pi$ which is contained totally in $\sigma$.

Altogether we have the theorem

(5.1) Let A be a quadratic algebra with rank n. Then $A_K$ can be embedded as a quadric (without radical) into the (n+1)-dimensional projective space over K. The chains are represented by the intersection of $\sigma$ with planes of $\pi$ having no line but at least two points in common with $\sigma$.

Vice versa the quadratic algebras are the only ones such that $A_K$ is representable in the described way. A nonempty subset $\mathcal{Q}$ of a projective space $\pi$ is called a quadratic set, if

(i) $|l \cap \mathcal{Q}| \leq 2$ or $l \subseteq \mathcal{Q}$ for all lines l of $\pi$

(ii) $\mathcal{Q}_P := \bigcup_{\substack{|l \cap \sigma| \neq 2 \\ P \in l}} l$ l a line, is the whole space $\pi$ or a hyperplane of $\pi$.

If $\mathcal{Q}_P = \pi$ we call P a singular point. The set of all singular points we denote by $R(\mathcal{Q})$. In his paper [ 12 ] Hotje proved the following theorem :

(5.2) Let m be the number of the maximal ideals of A, dim A = n, and $|K| > n+m$. If $A_K$ can be represented as a quadratic set $\mathcal{Q}$ (without $R(\mathcal{Q})$) of the (n+1)-dimensional projective space $\pi$ over K, such that the chains are plane sections, and the images of parallel points are lying on lines of $\mathcal{Q}$, then A is quadratic.

In the meantime Hotje's way of embedding was applied to some more general geometries $A_K$([ 24 ]).

A. Let $\tilde{A} := A \oplus K$, A a n-dimensional quadratic algebra. If $|K| > 2$ we have

(5.3) $\tilde{A}_K$ can be represented as the intersection $\tilde{\sigma}$ of $\frac{n^2+3n+6}{2}$ quadrics (without radicals) of the (2n+3)-dimensional projective space $\pi$ over K, the chains being the intersection of $\tilde{\sigma}$ with suitable 3-dimensional subspaces of $\pi$ .

More precisely: Let $\tilde{a} = (a, a_k) \in A \oplus K = \tilde{A}$. Then there is an injective mapping $\Delta : \tilde{A} \to \pi := (A\times A\times K\times K\times K\times K)^*/_{K^*}$ defined by $\left(\frac{\tilde{a}}{\tilde{b}}\right)^{\Delta} := [a\bar{b}a_k, a\bar{b}b_k, N(a)a_k, N(a)b_k, N(b)a_k, N(b)b_k]$

and it is $\tilde{\sigma} := \{[x_{o1},\dots,x_{on},x_{11},\dots,x_{1n},x_2,\dots,x_5] \mid$

$Q_1 = N(x_{o1},\dots,x_{on}) - x_2x_4 = 0,\ Q_2 = N(x_{11},\dots,x_{1n}) - x_3x_5 = 0,$

$x_2x_5 - x_3x_4 = 0,\ x_{oi}x_3 - x_{1i}x_2 = 0,\ x_{oi}x_5 - x_{1i}x_4 = 0,$

$x_{oi}x_{1j} - x_{oj}x_{1i} = 0$ for $i,j = 1,\dots,n\} \setminus (Rad_{Q_1} \cap Rad_{Q_2})$ .

By $Q_\nu$ we mean the first and the second quadratic form from the set above.

Because of this theorem the classical geometries $A_{\mathbb{R}}$ consisting of $A = \mathbb{C} \oplus \mathbb{R}$, $\mathbb{D} \oplus \mathbb{R}$, $\mathbb{A} \oplus \mathbb{R}$, respectively, can be embedded into a 7-dimensional projective space.

By the way, the algebras $\tilde{A}$ belong to the class of pseudo-quadratic (fast quadratische) algebras defined in [ 6 ].

B. In an analogous way we treat the geometries $A_K := K^n{}_K$, $n \in \mathbb{N} \setminus \{1\}$. The corresponding mapping $\Delta : \bar{A} \to \pi :=$ $:= (K^{2^n})^*/_{K^*}$ is defined by $\left(\frac{(a_1,\dots,a_n)}{(b_1,\dots,b_n)}\right)^{\Delta} =$

$= [a_1\cdot\dots\cdot a_n,\ a_1\cdot\dots\cdot a_{n-1}b_n, a_1\cdot\dots\cdot a_{n-2}b_{n-1}a_n,\dots,$

$b_1\cdot\dots\cdot b_n]$.

(5.4) The geometry $K^n{}_K$ can be embedded as a Segremanifold $S_{1,\dots,1} = S_{(1)^n}$ into the $(2^n-1)$-dimensional projective space, the chains being the Veronesian $V^1_n$ (Also see Halder [ 10 ])

C. In view of A and B it is obious that we are able to treat the geometries $\hat{A}_K$ in the same way where $\hat{A} := A \oplus K^n$, A a quadratic algebra and $n \in \mathbb{N}$.

D. For $A := K[x]/_{<x^3>}$, char $K \neq 2$, we consider

$$\alpha : \begin{cases} A \rightarrow A \\ (x_1, x_2, x_3) \mapsto (x_1^2, -x_1x_2, x_2^2 - x_1x_3) \end{cases}$$

(This is a specialization of the following, more general situation: Let $e_1, \ldots, e_n$ be elements of a basis of the n-dimensional algebra A, such that $x = (x_1, \ldots, x_n) = \sum_\upsilon x_\upsilon e_\upsilon$. Furthermore we define the multiplication in A by $e_i e_j := \sum_\upsilon \Gamma_{ij}^\upsilon e_\upsilon$. If we put $1 = \sum_\upsilon \varepsilon_\upsilon e_\upsilon$, then $\alpha$ is the following mapping

$$\alpha : \begin{cases} A \rightarrow A \\ x \mapsto \sum_{\sigma,\rho} (-1)^{\sigma+\rho} \varepsilon_\rho \, \| \sum_\upsilon \Gamma_{\upsilon j}^i x_\upsilon \|_{\sigma,\rho} \, e_\sigma \end{cases}$$

By $\| \cdots \|_{\sigma,\rho}$ we denote the determinant of the $(n-1)\times(n-1)$ - matrix, which we get from $(\sum_\upsilon \Gamma_{\upsilon j}^i x_\upsilon)$ dropping the $\sigma^{th}$ column and the $\rho^{th}$ row. For example, we have the properties $xx^\alpha \in K$, $xx^\alpha \neq 0 \Leftrightarrow x$ unit for all $x \in A$).

By $$\Delta : \begin{cases} \bar{A} \rightarrow \Pi_7 = (A \times A \times K \times K)^* /_{K^*} \\ \frac{a}{b} \mapsto [ab^\alpha, a^\alpha b, aa^\alpha, bb^\alpha] \end{cases}$$

$\bar{A}$ can be embedded into $\Pi_7$ (the image set is the intersection $\mathfrak{G}$ of some quadrics without radicals), chains are represented by the sections of $\mathfrak{G}$ with suitable 3-dimensional subspaces of $\Pi_7$.

## § 6 CROSS RATIOS AND CONFIGURATIONS

Within the first section of this paper the cross ratio $\begin{bmatrix} P & Q \\ S & R \end{bmatrix}$ was defined; we remind of $\begin{bmatrix} P & Q \\ S & R \end{bmatrix} = \begin{bmatrix} P^\gamma & Q^\gamma \\ S^\gamma & R^\gamma \end{bmatrix}$

for all $\gamma \in \Gamma$. Using the remarks about connexions between cross ratio and concircular points or the contact relation, the following theorem holds true (for a better understanding of the coming assumptions and statements we recommend to look at the cube model, fig 6 ).

(6.1) Let $k_o, k_1, k_2, k_3, k_4$ be chains and let A,B,C,D,

E,F,G,H be points such that $|\{A,B,E,F\}| = 4$, $E \nparallel F$, $G \nparallel H$. Furthermore we assume

$$(*)\; k_o \cap k_i = \begin{Bmatrix} A,C \\ B,C \\ B,D \\ A,D \end{Bmatrix},\quad k_i \cap k_{i+1} = \begin{Bmatrix} C,F \\ B,G \\ D,E \\ A,H \end{Bmatrix} \text{if} \quad i = \begin{cases} 1 \\ 2 \\ 3 \\ 4 \quad (5=1) \end{cases}$$

and $k_\nu P k_\mu$ if $|k_\nu \cap k_\mu| = 1$ in (*)

Then there exists a unique chain k such that

$$(**) \quad k \cap k_i = \begin{Bmatrix} F,H \\ F,G \\ E,G \\ E,H \end{Bmatrix} \quad \text{if} \quad i = \begin{cases} 1 \\ 2 \\ 3 \\ 4 \end{cases}$$

and $kPk_\nu$ if $|k \cap k_\nu| = 1$ in (**) .

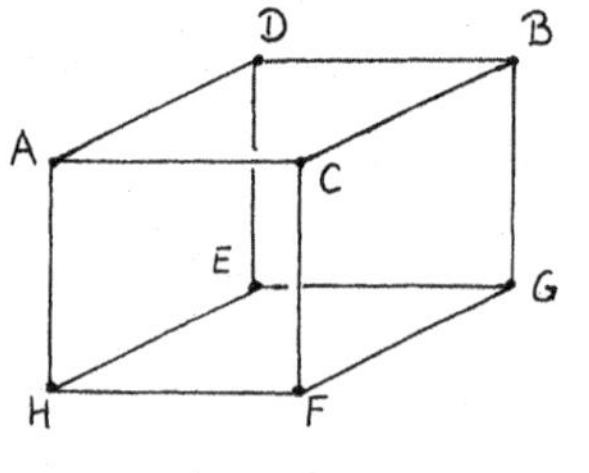


fig. 6 The faces of the cube represent the chains of the configuration

This theorem includes besides many other statements the theorem of Miquel ( $|\{A,...,H\}| = 8$) and the full theorem of Miquel ( $|\{A,...,G\}| = 7$, $H \notin \{B,...,G\}$).

The proof of (6.1) can be done by the following identities of cross ratios which occurred to [ 5 ],[ 20 ],[ 25 ].

(i) If $|\{A,\dots,H\}| = 8$, $\genfrac{[}{]}{0pt}{}{AB}{CD}\genfrac{[}{]}{0pt}{}{AF}{HC}\genfrac{[}{]}{0pt}{}{FB}{GC}\genfrac{[}{]}{0pt}{}{EB}{DG}\genfrac{[}{]}{0pt}{}{AE}{DH}\genfrac{[}{]}{0pt}{}{EF}{GH} = 1$

(ii) If $|\{A,\dots,H\}| = 7$, $\genfrac{[}{]}{0pt}{}{AB}{CD}\left(\genfrac{[}{]}{0pt}{}{AF}{DC} - \genfrac{[}{]}{0pt}{}{AF}{EC}\right)\genfrac{[}{]}{0pt}{}{FB}{GC}\genfrac{[}{]}{0pt}{}{EB}{GC}\genfrac{[}{]}{0pt}{}{EF}{GA} = 1$

such that A = H.

(iii) If $|\{A,\dots,H\}| = 6$:

α) $\genfrac{[}{]}{0pt}{}{AB}{CD}\left(\genfrac{[}{]}{0pt}{}{AF}{XC} - \genfrac{[}{]}{0pt}{}{AF}{DC}\right)\left(\genfrac{[}{]}{0pt}{}{DB}{XG} - \genfrac{[}{]}{0pt}{}{DB}{AG}\right)\genfrac{[}{]}{0pt}{}{FB}{GC}\genfrac{[}{]}{0pt}{}{DF}{GA} = 1$

such that A=H, D=E and $x \notin \{A,\dots,H\}$ another point lying on $k_4$.

β) $\genfrac{[}{]}{0pt}{}{AB}{CD}\left(\genfrac{[}{]}{0pt}{}{AF}{DC} - \genfrac{[}{]}{0pt}{}{AF}{EC}\right)\left(\genfrac{[}{]}{0pt}{}{BE}{CD} - \genfrac{[}{]}{0pt}{}{BE}{FD}\right)\genfrac{[}{]}{0pt}{}{EF}{BA} = 1$

such that A = H, B = G.

γ) $\left(\genfrac{[}{]}{0pt}{}{BA}{GC} - \genfrac{[}{]}{0pt}{}{BA}{EC}\right)\left(\genfrac{[}{]}{0pt}{}{AF}{BC} - \genfrac{[}{]}{0pt}{}{AF}{EC}\right)\genfrac{[}{]}{0pt}{}{FB}{GC}\genfrac{[}{]}{0pt}{}{EF}{GA} = 1$

such that A = H, B = D.

(iv) If $|\{A,\dots,H\}| = 5$:

α) $\genfrac{[}{]}{0pt}{}{FB}{EC}\left(\genfrac{[}{]}{0pt}{}{BA}{XC} - \genfrac{[}{]}{0pt}{}{BA}{EC}\right)\left(\genfrac{[}{]}{0pt}{}{AF}{BC} - \genfrac{[}{]}{0pt}{}{AF}{EC}\right)\left(\genfrac{[}{]}{0pt}{}{EA}{BF} - \genfrac{[}{]}{0pt}{}{EA}{XF}\right) = 1$

such that A = H, B = D, E = G and $X \notin \{A,\dots H\}$ lying on $k_3$.

β) $\left(\genfrac{[}{]}{0pt}{}{BF}{AE} - \genfrac{[}{]}{0pt}{}{BF}{CE}\right)\left(\genfrac{[}{]}{0pt}{}{FA}{BE} - \genfrac{[}{]}{0pt}{}{FA}{CE}\right) = \genfrac{[}{]}{0pt}{}{AF}{BC} - \genfrac{[}{]}{0pt}{}{AF}{EC}$

such that A=H, B=D, F=G.

(v) If $|\{A,\dots,H\}| = 4$:

α) $\left(\genfrac{[}{]}{0pt}{}{CE}{AB} - \genfrac{[}{]}{0pt}{}{CE}{XB}\right)\left(\genfrac{[}{]}{0pt}{}{BY}{AE} - \genfrac{[}{]}{0pt}{}{BY}{CE}\right)\left(\genfrac{[}{]}{0pt}{}{EY}{AB} - \genfrac{[}{]}{0pt}{}{EY}{CB}\right) = \genfrac{[}{]}{0pt}{}{AX}{BC} - \genfrac{[}{]}{0pt}{}{AX}{EC}$

such that A=H, B=D, C=F, E=G.

β) $\begin{bmatrix} AB \\ DC \end{bmatrix}^2 \left(\begin{bmatrix} DZ \\ YA \end{bmatrix} - \begin{bmatrix} DZ \\ BA \end{bmatrix}\right)\left(\begin{bmatrix} CX \\ AB \end{bmatrix} - \begin{bmatrix} CX \\ WB \end{bmatrix}\right)\left(\begin{bmatrix} BC \\ DX \end{bmatrix} - \begin{bmatrix} BC \\ YX \end{bmatrix}\right) =$

$= \begin{bmatrix} AW \\ DC \end{bmatrix} - \begin{bmatrix} AW \\ ZC \end{bmatrix}$ such that A=H, B=G, C=F, D=E and

$W \in k_1$, $X \in k_2$, $Y \in k_3$, $Z \in k_4$ further pairwise different points different from A,...,D.

We want to generalize the idea of chains in the plane case, i.e. $\dim_k A = 2$ (first done by Schröder [28])

Definition Let $P, Q \in \overline{A}$ non-parallel. If $A = \mathbb{D}$, $\mathbf{A}$ respectively then $k = k(P,Q) := \{X \in \overline{A} \mid X \parallel P \text{ or } X \parallel Q\}$

$k = k(P) := \{X \in \overline{A} \mid X \parallel P\}$ respectively

is called an irregular chain.

We want to abbreviate irregular chains by i-chain. By a g-chain (generalized chain) we denote either an i-chain or a usual chain.

In $\overline{\mathbb{D}_K}$ an i-chain either consists of two singular lines of the affin geometry Δ or of all improper points and one singular line. In K ⊕ K an i-chain either consists of two non-parallel singular lines and two special improper points or of one singular line and some special improper points, or of all improper points. We can connect cross ratios and g-chains.

(6.2) Let A,B,C,D be points such that $|\{A,\dots,D\}| \geq 3$, $A \nparallel D$, $B \nparallel C$. Then it is

(i) $\begin{bmatrix} AB \\ DC \end{bmatrix} \in \{o,1\} \Rightarrow \exists_1$ g-chain k such that $A,B,C,D \in k$

(ii) Let k be an i-chain such that $A,B,C,D \in k$

$\Rightarrow \begin{bmatrix} AB \\ DC \end{bmatrix} \in \{o,1\}$.

As a corollary, (6.2) implies

(6.3) Let A,B,C,D be pairwise different points such that $D \nparallel A \nparallel C \nparallel B \nparallel D$. Then it follows

$\begin{bmatrix} AB \\ DC \end{bmatrix} = 1 \Leftrightarrow \exists_1$ i-chain k such that $A,B,C,D \in k$

Therefore in plane geometries $A_K$ we are able to enlarge theorem 6.1.

(6.4) Let $k_o, k_1, k_2, k_3, k_4$ be g-chains and let A,B,C,D, E,F,G,H be points such that $E \neq F$. Assume (*) of (6.1). If $A_K$ is not the minimal Laguerre geometry $\mathbb{D}_{GF(2)}$, then there exists a unique g-chain k such that (**) of (6.1) holds true.

The proof is done by some simple reflexions of incidence and with the help from the mentioned cross ratio-identities. For more details see [ 23 ].

Now the so-called symmetrieaxiom s in [ 4 ],[ 5 ] is the special case $|\{A,\ldots,H\}| = 8$, $k_o$ a chain, $k_\nu$ i-chains for $\nu=1,2,3,4$ in (6.4). The condition of incidence $\Pi$ from [ 1 ] is another specialization of (6.4):

Put $A = C$, $F = H$, $|\{A,\ldots,H\}| = 6$, $k_3$ as an i-chain and all other $k_\nu$ as usual chains.

References

[ 1 ] Artzy,R.: A Symmetry Theorem for Laguerre planes, 1974, J.Geometry 5, pp.109-116

[ 2 ] Bartolone,C., di Franco, F.: A Remark on the Projectivities of the Projective Line over a Commutative Ring, 1979, Math.Z. 169,pp.23-29

[ 3 ] Benz, W.: Zur Geometrie der Körpererweiterungen, 1969, Canad.J.Math.21, pp.1097-1122.

[ 4 ] Benz, W.: Permutations and Plane Sections of a Ruled Quadric, 1970, Symposia Mathematica 5, pp.325-339

[ 5 ] Benz,W.: Vorlesungen über Geometrie der Algebren, Berlin - Heidelberg - New York, 1973

[ 6 ] Benz,W.: Zum Büschelsatz in der Geometrie der Algebren, 1973, Monatshefte f.Math.77,pp.1-9

[ 7 ] Blaschke, W.: Vorlesungen über Differentialgeometrie und geometrische Grundlagen von Einsteins Relativitätstheorie. III.Differentialgeometrie der Kreise und Kugeln. Bearbeitet von G.Thomsen, Berlin 1929.

[ 8 ] Burau, W.: Mehrdimensionale und höhere Geometrie, Berlin 1961

[ 9 ] Fladt, K., Baur,A.: Analytische Geometrie spezieller Flächen und Raumkurven,Braunschweig 1975

[10] Halder, H.R.: Produktbildung für Inzidenzstrukturen und Segresche Geometrien, Habilitationsschrift München 1979

[11] Hotje, H.: Einbettung gewisser Kettengeometrien in projektive Räume, 1974, J.Geometry 5/1,pp.85-94

[12] Hotje, H.: Die Algebren einbettbarer Berührstrukturen, 1978, Geometriae Dedicata 7,pp.355-362

[13] Laguerre, E.N.: Oeuvres de Laguerre. Tome II. Géomêtrie, Paris 1905

[14] Limaye, N.B.: Cross-Ratios and Projectivities of a Line, 1972, Math.Z.129, pp.49-53

[15] Limaye,B.V.,Limaye N.B.: The Fundamental Theorem for the Projective Line over Non-Commutative Local Rings, 1977, Arch.Math.28, pp.

[16] Limaye, B.V.,Limaye,N.B.: The Fundamental Theorem for the Projective Line over Commutative Rings, 1977, Aequationes Mathematica e 16, pp.275-281

[17] Mäurer, H. Metz,R.,Nolte,W.: Die Automorphismengruppe der Möbius-Geometrie einer Körpererweiterung, 1980,Aequationes Math. 21, pp.110-112

[18] Mäurer, H.: 1967, Möbius- und Laguerre-Geometrien über schwach konvexen Semiflächen, Math.Z.98, pp.355-386

[19] Möbius, F.: Gesammelte Werke (hier: Theorie der Kreisverwandtschaften), Leipzig 1886

[20] Peczar,L.: Über eine einheitliche Methode zum Beweis gewisser Schließungssätze, 1950, Monatshefte f.Mathematik 54, pp.210-220

[21] Samaga, H.-J.: Über nicht-ebene miquelsche Kurvensysteme, Dissertation Hamburg 1976

[22] Samaga, H.-J.: Dreidimensionale Kettengeometrien über $\mathbb{R}$, 1976, J. Geometry 8, pp.61-73

[23] Samaga, H.-J.: A Unified Approach to Miquel's Theorem and its Degenerations, 1979 in: Lecture Notes in Mathematics 792, pp.132-142

[24] Samaga, H.-J.: Zur Einbettung von Kettengeometrien in projektive Räume, erscheint demnächst

[25] Schaeffer,H.: Benz-Peczar-Doppelverhältnisidentitäten zum allgemeinen Satz von Miquel , 1974, Abh.Math.Sem.Univ.Hamburg 42, pp.228-235

[26] Schaeffer, H.: Das von Staudtsche Theorem in der Geometrie der Algebren, 1974, J. Reine Angew. Math. 267, pp.133-142

[27] Schaeffer,H.: Zur Möbius-Geometrie über Alternativkörper, Geometriae Dedicata 10, pp. 183-189.

[28] Schröder, E.: Eine allgemeine Fassung des Satzes von Miquel, 1978, (nicht veröffentlicht)

[29] Schröder, E.: Kreisgeometrische Darstellung metrischer Ebenen und verallgemeinerte Winkel- und Distanzfunktionen, 1974, Abh.Math.Sem. Univ.Hamburg 42, pp.154-186

[30] Staudt, G.K.C.von: Beiträge zur Geometrie der Lage, Bd.I, Nürnberg 1856

# PROJECTIVITIES IN FREE-LIKE GEOMETRIES

A. Barlotti

Università di Bologna, Bologna, Italy

## 1. Introduction.

In a projective plane all the projectivities of a line onto itself form a group $\Pi$ with respect to the composition of mappings. This group is an invariant for the plane since different lines have groups which are isomorphic (also as permutation groups).

In [28] K.G.Ch.v. Staudt proved that in the real projective plane the group $\Pi$ is sharply 3-transitive. This result led to the "fundamental theorem of projectivity": in a projective plane $\mathcal{P}$ the group $\Pi$ is sharply 3-transitive if and only if $\mathcal{P}$ is pappian.

The possibility to deduce from properties of under-structures of a given structure S , characterizations of S , is known as the v. Staudt principle [26]. The



*P. Plaumann and K. Strambach (eds.), Geometry – von Staudt's Point of View, 151–164.*

purpose of these lectures is to present some known results from the v. Staudt point of view in the free planar geometries and in other planar geometries obtained by certain extension processes.

Let $\mathcal{R} = (\mathcal{P}, \mathcal{L}, \mathrm{I})$ be a projective plane, $L, M \in \mathcal{L}$, $p \in \mathcal{P}$ and $p \not\mathrm{I}\, L$, $p \not\mathrm{I}\, M$. A perspectivity $\gamma = \{L, p, M\}$ of $L$ onto $M$ is the bijective mapping of $L$ onto $M$ defined in a natural way by the pencil $(p)$ of all lines incident with $p$. The point $p$ is called the centre of $\gamma$. A projectivity $\alpha$, of a line $L_o$ onto a line $L_n$ is the bijection defined as a product of perspectivities of the type:

$$(1) \qquad \bar{\alpha} = \prod_{i=1}^{n} \{L_{i-1}, p_i, L_i\} .$$

$\bar{\alpha}$ is called a representation of $\alpha$ and the integer $n$ is the lenght of $\bar{\alpha}$. The representation $\bar{\alpha}$ is called irreducible (or reduced) if $L_i \neq L_{i+1}$ and $p_i \neq p_{i+1}$. Every projectivity $\alpha \neq 1$ has at least one irreducible representation of the type (1). If $L_o = L_n$ and $\alpha = 1$ we assume the empty set to be an irreducible representation of $\alpha$.

For the definitions of projectivities in other geometries see, e.g., [5] or [11].

A permutation group is said to be n-regular if in

it every element with n fixed points is the identity, but there are elements different from the identity with n-1 fixed points.

A permutation group G on an infinite set M is called $\omega$-transitive if for every pair of n-subsets $S_1$ and $S_2$ of M (where $n \geq 1$ is any natural number) there is an element of G mapping $S_1$ onto $S_2$.

## 2. Basic notions.

For the definitions of projective and affine planes see, e.g. [16] or [20]; definitions of Benz-planes are given in [10], [12] and [14].

An incidence structure (following [9] ([1]) ) is a triple $S = (P,L,I)$ where P, L, I are sets with $P \cap L = \emptyset$ and $I \subseteq P \times L$ . The elements of P are called points, the elements of L are called lines (or, sometimes, blocks) and I defines the "incidence".

A partial plane is an incidence structure such that there is at most one line through any two distinct points ([2]) .

Consider a partial plane $\pi_o$ and define recursively a familiy of incidence structures using these two steps:

a) If $\pi_{2k}$ is given, for each pair of points not already joined by a line in $\pi_o$ create a new line, incident in $\pi_o$ exactly with the two points defining the new line. The structure $\pi_{2k+1}$ results by adding all these new lines to $\pi_{2k}$.

b) Consider in $\pi_{2k+1}$ all the pairs of lines that do not intersect. For each of these create a new point, incident in $\pi_{2k+1}$ exactly with the two lines defining the new point. By adding these new points to $\pi_{2k+1}$ we obtain $\pi_{2k+2}$. Then $\pi = \bigcup_i \pi_i$ (with the incidence relation induced by the $\pi_i$) is a projective plane which is called the free planar extension, $E(\pi_o)$, of the basic configuration $\pi_o$. If $\pi_o$ consists only of $n$ points ($n \geq 2$) on a line, together with two points without incidences, then $E(\pi_o) = F^{n+2}$ is called a <u>free plane</u>. An extensive study of projective planes which are free planar extension can be found in [13], [16], and [20]. In 1965 L.C. Siebenmann [25] has introduced the important notion of "extension process" leading to a definition of free planes which coincides with M. Hall's definition.

For the definitions of free extensions in the class of affine, Möbius and Laguerre planes see [23] (and also [4]) and for free extensions leading to Minkowski planes see [15]. For the class of Benz planes M. Funk [11] has given a procedure which unifies those of the preceding authors.

In a projective plane $E(\pi_0)$ which is a free extension of the proper partial plane $\pi_0$ there is in a natural way defined a map, $\alpha$, of the elements of the plane (points and lines) in $\mathbb{N}$ ($^3$), and $\alpha(x)$ is called the stage of the element $x$. Precisely: the elements of stage zero are those of $\pi_0$; the elements of stage $i$ are exactly those which belong to $\pi_i$ but not to $\pi_{i-1}$. For this map the two following properties hold.

i) Every element of stage $h$ ($> 0$) is incident with exactly two elements of preceding stages.

ii) Every element of stage $h$ ($> 0$) is incident with at least one element of stage $h - 1$.

<u>Definition (2.1).</u> Let $S$ be a linearly ordered system of numbers. A projective plane is called an S-plane if there exists a map $\alpha$ of its elements in $S$ (and $\alpha(x)$ will be called the stage of x), and no two elements of the same stage are incident.

We list some questions that naturally arise.

(2.2) Denote by $\mathbb{N}$ the naturals including the zero with the usual order. Let $\mathcal{R}$ be an $\mathbb{N}$-plane for which the properties i) and ii) hold. Determine whether it is true or not that $\mathcal{R}$ is a free extension of a proper partial plane.

(2.3) The same as before except that now we do not assume the validity of ii).

Denote by $\mathbb{Z}$ the integers with the usual order.

(2.4) Is it possible to prove that the set of the $\mathbb{Z}$ - planes ist not empty?

(2.5) If the previous question has a positive answer, how does one characterize the $\mathbb{Z}$ - planes?

(2.6) To formulate and study questions similar to the above for other kind of free extensions.

## 3. The group of projectivities in the projective NC-planes.

A projective plane $\mathcal{R}$ is called confined if every point of $\mathcal{R}$ belongs to some confined configuration. In a non confined plane (briefly an NC-plane) there exist points which do not belong to any confined configuration. Examples of NC-planes are the free planes, the open planes, the free extensions of proper partial planes (see [8], [13], [20], [22]).

The group of projectivities in the NC-planes has been studied in [22]. The main results are the following:

Theorem (3.1). In an NC-plane the only irreducible representation of the identity of $\Pi$ is given by the empty set.

Corollary (3.2). In a projective NC-plane every projectivity has exactly one irreducible representation.

Theorem (3.3). In an open plane $\Pi$ is 6-regular.

Theorem (3.4). In a projective NC-plane there is no projectivity of a line onto itself which has a non empty irreducible representation and is induced by an automorphism of the plane.

From the fact that in a free plane can be given a full order (see [17]) there follows:

Theorem (3.5). In an NC-plane the group $\Pi$ is 3-transitive.

## 4. The group of projectivities in other planes.

In an affine plane a projectivity is product of affine perspectivities. For a free affine plane theorems analogous to those listened in n.3 can be proven (see [5], [24]). We mention here only that in a free affine plane the group of projectivities of a line onto itself is 4-regular.

In the class of Benz planes it is convenient to define a projectivity as product of three types of perspectivities: "proper", "affine" and "free" perspectivities (see [11]). The main result in the study of the group $\Pi$ of projectivities of a circle onto itself is given by the following

Theorem (4.1) (Funk, [11]). In a free Benz plane $\Pi$ is 6-regular.

For other results analogous to some of those given in n.3, or concerning the groups generated by the affine and the proper perspectivities see [11].

We wish to present here an open problem. A k-confined configuration (k an integer $\geq 3$) is a finite incidence structure in which every element is incident with k elements of the structure. A projective plane $\mathcal{P}$ is k-open (k an integer $> 3$) if $\mathcal{P}$ does not contain any k-confined configuration but has at least a (k-1)-confined configuration (see [3]). The question of studying the group of projectivities of k-open planes has not yet been considered.

## 5. Groups of projectivities with certain prescribed properties.

In any projective plane the group $\Pi$ , as permutation group on the points of a line, is at least 3-transitive. The question arises whether or not, for infinite planes, there is an upper bound on the value of $n$ for which the group $\Pi$ is n-transitive. Analogous questions hold for affine or Benz planes. The answers are given by the following theorems (see [5]):

Theorem (5.1). There exist projective [affine] planes in which the group $\Pi$ of projectivities [affine projectivities] of a line onto itself is at least t-transitive for a chosen $t \geq 3$ [$t \geq 2$]. There are planes in which the group $\Pi$ is $\omega$-transitive.

Theorem (5.2). Chosen any integer $t \geq 3$ , there are Benz planes in which the group $\Pi$ of projectivities of a circle onto itself is at least t-transitive. There are Benz planes in which the group $\Pi$ is $\omega$-transitive.

It is certainly difficult to make a comparison between the classical geometries, classes of geometries with closed configurations and the free geometries.

Among the many results that can be used at this end, we can list also those expressed by the following theorems (see [5]):

Theorem (5.3). For every $n \geq 5$ $[n \geq 3]$ there exists a projective [affine] plane such that the group $\Pi$ of projectivities [affine projectivities] of a line L has the following properties:

The stabilizer in $\Pi$ of $n + 1$ distinct points consists of the identity only, but there are $n$ different points such that their pointwise stabilizer contains elements different from the identity.

Theorem (5.4). For every $n \geq 5$ there are Möbius, Laguerre and Minkowski planes such that the group of projectivities $\Pi$ of a circle onto itself has the following properties:

The stabilizer in $\Pi$ of $n + 1$ different points consists of the identity only but there are $n$ distinct points such that their pointwise stabilizer contains elements different from the identity.

## 6. The group of projectivities in binary systems.

In [6] has been developed the study of the group of projectivities of a line onto itself in 3-nets and in quasigroups. We mention here only the following two theorems.

Theorem (6.1). In a free net the pointwise stabilizer of every four distinct points consists only of the identity.

Theorem (6.2). (A v. Staudt theorem for quasigroups). A quasigroup Q is an abelian group without involutions if and only if the pointwise stabilizer of every two elements in the group of projectivities of Q consists only of the identity.

Notes.

([1]) This notion should not be confused with the "Inzidenzstruktur" defined in [20].

([2]) This is equivalent to the fact that there is at most one point common to two distinct lines.

([3]) We assume that $\mathbb{N}$ includes the zero.

REFERENCES

[1] A. Barlotti; La determinazione del gruppo delle proiettività di una retta in sè in alcuni particolari piani grafici finiti non desarguesiani. Boll. Un. Mat. Ital. 14 (1959), pp. 182 - 187.

[2] A. Barlotti; Sul gruppo delle proiettività di una retta in sè nei piani liberi e nei piani aperti; Rend. Sem. Mat. Padova, 34 (1964), 135 - 159.

[3] A. Barlotti; Configurazioni k-chiuse e piani k-aperti; Rend. Sem. Mat. Padova, 35 (1965), 56 - 64.

[4] A. Barlotti; Sulle m-strutture di Möbius; Rend. Ist. Mat. Univ. Trieste 1 (1969), 35 - 46

[5] A. Barlotti, E. Schreiber, K. Strambach; The group of projectivities in free-like geometries; Rend. Sem. Mat. Univ. Padova 60 (1978), 183 - 200.

[6] A. Barlotti, K. Strambach; The geometry of binary systems, to appear.

[7] W. Benz; Vorlesungen über Geometrie der Algebren; Berlin-Heidelberg-New York, Springer Verlag, 1973.

[8] P. Dembowski; Freie und offene projektive Ebenen; Math. Z., 72 (1960), 410 - 438.

[9] P. Dembowski; Finite Geometries; Berlin-Heidelberg-New York, Springer Verlag, 1968.

[10] H. Freudenthal, K. Strambach; Schließungssätze und Projektivitäten in der Möbius- und Laguerregeometrie; Math. Z., 143 (1975), 213 - 234.

[11] M. Funk; Regularität in Benz-Ebenen; Ph. D. Thesis, Erlangen (1980).

[12] H.R. Halder, W. Heise; Einführung in die Kombinatorik; München-Wien, Hanser-Verlag (1976).

[13] M. Hall Jr.; Projective planes; Trans. Amer. Math. Soc., 54 (1943), 229 - 277.

[14] W. Heise, H. Seybold; Das Existenzproblem der Möbius-, Laguerre- und Minkowski-Erweiterungen endlicher affiner Ebenen; Sitz. Ber. Bayer. Akad. Wiss., Math. Nat. Kl. 1975, 43 - 58.

[15] W. Heise, K. Sörensen; Freie Minkowski-Ebenenerweiterungen, J. Geom. 3, (1973) 1 - 4.

[16] R.D. Hughes, F.C. Piper; Projective planes; New York- Heidelberg-Berlin, Springer Verlag, 1973.

[17] J. Joussen; Die Anordnungsfähigkeit der freien Ebenen; Abh. Math. Sem. Univ. Hamburg 29 (1966), 137 - 184.

[18] H.J. Kroll; Die Gruppe der eigentlichen Projektivitäten in Benz-Ebenen; Geometriae Dedicata 6, 407 - 413 (1977).

[19] H.J. Kroll; Perspektivitäten in Benz-Ebenen; in Beiträge zur geometrischen Algebra, 203 - 207, Basel-Stuttgart, Birkhäuser 1977.

[20] G. Pickert; Projektive Ebenen; Berlin-Heidelberg-New York, zweite Auflage, Springer Verlag, 1975.

[21] A. Schleiermacher; Bemerkungen zum Fundamentalsatz der projektiven Geometrie; Math. Z. 99, 299 - 304 (1976).

[22] A. Schleiermacher, K. Strambach; Über die Gruppe der Projektivitäten in nichtgeschlossenen Ebenen; Arch. Math. 18, 299 - 307 (1967).

[23] A. Schleiermacher, K. Strambach; Freie Erweiterungen in der affinen Geometrie und der Geometrie der Kreise (I u. II); Abh. Math. Sem. Univ. Hamburg 34, 22 - 37 and 209 - 226 (1969-70).

[24] E. Schreiber; Freie Strukturen und die Gruppe der affinen Projektivitäten; Dissertation, Erlangen, 1979.

[25] E.C. Siebenmann; A characterization of free projective planes; Pac. Journ. 15 (1965) pp. 293 - 298.

[26] K. Strambach; Die Gruppe der Projektivitäten in projektiven und affinen Ebenen; Erlangen 1976.

[27] K. Strambach; unpublished lectures noted by H. Krauß; Erlangen (1978); Vorlesungsausarbeitung.

[28] K.G.Ch. v. Staudt; Geometrie der Lage; Nürnberg, Verlag von Bauer und Raspe 1847.

# EXISTENTIALLY CLOSED PROJECTIVE PLANES

Otto H. KEGEL

Mathematisches Institut der Universität
Freiburg i.Br., Bundesrepublik Deutschland

As an illustration of existentially closed models in certain classes of geometries the notion of existentially closed projective plane in the class of all projective planes is discussed. It is shown that the group of projectivities of such a plane with respect to a line acts highly transitively on the points of that line. In general, however, such a plane is not homogeneous. If one considers instead the class of projective planes with attached group of collineations, the existentially closed models of this class are existentially closed projective planes with an existentially closed group attached as a group of collineations which acts transitively on each isomorphism class of finitely generated subplanes.

Model theory has evolved - at least in part - out of



*P. Plaumann and K. Strambach (eds.), Geometry - von Staudt's Point of View, 165–174.*

the axiomatic foundations of geometry. Thus it seems reasonable to investigate what contribution present day model theory with its results, methods, and problems can bring to the area of geometry. This is a vast and - in my eyes - rather promising project. In this lecture, meant to stimulate interest in this direction, I would like to draw the attention of geometers to a tiny part of this project: the notion of existentially closed model in a given class of models. This notion generalises the notion of algebraically closed field (in the class of all fields); it has attracted considerable interest by logicians (Chapter 1 of [2] is a pertinent reference, cf. also [1] and [6]).

Let $\mathcal{L}$ be a first-order language describing a certain class of structures, e.g. (two sorted) incidence geometry. Let $\mathcal{C}$ be a class of models for these structures, e.g. affine or projective planes. The model A for $\mathcal{L}$ is <u>existentially closed in the model</u> $B \in \mathcal{C}$ if $A \in \mathcal{C}$, $A \subseteq B$, and every finite set of existential sentences of $\mathcal{L}$ with constants in A which may be satisfied in B may already be satisfied in A. The model A for $\mathcal{L}$ is <u>existentially closed in the class</u> $\mathcal{C}$ if one has that $A \in \mathcal{C}$ and A is existentially closed in every model $B \in \mathcal{C}$ containing it. - If the class $\mathcal{C}$ is inductive, i.e. closed with respect to unions of ascending chains of models in $\mathcal{C}$, then every model $A \in \mathcal{C}$ is contained in an existentially closed model of the class $\mathcal{C}$. In order to obtain information on the structure of a model A existentially closed in the class $\mathcal{C}$, one must search for constructions of models $B \in \mathcal{C}$ containing A and satisfying certain existential sentences of $\mathcal{L}$ with constants in A. Now, let $\mathcal{L}$ be the language of two sorted incidence geometry the elements of which will be called points and lines and between which an incidence relation (symbol I) is given. In order to specify classes of incidence geometries we shall use sets of sentences of $\mathcal{L}$, the axioms, and require every model in the class to satisfy all the axioms of the corresponding set. Thus, one has the class of partial planes if - informally - in each of these geometries there is at most one line incident with any two given distinct points, and if there is at most one point incident with any two given distinct lines. Obviously, the class of all partial planes is inductive. Let E be existentially closed in the class of all partial planes, then E is in fact existentially closed in the class of all projective planes. - To see this, we have first to convince ourselves that E is a projective plane. But this is clear, since by the procedure of free extension (or free closure), cf. [3], chapter XI, every partial plane may be embedded into a projective plane. Thus the axioms of projective planes will be true statements for E . Every existential sentence in $\mathcal{L}$ with constants in E which can be satisfied in a projective plane containing E can also be satisfied in E .

Now let us consider the class of all pappian projective planes, that of all desarguesian projective planes, and that of all projective planes. Each of these three classes is inductive. Let E be existentially closed in any of these classes. Since algebraic properties of any planar ternary ring coordinatising a projective plane may be translated - if artificially - into geometric properties, one obtains that E is a projective plane over an algebraically closed field, or over an existentially closed skew field, or over an existentially closed planar ternary ring. And conversely, projective planes coordinatised in this way are existentially closed in the class of all pappian projective planes, of all desarguesian, or the class of all projective planes. Of course, the last of these characterisations is scarcely more than a definition of existentially closed planar ternary rings, since one does not know how to handle these objects in all generality, except by means of the geometries they coordinatise. - For the class of all translation planes the study of the existentially closed models might be rewarding as quasifields have a structure which is much easier to handle than that of planar ternary rings.

If $E$ is existentially closed in the class of all projective planes, for every finite partial plane $P$ there is a finite partial subplane $P'$ of $E$ isomorphic to $P$ ; in fact, if $P$ contains a triangle, i.e. three points which are not collinear in $P$ , then one may prescribe the image of that triangle in $E$ . This result follows from the embedding results of [4] and the fact that $E$ was assumed existentially closed.

It is the theme of this conference to say something on the von Staudt group $G_\chi$ of a geometry. In our case we shall consider $G_\chi$ , the group of all projectivities of the projective plane $E$ leaving a fixed line invariant. For every natural number $n$ the group $G_\chi$ acts transitively on the set of $n$-tuplets of points incident with the line $\chi$ . To see this, let $p_1,\dots,p_n$ and $q_1,\dots,q_n$ be two disjoint $n$-tuplets of points incident with the line $\chi$ . Enlarge $E$ by adding to it a line $y$ with which further $n$ new points $r_1,\dots,r_n$ are incident; further add lines $x_i$

and $z_i$ the only points incident with these are $p_i$ and $r_i$ and $r_i$ and $q_i$, respectively. If we now decree points $s$ and $t$ such that $s$ is incident with all the lines $x_i$ and $t$ with all the lines $z_i$, then the structure obtained from adding to $E$ the points $r_1,r_2,\ldots,r_2$, $s$, $t$, and the lines $y$, $x_1,\ldots,x_n$, $z_1,\ldots,z_n$ with the incidences described is a partial plane. Thus, in the free extension of this partial plane there is a line $y$ and there are two points $s,t$ so that the projectivity of $\ell$ defined by projecting the points of $\ell$ onto $y$ with focus $s$ and those of $y$ back to $\ell$ with focus $t$ will just map the point $p_i$ to the point $q_i$ for every $i = 1,\ldots,n$. As $E$ is existentially closed, there must be such points and such lines already in $E$. If the two $n$-tuplets to be considered are not disjoint, then consider an auxiliary $n$-tuplet of points on $\ell$, disjoint from both the others. The preceding construction then yields projectivities $\pi$ and $\pi'$ of $E$ leaving $\ell$ invariant and mapping the $n$-tuplet $p_1,\ldots,p_n$ to the auxiliary $n$-tuplet and the auxiliary $n$-tuplet to $q_1,\ldots,q_n$ such that the composed projectivity $\pi \circ \pi'$ maps $p_i$ to $q_i$, $i = 1,\ldots,n$.

Essentially the same construction as the one presented here of a projective plane $E$ such that the group

$G_{\mathcal{X}}$ of projectivities leaving $\mathcal{X}$ invariant acts highly transitively on the points incident with $\mathcal{X}$ appears in [0]; the limit procedure in the construction of [0] has here been transferred to the definition of an existentially closed projective plane.

The projective plane P is called homogeneous if the group of all collineations of P acts transitively on each isomorphism class of finitely generated subplanes of P. By [4] every infinite projective plane may be embedded into a homogeneous projective plane of the same cardinality. For countable projective planes Cantor's standard back-and-forth argument yields:

A countable homogeneous projective plane is determined up to isomorphism by the set of isomorphism types of its finitely generated subplanes.

Are existentially closed planes in certain classes of projective planes homogeneous? This is so in the class of all pappian and in the class of all desarguesian projective planes. This fact is due to the classical close connection between projectivities and collineations in desarguesian projective planes. For E existentially closed in the class of all projective planes one cannot in general prove homogeneity. An example of a countable, but not homogeneous plane E which is existentially closed in the class of all projective planes was shown to me by M. ZIEGLER 1978. The reason for this defect is to be sought in the fact that collineations are not part of the language $\mathcal{L}$ of two sorted incidence geometry and hence they cannot openly appear in the system of axioms for any class of geometries considered.

That this shortcoming may be mended is suggested by the results of [4] mentioned above. Denote by $\mathcal{L}'$ the language $\mathcal{L}$ augmented by a set $\mathcal{F}$ of functions mapping points to points, lines to lines, and preserving incidence. Restricting the class of all models of $\mathcal{L}'$ somewhat, we shall consider the class of all pairs $(E,G)$ where $E$ is an incidence geometry and $G$ is a group acting as group of collineations on $E$. Define $(E,G)$ to be a subpair of the pair $(E_1,G_1)$ if $E \subseteq E_1$, i.e. $E$ is a subgeometry of $E_1$, $G$ is a subgroup of $G_1$, and if the subgroup $G$ of $G_1$ acts in such a way on $E_1$ that the subgeometry $E$ remains invariant. With this definition the class of all such pairs $(E,G)$ is an inductive class and hence, every such pair is a subpair of one that is existentially closed in this class. What can one say about such a pair?

If we restrict the geometries to be partial planes and denote by $\mathcal{C}$ the class of all such pairs (P,G) such that P is a partial plane and the group G acts on P as a group of collineations then the methods of [4] (and [5]) make information on the structure of pairs (E,G) existentially closed in $\mathcal{C}$ available. In fact, one finds that G _is_ a group of collineations of E, E is existentially closed in the class of all projective planes, and G is existentially closed in the class of all groups. Furthermore, the constructions of [4] show that the group G acts transitively on each isomorphism class of finitely generated subplanes of E, thus E is in fact homogeneous.

Also for other sorts of geometries for which the pasting and gluing techniques of [4] are available one may get information of the above sort on existentially closed models in the class of all such geometries or in the class of all pairs (E,G), where E is a geometry of our class and G is a group acting as group of automorphism (collineations) on E. One such class, not too far removed from the class of projective planes is the class of Möbius planes.

BIBLIOGRAPHY

[0] A. BARLOTTI, E. SCHREIBER, K. STRAMBACH : The group of projectivities in free-like geometries. Rend. Sem. Univ. Padova, 60 (1978) 183 - 200

[1] P. C. EKLOF : Ultraproducts for algebraists. pp. 105 - 137 in Handbook of Mathematical Logic, ed. J. Barwise, North-Holland 1977

[2] J. HIRSCHFELD, W.H. WHEELER : Forcing, Arithmetic, and Division Rings, Lecture Notes in Mathematics, Vol. 454, Springer 1975

[3] D. R. HUGHES, F.C. PIPER : Projective Planes, Graduate Texts in Mathematics, Vol. 6, Springer 1973

[4] O.H. KEGEL, A. SCHLEIERMACHER : Amalgams and embeddings of projective planes. Geometriae Dedicata 2 (1973) 379 - 395

[5] O.H. KEGEL : Examples of highly transitive permutation groups. Rend. Sem. Univ. Padova, 63 (1980)

[6] A. MACINTYRE : Model completeness. pp. 139 - 180 in Handbook of Mathematical Logic, ed. J. Barwise, North-Holland 1977

# CONICOIDS: CONIC-LIKE FIGURES IN NON-PAPPIAN PLANES

T. G. Ostrom

Washington State University,
Pullman, Washington

Abstract.
In a Pappian plane of odd characteristic (or odd order) a conic may be defined by a polarity, a projectivity or a second degree equation; the conic (if non-degenerate) is an oval. The extensions of these concepts to non-Pappian planes are not equivalent; we look at the differences. The curve defined by the equation $y = x^2$ over a commutative semifield does satisfy restricted versions of all these definitions. In many other cases, the restriction of the "concoid" to a Pappian subplane is a conic.

In a Pappian plane, several important properties of a conic are equivalent and each can be taken as a definition. This is not the case in a non-Pappian plane. The word "conic" is used in different places by different authors with different meanings. Perhaps it would be better to reserve the term for reference to Pappian planes and to use the word "conicoid" for the various generalizations to non-Pappian planes.We



*P. Plaumann and K. Strambach (eds.), Geometry - von Staudt's Point of View, 175–196.*

intend to follow this usage in the present paper; we hope that this does not lead to further confusion.

We shall be interested in four types of conicoids and especially in the relations between them in various cases. The flavor of what we intend to look at should be apparent from the following list, even though we do not give definitions:

1. Conicoids defined by polarities.
2. Conicoids defined by projectivities.
3. Conicoids defined by second degree equations over a coordinate system.
4. Ovals.

We may refer to conicoids of type 1, 2, 3 or 4 but in general we intend to preserve enough ambiguity in our use of the term so that we can use it when we feel like it! We shall be more inclined to feel like it when we encounter something that is a conicoid in more than one of the above senses. We shall examine these notions in four different contexts. The reader may also wish to consult Segre's book [12].

I. Skewfields. Let K be a division ring (skewfield) which is not necessarily commutative. Let V be a 3-dimensional left vector space over K with its elements represented by row matrices $X = (x_1, x_2, x_3)$ so that $(cx_1, cx_2, cx_3)$ is a scalar multiple of X and

$x_1$, $x_2$, $x_3$, c all belong to K.

Throughout this paper we shall assume that none of our algebraic structures has characteristic 2 and that finite planes have odd order. Let $V^*$ be a 3-dimensional right vector space over K, with its elements represented by column matrices L. The points of the projective plane $\pi$ are one-dimensional subspaces of V and the lines of $\pi$ are one-dimensional subspaces of $V^*$. The point X is incident with the line L (more properly: the point whose coordinates are given by X and the line whose coordinates are given by L) iff $XL = 0$.

Suppose that K admits an anti-automorphism of order two: $a \to \bar{a}$. Let $\bar{X} = (\bar{x}_1, \bar{x}_2, \bar{x}_3)$.

If A is a matrix representing a non-singular linear transformation on $V^*$ then the mapping $X \to A\bar{X}^T$ induces a duality of $\pi$. The duality will be a polarity if A is hermitian. The absolute points of the polarity satisfy the equation $XA\bar{X}^T=0$. In terms of coordinates we get a scalar equation

$$(1)\quad x_1 a_{11} \bar{x}_1 + x_2 a_{22} \bar{x}_2 + x_3 a_{33} \bar{x}_3 + x_1 a_{12} \bar{x}_2 + x_2 \bar{a}_{12} \bar{x}_1 + x_2 a_{23} \bar{x}_3 + x_3 \bar{a}_{23} \bar{x}_2 + x_1 a_{13} \bar{x}_3 + x_3 \bar{a}_{13} \bar{x}_1 = 0.$$

Thus the set of points satisfying (1) form a conicoid in sense 1. Now let us look at projectivities. As

Pickert points out in his book [10], pp. 113-114, each projectivity in a Desarguesian plane is induced by a projective collineation and the projective collineations are precisely those represented by non-singular linear transformations.

Without loss of generality, suppose that the transformation carries (1,0,0) into (0,1,0) and is represented by the matrix A, where

$$A = \begin{pmatrix} 0 & 1 & 0 \\ a_{21} & a_{22} & a_{23} \\ a_{31} & a_{32} & a_{33} \end{pmatrix},$$

Here it will be convenient to identify a line L with the two-dimensional subspace of V consisting of vectors which are incident with L. The two points (0, 1, w) and (1, 1, w) generate a line L which contains (1, 0, 0). The image of L is generated by $(a_{21} + w\,a_{31}, a_{22} + w\,a_{32}, a_{23} + w\,a_{33})$ and $(a_{21} + w\,a_{31}, 1 + a_{22} + w\,a_{32}, a_{23} + w\,a_{33})$. The point $X = (x_1, x_2, x_3)$ is on the intersection of L with its image iff X is a linear combination of generators for L and a linear combination of generators for its image. Hence

$$x_1 = \beta_2 = (\alpha_1+\alpha_2)(a_{21} + w\,a_{31})$$
$$x_2 = \beta_1 + \beta_2 = (\alpha_1+\alpha_2)(a_{22} + w\,a_{32}) + \alpha_2$$
$$x_3 = (\beta_1+\beta_2)\,w = (\alpha_1+\alpha_2)(a_{23} + w\,a_{33})$$

for some $\alpha_1$, $\alpha_2$, $\beta_1$, $\beta_2$ in K.

If $\beta_1 + \beta_2 = 0$ then $x_2 = x_3 = 0$; without loss of generality $x = (1,0,0)$ in this case. Otherwise $x_3 = x_2 w$ and $x_2^{-1} x_3 = w$. If $a_{23} + w\, a_{33} = 0$ and $\beta_1 + \beta_2 \neq 0$ then $w=0$. In this case L has the equation $x_3 = 0$ (more properly $L = \begin{pmatrix}0\\0\\1\end{pmatrix}$) and L is invariant under the transformation. The induced projectivity is a perspectivity in this case.

Thus suppose that $x_2^{-1} x_3 = w$ and $a_{23} + w\, a_{33} \neq 0$. Then $x_3\ (a_{23} + w\, a_{33})^{-1} = \alpha_1 + \alpha_2$ and $x_1 = x_3\ (a_{23} + w\, a_{33})^{-1}\ (a_{21} + w\, a_{31})$.

(2) $\quad X_1 = x_3 (a_{23} + x_2^{-1}\ x_3\ a_{33})^{-1}\ (a_{21} + x_2^{-1}\ x_3\ a_{31})$.

With the exception noted, X is on the intersection of a line through (1,0,0) with its image iff the coordinates of X satisfy (2).

If we multiply (2) on the left in succession by $x_3^{-1}$, $a_{23} + x_2^{-1}\ x_3\ a_{33}$, $x_2$ and $x_3^{-1}$ again we obtain

(3) $\quad x_3^{-1}\ x_2\ a_{23}\ x_3^{-1}\ x_1 + a_{33}\ x_3^{-1}\ x_1 = x_3^{-1}\ x_2\ a_{21} + a_{31}$.

If we set $x_3^{-1}\ x_1 = x$ and $x_3^{-1}\ x_2 = y$ we obtain the non-homogeneous equation

(4) $\quad y\ a_{23}\ x + a_{33}\ x - y\ a_{21} - a_{31} = 0$.

Thus, together with the "points at infinity" (1,0,0) and (0,1,0), the set of points whose non-homogeneous coordinates satisfy (4) form a conicoid in senses 2 and 3.

However, we have a different type of second degree equation from what we had in (1). That is (3) is not equivalent to a special case of (1) unless K is commutative. To put it another way, some conicoids of type 3 are also of type 1 and some are of type 2. There seem to be none in planes over proper skewfields which are of both type 1 and type 2.

Furthermore even in the special case of (4) given by $yx - 1 = 0$ or the special case of (1) given by $x_1 x_2 + x_2 x_1 - x_3^2 = 0$ the conicoid may contain three collinear points (at least in the case where K is the quaternion skewfield). Thus take the line $y = -x$ or $x_2 = -x_1$. The fact that $-1$ has more than two square roots implies that in either case the conicoid contains at least three collinear points.

II. Commutative Semi-fields. Perhaps it should not be surprising that commutativity turns out to be more important than associativity. Let K be a commutative semi-field. (Semi-fields used to be called non-associative division rings. Both distributive laws hold in a semi-field and the non-zero elements form a loop under multiplication.) Let k be an element $\neq 0$ in the nucleus of K; that is, k associates with every pair of elements of K in all positions. Then it turns out that the set of affine points (x,y) satisfying

$y = k x^2$ together with $(\infty)$ form a conicoid in all four senses. If the term "conic" is going to be used for a configuration in non-Pappian planes, this would seem to be the place. Without loss of generality, we can take $k = 1$ or $k = \frac{1}{2}$ . (Recall that we are avoiding characteristic 2.)

If $\pi$ is a plane coordinatized by a commutative semi-field K, then $\pi$ admits a polarity such that point with coordinates (a,b) corresponds to the line with equation $y = xa - b$. The line $x=a$ and the point (a) on $\ell_\infty$ correspond. (We like to use the symbol $\infty$ by analogy with the case of the extended real plane. The line $\ell_\infty$ is not necessarily "at infinity". It is also not "improper" and calling it the "ideal line" seems to imply that it doesn't really exist. In this spirit, we frequently refer to the symbol "$\infty$" as "lazy eight".) The absolute points of this polarity are the affine points satisfying $y = \frac{1}{2} x^2$ together with $(\infty)$.

Proposition. Let K be a commutative semi-field of odd characteristic. Let a, b, c be elements of K which are not all zero such that a is in the nucleus of K. Then the equation $x^2 a + x b + c = 0$ has at most two solutions in K.

Proof. Let $x_1$ be a solution in K so that $x_1^2 a + x_1 b + c = 0$. Then, by subtraction,

$(x^2 - x_1{}^2)a + (x - x_1)b = 0$ or

$(x - x_1)[(x + x_1)a + b] = 0$. Thus either $x = x_1$ or $x$ is the unique value of K such that $(x + x_1)a = -b$.

Theorem. Over a commutative semi-field no three affine points of the curves $yx = 1$ and $y = x^2$ are collinear. By adjoining the points $(0)$, $(\infty)$ or $(\infty)$ respectively one gets a "curve" in which no three points are collinear. The proof is left to the reader.

Thus the curve $y = \frac{1}{2} x^2 \cup (\infty)$ is a conicoid in senses 1, 3 and 4.

The next result does not require a commutative semi-field.

Theorem. Let $\pi$ be a projective plane coordinatized by a ternary such that (a) Addition is associative and commutative. (b) There is a point $(-1)$ on $\ell_\infty$ such that all affine lines thru $(-1)$ are represented by equations of the form $y = -x + b$. Then $\pi$ admits a projectivity such that an affine point $(x,y)$ is the point of intersection of corresponding lines iff $y = x^2$.

Proof. Let $\rho$ be a perspectivity with axis $y=1$ between the lines through $(\infty)$ and through $(-1)$. Let $\sigma$ be the perspectivity with axis $x=1$ between the lines through $(-1)$ and the lines through $(0,0)$. Then

$$(x = a) \xrightarrow{\rho} (y = -x+a+1) \xrightarrow{\sigma} (y = xa).$$

The point of intersection of the lines $x = a$ and

$y = xa$ has coordinates $(a, a^2)$.

Note that if $k \neq 0$ is in the nucleus of a commutative semi-field the mappings $(x,y) \to (x\,k,\, y)$ and $(x,y) \to (x,\, y\,k)$ are collineations so that all of the conicoids $y = kx^2$ are equivalent for k in the nucleus.

III. Not necessarily commutative semi-fields.

Note that skewfield planes and Moufang planes are included.

Berz [3] and Krueger [8] have studied the curve $yx = 1$ over skewfields (particularly the quaternions) and in Moufang planes. Krueger shows that in an arbitrary plane with an arbitrary ternary there is a projectivity leading to the curve $yx = 1$. Artzy [1] has examined projectivities in Moufang planes leading to $y = x^2$. He shows that, in a Moufang plane coordinatized by an alternative field the curve $yx = 1$ can be carried into $y = x^2$ by a collineation. Note also the projectivity we used in the last section.

Even these simple equations don't tell us much unless we assume something about the coordinatizing algebra. In general it appears clear that conicoids of type three need not be ovals. What might perhaps be a tangent line turns out sometimes not to be a tangent line. For more details, see the above mentioned papers by Artzy, Berz and Krueger.

For the moment, let us return to the case where K is a skewfield. Let F be a field in the center of K. Then the plane coordinatized by K has a subplane $\pi_0$ coordinatized by F. Let $\mathcal{C}$ be a conicoid of type 3 with its coefficients in F. In particular, suppose that $\mathcal{C}$ is represented by an equation which is a special case of either equation (1) or (4) with the $a_{ij}$ in F. Then the intersection of $\mathcal{C}$ with $\pi_0$ is a conic in the Pappian plane $\pi_0$ (unless, perchance, this intersection is trivial or degenerate).

We wish to generalize this notion to semi-fields, so we will use non-homogeneous coordinates, but the argument can be modified to apply to homogeneous co-ordinates.

Let K be a semi-field which is a finite dimensional vector space over some field F in its center. Let $t_1, t_2, \ldots, t_d$ be a basis for K over F so that for each x,y in K we may write

$$x = \alpha_1 t_1 + \ldots + \alpha_d t_d$$
$$y = \beta_1 t_1 + \ldots + \beta_d t_d$$

where $\alpha_1,\ldots,\alpha_d, \beta_1,\ldots,\beta_d$ are in F and $t_1 = 1$.

Let Q(x,y) be a quadratic polynomial. We do not need to require that the coefficients be in F. If K is a skewfield, Q(x,y) will look something like

$$Q(x,y) = a_1xb_1xc_1 + a_2yb_2yc_2 + a_3xb_3yc_3 + a_4yb_4xc_4 + a_5xb_5 + a_6yb_6+b_7 .$$

If we do not have associativity, parentheses may be inserted in various ways.

Under quite general circumstances,

$Q(x,y) = \sum_{i=1}^{d} Q_i(\alpha_1,\ldots,\alpha_d ; \beta_1,\ldots,\beta_d)t_i$ where $Q_i$ is a polynomial of degree at most two over F. Then $\{(x,y) \mid Q(x,y) = 0\}$ is defined by the simultaneous solution of the d equations

$$Q_i(\alpha_1,\ldots,\alpha_d; \beta_1,\ldots,\beta_d) = 0 \quad i=1,\ldots,d.$$

We may identify the points of $\pi$ with the points of an affine space $\mathcal{A}$ of dimension 2d over F. If $Q_i$ has degree two, then the equation $Q_i = 0$ represents a (possibly degenerate) quadratic hypersurface in $\mathcal{A}$. It could also happen that the $Q_i$ is identically zero or is constant or determines a hyperplane in $\mathcal{A}$ .

Let $\pi_1$ be a plane of $\mathcal{A}$ . Thus the points of $\pi_1$ may form a two-dimensional vector space over F or may be gotten by taking a translate of such a vector subspace. Then the intersection of $\pi_1$ with all of the surfaces $Q_i = 0$ must come under one of the following cases:

(a) The empty set.

(b) A single point.

(c) A conic in $\pi_1$.

(d) A line of $\pi_1$.

(e) A set of 2, 3, 4 points of intersections of conics in $\pi_1$.

In particular, this applies to the case where $\pi_1$ consists of all points in which $\alpha_1$ and $\beta_1$ are the only nonzero coordinates over F. In this case, $\pi_1$ can be identified with a subplane of $\pi$ that is coordinatized by F.

Thus a conicoid over K will contain subsets which are isomorphic to conics in subplanes coordinatized by F. Various types of degeneracy are of course possible.

It may be worthwhile to look at these type three conicoids from a closely related, but slightly different, point of view. To keep everything relatively explicit, we will restrict our discussion to the case $Q(x,y) = yx - 1$. Recall that x,y are elements of a (not-necessarily commutative) semi-field coordinatizing an affine plane $\pi$.

Let $(\bar{x},\bar{y})$ and $(\hat{x},\hat{y})$ be fixed points of $\pi$ with $(\hat{x},\hat{y}) \neq (0,0)$.

Then $\{(x,y) \mid y = \bar{y} + \alpha\hat{y},\ x = \bar{x} + \alpha\bar{x}\ \alpha\varepsilon F\}$ is a line $\ell$ of $\mathcal{Q}$. Suppose that $\bar{y}\,\bar{x} = 1$. The intersections of $\ell$ with the conic $\mathcal{C}$ : $yx = 1$ correspond to choises of $\alpha$ such that $(\bar{y} + \alpha\hat{y})(\bar{x} + \alpha\hat{x}) = 1$, i.e.

$$\alpha(\hat{y}\,\bar{x} + \bar{y}\,\hat{x}) + \alpha^2\,\hat{y}\,\hat{x} = 0.$$

Thus $\ell$ intersects $\mathcal{C}$ in one or two points if the above solution has one or two solutions for $\alpha$ in F. The other possibility might be that $\ell$ lies completely in $\mathcal{C}$ but we must have $\hat{y}\,\hat{x} = 0$ for this to happen. Since K has no divisors of zero, we must have $\hat{y} = 0$ or $\hat{x} = 0$. But we must also have $\hat{y}\,\bar{x} + \bar{y}\,\hat{x} = 0$ and $\bar{y}\,\bar{x} \neq 0$. We conclude that the number of intersections of $\ell$ with $\mathcal{C}$ must be 1 or 2 and hence that no line of $\mathcal{Q}$ intersects $\mathcal{C}$ in more than two points.

A line of $\pi$, of course, is a union of lines of $\mathcal{Q}$ so that a line of $\pi$ can intersect $\mathcal{C}$ in more than two points as previously pointed out. Thus $\ell$ above will lie in $y = -x$ if $\bar{y} = -\bar{x}$ and $\hat{y} = -\hat{x}$. It follows that $\ell = \ell(\bar{x},\hat{x})$ will intersect $\mathcal{C}$ in two points if there is an $\alpha \neq 0$ in F such that $-[\hat{x}\,\bar{x} + \bar{x}\,\hat{x}] = \alpha(\hat{x})^2$. Over the quaternions if we take $(\bar{x},\bar{y}) = (i,-i)$ (where $i^2 = j^2 = k^2 = -1$) it suffices to let $\hat{x} = \lambda_1 i + \lambda_2 j + \lambda_3 k$ $\lambda_1 \neq 0$. Thus the line $y = -x$ of $\pi$ contains more than one line of $\mathcal{Q}$ which intersects $\mathcal{C}$ in more than one point.

IV. Ovals. We have been using the term "oval" without saying exactly what we mean.

Definition. Let $\mathcal{G}$ be a non-empty set of points in a projective plane $\pi$ such that no three points of $\mathcal{G}$ are collinear. A line of $\pi$ which is incident with exactly one point of $\mathcal{G}$ will be called a tangent to $\mathcal{G}$. We shall refer to $\mathcal{G}$ as an <u>oval</u> if either of the following conditions are also satisfied:

(a) No point of $\pi$ is incident with three tangents.

(b) Some point is incident with all of the tangents to $\mathcal{G}$.

In this paper we are primarily interested in ovals satisfying condition (a). Unless we give a specific indication to the contrary further references to ovals will be to ovals of type (a).

It is well known -- going back to Qvist [11] that if $\pi$ is a finite projective plane of odd order n then a set of n+1 points, no three of which are collinear, is an oval in the above sense.

To see this, let $\mathcal{G}$ be a set of n + 1 points no three of which are collinear and let $\ell$ be a tangent line intersecting $\mathcal{G}$ in the point P. Let Q be any point $\neq$ P on $\ell$. Since $\mathcal{G}$ has an odd number n of points different from P and each line through Q intersects $\mathcal{G}$ in 0, 1 or 2 points, the number of tangents through Q must be even. Hence the n tangents at $\mathcal{G}$ at points $\neq$ P must intersect $\ell$ in n distinct points so each point $\neq$ P on $\ell$ lies on exactly two tangents.

In a Pappian plane of characteristic different from 2 every conic is an oval unless it degenerates into a line, pair of lines, or a single point. Segre [13] has shown that in a finite Desarguesian plane of odd order every oval is a conic.

Baer [2] has shown that in a finite projective plane of odd non-square order the absolute points of a polarity always form an oval. Here the plane need not be Desarguesian and hence need not be Pappian.

Even in finite planes of odd order, ovals do not necessarily determine polarities. However, each oval does determine a natural correspondence between certain points and certain lines.

Definition. Let $\mathcal{O}$ be an oval of type (a) in a projective plane $\pi$ not necessarily of odd order. A point P will be called an <u>exterior point</u> if P lines on two tangents to $\mathcal{O}$ ; P will be called an <u>interior point</u> if P lies on no tangents to $\mathcal{O}$ . A line $\ell$ will be called a <u>secant line</u> if $\ell$ is incident with two points of $\mathcal{O}$ , $\ell$ will be called an <u>exterior line</u> if $\ell$ is incident with no points of $\mathcal{O}$ .

There is a natural correspondence between secant lines and exterior points. Let $\ell$ be a secant line intersecting $\mathcal{O}$ in the points A and B. Then the tangents at A and B intersect in an exterior point P. Conversely if P is an exterior point, the two tangents at P determine two points of tangency to and hence a line $\ell$ joining these two points of tangency. We find it convenient to use the words "pole" and "polar" for this correspondence even if it does not extend to a polarity.

If $\mathcal{G}$ is an oval in $\pi$ and if $\pi$ admits a homology $\sigma$ leaving $\mathcal{G}$ invariant, then $\sigma$ must be an involution. In [9] we showed that if every point U not on $\mathcal{G}$ is the center of a central collineation $\sigma_u$ leaving $\mathcal{G}$ invariant then $\mathcal{G}$ determines a polarity. (Actually $\sigma_u$ will be an involution whose axis is the polar of U.) We should mention that although [9] was written for the finite case, most of the arguments work as well for the infinite case if we use our present definition of an oval. One of the big differences is that in the finite case every exterior line contains both interior and exterior points.

When we wrote [9], we were interested in generalizing the notion of harmonic sets. The following now seems to be a more natural formulation of one of the assumptions in [9]:

Assumption A1. Let U be an arbitrary exterior point of $\mathcal{G}$ and let $\ell$, $\ell_1$ be two secant lines through U intersecting $\mathcal{G}$ at A, B and $A_1$, $B_1$ respectively. Then $AA_1 \cap BB_1$ and $AB_1 \cap BA_1$ are on the polar of U. Here A, B, $A_1$, $B_1$ are assumed to be distinct.

Note that Assumption A1 holds for any exterior point U which is the center of an involutory homology which leave $\mathcal{O}$ invariant. We refered in [9] to Hilbert's [7] example of a non-Desarguesian plane obtained by modifying the interior of a conic in the real projective plane. We remarked that this gave an example of a non-Desarguesian plane in which Assumption $A_1$ was satisfied. On re-examination this seems unlikely. We probably made the tacit assumption that both $AA_1 \cap BB_1$ and $AB_1 \cap A_1B$ were exterior points.

If $\mathcal{O}$ is the oval in Hilbert's construction; if $AA_1 \cap BB_1 = W$ is an exterior point then W will be on the polar of U since the exterior parts of lines of $\mathcal{O}$ are unmodified. In passing, we note that $\mathcal{O}$ will be an oval in Hilbert's plane and that $\mathcal{O}$ will determine a polarity on the partial plane left if all of the interior points of $\mathcal{O}$ are deleted.

Definition. If Assumption A1 is satisfied for every exterior point of $\mathcal{O}$, then we call $\mathcal{O}$ an harmonic oval. We were able to show that if A, B, C, $A_1$, $B_1$, $C_1$ are six distinct points of an harmonic oval and if $AA_1 \cap BB_1 \cap CC_1$ is an exterior point then the hexagon A, B, C, $A_1$, $B_1$, $C_1$ satisfies Pappos' Theorem.

By making further assumptions of a similar nature to that of Assumption A1, we were able to show that $\mathcal{O}$ determined a polarity.

So far as I know it is still an open question as

to whether harmonic ovals (or ovals satisfying the additional assumptions A2 and B) exist that are not conics in Pappian planes. Buekenhout's work may be pertinent here. It is my understanding that this work will be discussed in other lectures. The conicoid $y = x^2$ in a commutative semi-field plane turns out to satisfy A1 if $(\infty)$ is one of the points A, B, $A_1$, $B_1$.

It should be pointed out that we used the word "conic" in [9] to denote an oval which did consist of the absolute points of a polarity and which also satisfied what amounts to a strengthening of Assumption A1 by allowing the point U to be either exterior or interior. This appears to have led to some confusion with the work of other authors who used the term in a different sense.

Note that if we have a polarity in a plane (finite or infinite) such that no three absolute points are collinear then, by duality, no three tangents are concurrent since the tangents are absolute lines. Thus (unless the set of absolute points is empty) we have an oval in this case.

Ganley [6] has shown that if a finite semi-field plane $\pi$ admits a polarity whose absolute points form an oval, then $\pi$ can be coordinatized by a commutative semi-field. We wonder whether this is not also true for infinite planes, at least in the case where the dimension over the kernel is finite.

Hilbert's construction [7] mentioned earlier indicates that, at least in the infinite case, one cannot deduce much about the interior of an oval from knowledge of the exterior. In the finite case every line that is not a tangent line contains both interior and exterior points.

Jha has raised the following question in conversation with the author: Let $\pi$ be the Hilbert non-Desarguesian plane. Does $\pi$ contain some other oval to which Hilbert's construction can be applied to get a plane $\pi'$? Thus $\pi'$ would contain two singular ovals $\mathcal{O}_1$ and $\mathcal{O}_2$ such that the intersection of the exteriors of $\mathcal{O}_1$ and $\mathcal{O}_2$ can be identified with a portion of the real projective plane. The answer is "yes". I understand that Salzmann has investigated this problem.

With current methods of investigation, the mere knowledge that a projective plane contains an oval does not seem to help much in determining the nature of the plane. We are not aware of much activity concerned with collineation groups leaving invariant a conicoid in a non-Pappian plane.

We are aware of one context in which polarities

determining ovals arise naturally from collineation groups.

Let $\pi$ be a projective plane admitting a group G of collineations. Suppose that G is abelian and sharply transitive on both points and lines. Let P be any point chosen as a reference point and let $\ell$ be a reference line. Let the point $P\lambda$ correspond to the line $\ell\lambda^{-1}$ for each $\lambda$ in G. Note that $P\sigma$ belongs to $\ell\lambda^{-1}$ (for $\sigma$, $\lambda \in G$) iff $P\sigma\lambda$ belongs to $\ell$ and $P\lambda$ belongs to $\ell\lambda^{-1}$. Thus the correspondence is a polarity and $P\lambda$ is an absolute point iff $P\lambda^2 \in \ell$.

If $\pi$ is finite, G has odd order (since the number of points in a finite projective plane is odd) and the number of absolute points is equal to the number of points on a line so that the absolute points form an oval. The case where $\pi$ is finite and G is cyclic is the well known difference set problem. The only known finite cyclic planes are Desarguesian.

Essentially the same argument works if $\pi$ is an affine plane with a special point $(\infty)$ on $\ell_\infty$ and G is sharply transitive on affine points and on lines not through $(\infty)$. In this case the above argument does not give a polar for points on $\ell_\infty$ nor a pole for lines through $(\infty)$.

Dembowski and the author [5], Theorem 2 showed that

in the finite case the projective version of $\pi$ does indeed admit a polarity in which the absolute points form an oval. An earlier result of Dembowski [4] implies that $\pi$ is $((\infty), \ell_\infty)$ transitive.

Theorem 3 of our joint paper says, in part, that under these circumstances $\pi$ can be coordinatized by a cartesian group with commutative addition and multiplication. The only known case occurs when $\pi$ is a commutative semi-field plane. In this case, G consists of mappings of the form $(x,y) \to (x + a, y + xa + b)$. The conicoid $y = \frac{1}{2} x^2$ is invariant under the subgroup consisting of mappings of the form $(x,y) \to (x + a, y + xa + \frac{a^2}{2})$.

Ovals seem to arise naturally in coding theory. If one interprets the rows of an incidence matrix as elements of a vector space over GF(2) the code words are elements of row space. Sometimes one can show that certain code words define ovals. We have been avoiding planes of even order but we should mention that this has been the basis for the most serious attempts on the plane of order 10 problem. A not necessarily complete list of people who have worked on this problem: John Thompson, Marshall Hall, and David Erbach.

This research was supported in part by a grant from the National Science Foundation.

References

1. Artzy, R.: The conic $y = x^2$ in Moufang planes. Aeq. Math. 6, 30 - 35 (1971).
2. Baer, R.: Polarities in finite projective planes. Bull. Am. Math. Soc. 52, 77 - 93 (1946).
3. Berz, E.: Kegelschnitte in desarguesschen Ebenen. Math. Z. 78, 55 - 85 (1962).
4. Dembowski, P.: Gruppentheoretische Kennzeichnungen der endlichen desarguesschen Ebenen. Abh. Math. Sem. Hamb. 29, 92 - 106 (1965).
5. _________ and Ostrom, T.G.: Planes of order n with collineation groups of order $n^2$. Math. Z. 103, 239 - 258 (1968).
6. Ganly, M.: Polarities in translation planes. Geom. Ded. 1, 103 - 116 (1972).
7. Hilbert, D.: The foundations of geometry. Open Court Publishing Co., LaSalle, Illinois (1938). (Published in German as "Grundlagen der Geometrie".)
8. Krueger, W.: Kegelschnitte in Moufangebenen. Math. Z. 120, 41 - 60 (1971).
9. Ostrom, T.G.: Ovals, dualities and Desargues's Theorems.Can. J. Math. 7, 417 - 431 (1955).
10. Pickert, G.: Projective Ebenen. Springer Verlag, Berlin (1955).
11 Qvist, B.: Some remarks concerning curves of the second degree in a finite plane. Ann. Ac. Fennicae 134, 1 - 27 (1952).
12. Segre, B.: Lectures on Modern Geometry. Cremonese, Rome (1961).
13. _________: Ovals in a finite projective plane. Can. J. Math. 7, 414 - 416 (1955).

SYMMETRIES OF QUADRICS

Helmut Mäurer

Darmstadt

## 1. QUADRICS, QUADRATIC SETS, OVOIDS AND OVALS

Quadrics in projective geometry are usually introduced in an algebraic way: Starting with an at least 3-dimensional vectorspace V over a commutative field K, we consider a quadratic form $q : V \longrightarrow K$ i.e. $q(kv) = k^2 q(v)$ for all $(k,v) \in K \times V$ and

$$\langle\ ,\ \rangle : \begin{array}{l} V \times V \longrightarrow K \\ (x,y) \longrightarrow \langle x,y\rangle := q(x+y) - q(x) - q(y) \end{array} \quad \text{is bilinear.}$$

The projective geometry *PG(V)* is the lattice of all linear subspaces of V . The set of all (i+1)-dimensional subspaces of V will be denoted by $PG_i(V)$ . As usual, the elements of $PG_0(V)$, $PG_1(V)$ and $PG_2(V)$ are called *points*, *lines* and planes *respectively*.

A quadric is defined as the set of solutions of the equation $q = 0$ , more precisely:



*P. Plaumann and K. Strambach (eds.), Geometry – von Staudt's Point of View, 197–229.*

$$\mathcal{Q}(V,q) := \{P \in PG_0(V) \mid q(P) = 0\}$$

is called the *quadric* in PG(V), determined by $q$. In the case of $\dim V = 3$ we speak of a *conic* in the projective plane PG(V). Coordinatizing $V$ relative to a basis $B = \{b_i \mid i \in I\}$, we see that the point $Kx$ with

$$x = \sum_{i \in I} x_i b_i \neq 0$$

lies in $\mathcal{Q}(V,q)$ if and only if

$$\sum_{(i,j) \in I^2} a_{ij} x_i x_j = 0 .$$

Here $a_{ii} := q(b_i)$ and for $i \neq j$ the elements $a_{ij}, a_{ji}$ are chosen in such a way that $a_{ij} + a_{ji} = \langle b_i, b_j \rangle$.

As far as the incidence-properties of a quadric are concerned, we are mainly interested in the behaviour of the intersection of lines with $\mathcal{Q} := \mathcal{Q}(V,q)$. Let $l = Ka + Kx$ be a line where $A = Ka$ is a point of $\mathcal{Q}$. Any point of $l$ distinct from $A$ has the form $K(x + ka)$ with $k \in K$. This point is on $\mathcal{Q}$ iff $q(x) + k\langle x,a \rangle = 0$. From this we deduce that in the case $\langle x,a \rangle \neq 0$ the line $l$ intersects $\mathcal{Q}$ in exactly 2 points. Such a line is called a *secant*. In case $\langle x,a \rangle = 0$ the line $l$ is contained in $\mathcal{Q}$ (iff $q(x) = 0$) or $\mathcal{Q} \cap l = \{A\}$ (iff $q(x) \neq 0$). In this situation the line $l$ is called a *tangent* to $\mathcal{Q}$ at $A$.

The equation $\langle x,a \rangle = 0$ for tangents at A shows the validity of

(1) For all points $A \in \mathcal{Q}$, the union $\mathcal{Q}_A$ of all tangents to $\mathcal{Q}$ at A is either a hyperplane or the whole space.

In the second alternative of (1) A is said to be a *double point*. We observe, that $\mathcal{Q}$ is a *quadratic set* in the sense of F. BUEKENHOUT [4], i. e. $\mathcal{Q}$ is a set of points in a projective geometry - projective planes included - for which the concepts of secants and tangents are defined as above and for which the conditions (1) and (2) are satisfied. Here (2) means:

(2) Any line l intersecting $\mathcal{Q}$ in at least 3 points is entirely contained in $\mathcal{Q}$.

In [4] BUEKENHOUT has proved the following results: A quadratic set $\mathcal{Q}$ is a quadric, if $\mathcal{Q}$ contains a line l, such that no point of l is a double point. The set D of double points is a subspace and $\mathcal{Q}$ is the union of D and all lines connecting a point of D and a point of $\mathcal{Q} \cap C$, where C is a complement of D, i. e. $V = D \oplus C$.

In view of these results, for the characterization of the quadrics among the quadratic sets, we have only to consider *ovoids*, which means non empty quadratic sets containing no double point and no line. For an ovoid in a projective plane we usually use the name *oval*.

A further useful reduction of questions about quadrics or quadratic sets to the planar situation can be obtained as a consequence of the following results of BUEKENHOUT: A set $\mathcal{Q}$ of points is a quadratic set respectively a quadric iff every plane section of $\mathcal{Q}$ is a quadratic set respectively a conic. ([4], 2.2 and 4.4).

## 2. PERSPECTIVITIES STABILIZING A QUADRATIC SET

Let $\sigma \neq 1$ be a perspectivity with center $C$ stabilizing $\mathcal{Q}$, i. e. $\sigma(\mathcal{Q}) = \mathcal{Q}$. If $C$ is on $\mathcal{Q}$, then each point $Q \in \mathcal{Q} \setminus \mathcal{Q}_C$ is fixed by $\sigma$, because $C + Q$ is a fixed secant. Therefore $\mathcal{Q}$ is contained in the union of $\mathcal{Q}_C$ and the axis of $\sigma$. Thus either $C$ is a double point or $\mathcal{Q}$ is the union of 2 hyperplanes. In the second case the intersection of these hyperplane consists of double points. This proves a part of

LEMMA 1: If $\mathcal{Q}$ is a *regular* quadratic set, i. e. $\mathcal{Q}$ has no double points, then every perspectivity $\sigma \neq 1$ stabilizing $\mathcal{Q}$ has order 2 and its center $C$ is outside of $\mathcal{Q}$. Furthermore $\sigma$ is uniquely determined by the center $C$.

PROOF: It has been shown already that C is not on $\mathcal{Q}$. We assert that for any secant l through C with $l \cap \mathcal{Q} = \{A,B\}$, the points A,B are interchanged by $\sigma$. If we assume this as being shown already, then $\sigma^2$ and $\sigma\sigma'$ are equal to the identity on $\mathcal{Q}$, for any perspectivity $\sigma'$ with center C and $\sigma'(\mathcal{Q}) = \mathcal{Q}$. Therefore $\sigma^2 = 1$ and $\sigma\sigma' = 1$ or $\mathcal{Q}$ is contained in the axis of $\sigma^2$ respectively of $\sigma\sigma'$. This last alternative is impossible, because $\mathcal{Q}$ is not contained in a hyperplane. Thus it remains to show $\sigma(A) = B$. By way of contradiction we now assume $\sigma(A) = A$ and $\sigma(B) = B$. The two hyperplanes $\mathcal{Q}_A$, $\mathcal{Q}_B$ which both do not pass through C are then fixed by $\sigma$, which yields a contradiction.

NOTATION: A perspectivity $\sigma \neq 1$ with center C fixing $\mathcal{Q}$ is called a *reflection* of $\mathcal{Q}$ at the point C. If such a reflection exists at all, then $\mathcal{Q}$ is said to be *symmetric* to the point C.

LEMMA 2: Let $\mathcal{Q}$ be a quadric and C = Kc a point not in $\mathcal{Q}$, then

$$\lambda_C : \begin{array}{l} V \longrightarrow V \\ x \longmapsto x - \frac{\langle x,c\rangle}{q(c)} \cdot c \end{array}$$

is a bijection of order 1 or 2, leaving invariant

<u>q and $\mathfrak{Q}$. $\lambda_C = 1$ or $\lambda_C$ induces in PG(V) a reflection of $\mathfrak{Q}$ at the point C.</u>

<u>PROOF:</u> By direct computation we observe that $\lambda_C$ preserves the form q and that $\lambda_C$ has order 1 or 2. Furthermore C is fixed by $\lambda_C$ and $\lambda_C$ induces the identity in the quotient space $V/_C$. Therefore $\lambda_C$ acts on PG(V) as a perspectivity with center C.

<u>LEMMA 3: Let A,B,C be collinear points outside of the quadric $\mathfrak{Q}$ in PG(V). Then there exists a point $D \notin \mathfrak{Q}$ collinear with A,B,C, such that $\lambda_C\lambda_B\lambda_A = \lambda_D$.</u>

<u>PROOF:</u> Because $\lambda_C\lambda_A\lambda_A = \lambda_C$, we may assume $A \neq B$. A simple, but lengthy computation shows, that for $A = Ka$, $B = Kb$, $C = K(\alpha a + \beta b)$ we get $\lambda_C\lambda_B\lambda_A = \lambda_D$ where $D = Kd$,

$$d = (\beta q(b) + \alpha\langle a,b\rangle)\, a - (\alpha q(a))\, b$$

and $q(d) = q(a) \cdot q(b) \cdot q(\alpha a + \beta b)$.

## 3. PASCAL'S THEOREM AND ITS CONVERSE

Let $\mathfrak{Q}$ be an oval in a projective plane. $\mathfrak{Q}$ is called *pascalian* if for each sixtuple $A_1,A_2,A_3,B_1,B_2,B_3$ of not necessarily distinct points of $\mathfrak{Q}$ with $A_i \neq A_j$, $B_i \neq B_j$, $A_i + B_j \neq A_j + B_i$ for all $i \neq j$, the 3 distinct points $D_1 := (A_2 + B_3) \cap (A_3 + B_2)$, $D_2 := (A_1 + B_3) \cap (A_3 + B_1)$, $D_3 := (A_1 + B_2) \cap (A_2 + B_1)$ are collinear. (When $A_i$

coincides with $B_j$ the sum $A_i + B_j$ denotes the tangent to $\mathfrak{q}$ at $A_i = B_j$ .)

THEOREM 1 (PASCALS THEOREM): Every conic is pascalian.

PROOF: Usually this theorem is proved by using perspectivities from a conic onto itself. Another proof can be obtained by lemma 3 in section 2: Let $D_1, D_2$ be the points as defined above, and set $D_3 := (D_1 + D_2) \cap (A_2 + B_1)$ . The product of the factors $\lambda_{D_3}, \lambda_{D_2}, \lambda_{D_1}, \lambda_{D_3}, \lambda_{D_2}, \lambda_{D_1}$ is the identity and the point $B_2$ is moved succesively by these factors into the points $A_3, B_1, A_2, B_3, A_1$ , and finally back into $B_2$ . Therefore $\sigma_{D_3}(A_1) = B_2$ . Consequently $D_3 = (A_2 + B_1) \cap (A_1 \cap B_2)$ and Pascals theorem is proved.

In 1966 BUEKENHOUT [3] proved the converse, namely the following

THEOREM 2: Every pascalian oval in a projective plane is a conic.

Beside Buekenhout's proof, using a theorem of TITS about sharply 3-fold transitive permutation groups, other proofs have been given by several authors, see [1] , [8] and [13] . Here we present a slight modification of the original proof in order to avoid the application of TITS' theorem.

<u>PROOF OF THEOREM 2:</u> Let $\mathfrak{O}$ be a pascalian oval in the projective plane $\mathfrak{P}$. For any point $P \notin \mathfrak{O}$ we define an involutory bijection $\widetilde{P}$ from $\mathfrak{O}$ onto itself by the equation $(P + X) \cap \mathfrak{O} = \{X, \widetilde{P}(X)\}$, for all $X \in \mathfrak{O}$. First we remark that the mapping $P \longrightarrow \widetilde{P}$ is injectiv. A basic property used in the following is

(1): The points $P_1, P_2, P_3$ not on the oval $\mathfrak{O}$ are collinear if and only if $(\widetilde{P}_1 \widetilde{P}_2 \widetilde{P}_3)^2 = 1$.

If the points $P_1, P_2, P_3$ are not distinct, then (1) is obvious. In the other case one can see similary as in the proof of theorem 1 that (1) is a consequence of the Pascal-property of $\mathfrak{O}$.

For a line $l$ we consider $\widetilde{l} := \{\widetilde{Q} \circ \widetilde{P} \mid P, Q \in l \setminus \mathfrak{O}\}$.

(2): $\widetilde{l}$ is an abelian group, acting regularly on $\mathfrak{O} \setminus (l \cap \mathfrak{O})$. Furthermore $\widetilde{l} = \{\widetilde{X} \circ \widetilde{P} \mid X \in l \setminus \mathfrak{O}\}$ for any point $P \in l \setminus \mathfrak{O}$.

From (1) we deduce that the group $L$ generated by the products $\widetilde{Q} \circ \widetilde{P}$ with $P, Q \in l \setminus \mathfrak{O}$, is a commutative group. Let $A, B, P$ be points with $A, B \in \mathfrak{O} \setminus l$ and $P \in l \setminus \mathfrak{O}$. Then $D := l \cap (B + \widetilde{P}(A))$ is the unique point on $l \setminus \mathfrak{O}$ with $\widetilde{D}\widetilde{P}(A) = B$. Therefore the subset $\{\widetilde{X} \circ \widetilde{P} \mid X \in l \setminus \mathfrak{O}\}$ of $L$ acts regularly on $\mathfrak{O} \setminus l$. Now $\mathfrak{O} \setminus l$ and each single point of $\mathfrak{O} \setminus l$ form the orbits of the abelian group $G$. As a commutative group $L$ acts regularly on $\mathfrak{q} \setminus l$ and therefore $\{\tilde{X} \subset \tilde{P} \mid X \in \mathfrak{q} \setminus l\} = \check{l} = L$.

(3) Let $P_1, P_2, P_3$ be points of $l \setminus \mathfrak{q}$, where $l$ is a given line. Then there exists a unique point $P_4 \in l \setminus \mathfrak{q}$ with $\tilde{P}_3\tilde{P}_2\tilde{P}_1 = \tilde{P}_4$

This is corollary of (2).

Our next aim is to extend $\tilde{P}$ to a perspectivity of the projective plane $\mathfrak{P}$. For doing this, we first remark that for each point $X \in \mathfrak{P} \setminus \mathfrak{q}$ the product $\tilde{P}\tilde{X}\tilde{P}$ is - by (3) - of the form $\tilde{Y}$. Defining $\tilde{P}(X) = Y$ we get an extension, also denoted by $\tilde{P}$. This extension is a mapping from $\mathfrak{P}$ into itself. Because $\tilde{P}\tilde{X}\tilde{P} = \tilde{Y}$ is equivalent to $\tilde{P}\tilde{Y}\tilde{P} = \tilde{X}$, the extended mapping $\tilde{P}$ is an involution, in particular a bijection. Let $P_1, P_2, P_3$ be points of a line $l$. If $l \cap \mathfrak{q} = \emptyset$ then $(\tilde{P}_1\tilde{P}_2\tilde{P}_3)^2 = 1$ by (1). Conjugation with $\tilde{P}$ yields the collinearity of the points $\tilde{P}(P_i)$ $(i = 1,2,3)$. In the case $l \cap \mathfrak{q} = \{A,B\}$, where $A,B$ are not necessarily distinct points, we have $\tilde{P}_i(A) = B$ if $P_i \notin \mathfrak{q}$. By conjugation with $\tilde{P}$, we get $\tilde{P}\tilde{P}_i\tilde{P}(\tilde{P}(A)) = \tilde{P}(B)$. Therefore the point $\tilde{P}(P_i)$ is on the line $\tilde{P}(A) + \tilde{P}(B)$. Now we get easily the collinearity of the points $\tilde{P}(P_1), \tilde{P}(P_2), \tilde{P}(P_3)$ and $\tilde{P}$ is seen to be a collineation. By (3) each line through $P$ is fixed.

We summarize:

(4) The extension $\tilde{P} : \mathfrak{Z} \longrightarrow \mathfrak{Z}$ is a perspectivity of $\mathfrak{Z}$ with center P and $\tilde{P}(\mathfrak{O}) = \mathfrak{O}$ . If there is a secant through P , then $\tilde{P}$ has order 2 . Thus the oval $\mathfrak{O}$ is symmetric to the point P .

In the following $\tilde{P}$ is never used in the original meaning, but always as perspectivity of $\mathfrak{Z}$ . Thus $\tilde{I}$ in (2) becames a collineation group of $\mathfrak{Z}$ .

Now we coordinatize $\mathfrak{O}$ and distinguish 3 distinct points 0,E and $\infty$ on $\mathfrak{O}$ . The group $\tilde{1}_\infty$ , where $1_\infty$ is the tangent to $\mathfrak{O}$ at $\infty$ , is sharply transitive on $K := \mathfrak{O} \setminus \{\infty\}$ and fixes $\infty$ . The group $\widetilde{1_{0\infty}}$ ($1_{0\infty}$ the line through 0 and $\infty$) acts sharply transitively on $K \setminus \{0\} = \mathfrak{O} \setminus \{0,\infty\}$ and fixes the points 0 and $\infty$ . Both groups are abelian and they provide the set K with the structure of a field by defining:

$$\tau_2(0) + \tau_1(0) := \tau_2\tau_1(0) \quad \text{for all } \tau_1,\tau_2 \in \tilde{1}_\infty$$

$$\delta(E) \cdot X := \delta(X) \quad \text{for all } X \in K \text{ and all } \delta \in \widetilde{1_{0\infty}}$$

$$0 \cdot X = 0 \quad \text{for all } X \in K .$$

Indeed (K,+) is a group isomorphic to $1_\infty$ and 0 is the zero of this group. For $\delta_1,\delta_2 \in 1_{0\infty}$ we have $\delta_2(E) \cdot \delta_1(E) = \delta_2\delta_1(E)$ and therefore $(K \setminus \{0\}, \cdot)$ is a group isomorphic to $\widetilde{1_{0\infty}}$ and E is the unit of this group. $\widetilde{1_{0\infty}}$ is in the normalizer of $\widetilde{1_\infty}$, because $\infty$ is fixed by $\widetilde{1_{0\infty}}$. From this fact and from the commutative law one easily deduces that multiplication satisfies the distributive law.

(5) Let $P$ be a point on the line $1_\infty$. Then $\infty$ is fixed by $\tilde{P}$ and the restriction $\tilde{P}|_K$ of $\tilde{P}$ to $K = \mathfrak{Q}\setminus\{\infty\}$ is of the form $X \longrightarrow A - X$, where $A := \tilde{P}(0)$.

By (1) conjugation with $\tilde{P}$ induces in the group $\widetilde{1_\infty}$ the mapping $\tau \longrightarrow \tau^{-1}$. For $X := \tau(0)$ we now get $\tilde{P}(X) = \tilde{P}\tau(0) = \tau^{-1}\tilde{P}(0) = \tau^{-1}(A) = -X+A$.

(6) Let $P$ be a point on the line $1_{0\infty}$. Then the points $0,\infty$ are interchanged by $\tilde{P}$ and the restriction $\tilde{P}|_{K^\times}$ of $\tilde{P}$ to $K^\times := K\setminus\{0\}$ is of the form $X \longrightarrow A \cdot X^{-1}$, where $A := \tilde{P}(E)$.

Again, $\tilde{P}\delta = \delta^{-1}\tilde{P}$ for all $\delta \in \widetilde{1_{0\infty}}$. Setting $X = \delta(E)$, we get $\tilde{P}(X) = \tilde{P}\delta\ (E) = \delta^{-1}\tilde{P}(E) = X^{-1} \cdot A$.

(7) The group $G$ generated by the bijections $\tilde{P}|\mathfrak{Q}$ is the sharply 3-fold transitive permutation group $PGL(2,K)$, acting on $\mathfrak{Q} = K \cup \{\infty\}$.

Because the reflections considered in (5), (6) generate $PGL(2,K)$, we have $PGL(2,K) \subset G$. Conversely, for any point $Y \in \mathfrak{P}\setminus\mathfrak{Q}$ the image $\tilde{Y}(\infty)$ is on $\mathfrak{Q}$. If $\tilde{Y}(\infty) = \infty$, then $Y \in 1_\infty$ and by (5)

$\tilde{Y}|\mathcal{O}$ is an element of $PGL(2,K)$. If $\tilde{Y}(\infty) \neq \infty$ , then (2) guarantees the existence of a mapping $\tau \in \widetilde{1_\infty}$ with $\tau(\tilde{Y}(\infty)) = 0$. Now $\tau\tilde{Y}\tau^{-1} = \widetilde{\tau(Y)}$ interchanges the points $0,\infty$ and therefore $\tau(Y)$ is on $1_{0\infty}$ . By (6) $\widetilde{\tau(Y)}|\mathcal{O}$ is in $PGL(2,K)$ . With $\tau|\mathcal{O} \in 1_\infty|\mathcal{O} \subset PGL(2,K)$ we get $\tilde{Y}|\mathcal{O} \in PGL(2,K)$ and from this also the assertion $G = PGL(2,K)$ .

All generators $\tilde{P}|\mathcal{O}$ of $G$ can be extended to collineations of $\mathcal{P}$ . Therefore the same is true for the elements in $G$ . The uniqueness of the extension is obvious.

(8) To every element $\sigma$ of order 2 in $PGL(2,K)$ , there exists exactly one point $S \in \mathcal{P}\setminus\mathcal{O}$ such that $\tilde{S}|\mathcal{O} = \sigma$ . The oval $\mathcal{O}$ has a knot, i. e. a point lying on all tangents to $\mathcal{O}$ if and only if $K$ has characteristic 2 .

Because $\sigma$ has order 2 , there are distinct points $A,B \in \mathcal{O}$ interchanged by $\sigma$ . Choose $\alpha \in PGL(2,K)$ with $\alpha(A) = 0$ and $\alpha(B) = \infty$. Then $\alpha\sigma\alpha^{-1} \in PGL(2,K)$ interchanges the points $0,\infty$ . The restriction of this mapping to $K^\times$ is of the form $X \longrightarrow X^{-1}\cdot A$ with some $A \in K^\times$ . By (6) there exists some point $P \in 1_{0\infty}$ such that $\tilde{P}|\mathcal{O} = \alpha\sigma\alpha^{-1}$. If $\alpha^*$ denotes the extension of $\alpha$ to a collineation, then $\sigma = \tilde{S}|\mathcal{O}$ for $S = \alpha^{*-1}(P)$ . A point $D$ is a knot, if and only if

$D \notin \mathcal{Q}$ and $\tilde{D}$ is the identity. These conditions are also equivalent to $\infty \neq D \in l_\infty$ and $D(X) = X$ for all $X \in K$. With (6) we conclude that this occurs iff the characteristic of the field $K$ is $2$.

Within our considerations we have constructed a field $K$ and have determined the action of the group $G$ on $\mathcal{Q}$. Our next aim is to obtain some informations about the structure of the projective plane $\mathcal{P}$. For this purpose, we set

$$q : \begin{array}{ccc} K^3 & \longrightarrow & K \\ (x_o, x_1, x_2) & \longrightarrow & -x_o^2 + x_1 x_2 \end{array}$$

$q$ is a quadratic form on the 3-dimensional vector-space $K^3$ and defines a conic $\mathcal{Q}' := \{X \in PG_o(K^3) \mid q(X) = 0\}$ in the projective plane $\mathcal{P}' := PG(K^3)$. Because $\mathcal{Q}'$ is a pascalian oval in $\mathcal{P}'$ all the preceding considerations about $\mathcal{Q}$ and $\mathcal{P}$ can also be done for $\mathcal{Q}'$ and $\mathcal{P}'$.

There is a bijection $\gamma : \mathcal{Q} \longrightarrow \mathcal{Q}'$, in the following often denoted by a dash $'$, given by $\infty' = K(0,1,0)$, $0' = K(0,0,1)$ and $k' = K(1,k,k^{-1})$ for all $k \in K^\times$.

If $P$ is the point $K(1,a,0)$ with $a \in K$, then $P$ is a point on the tangent to $\mathcal{Q}'$ at $\infty'$. Because $P \notin \mathcal{Q}'$ there exists a perspectivity $\tilde{P}$ of $\mathcal{P}'$ with center $P$ leaving invariant the conic $\mathcal{Q}'$. It is easily seen that $\tilde{P}$ maps $0' = K(0,0,1)$ on the point $a'$. For $x \in K \setminus \{0,a\}$ we compute that $\mathfrak{Q}'$ intersects the line joining $x'$ and $P$ in the set $\{x',(a-x)'\}$. Therefore $\tilde{P}$ fixes $\infty'$ and maps $x'$ on $(a-x)'$ for each $x \in K$.

If $P$ is the point $K(0,1,-a^{-1})$ with $a \in K^{\times}$ then $P \notin \mathfrak{Q}'$ and $\tilde{P}$ interchanges the points $0',\infty'$. For $x \in K^{\times}$ we compute that $\mathfrak{Q}'$ intersects the line joining $x'$ and $P$ in the set $\{x',(ax^{-1})'\}$. Therefore $\tilde{P}(x') = (ax^{-1})'$ for all $x \in K^{\times}$.

In part (7) we have proved that the group $G'$ generated by the bijections $\tilde{P}|_{\mathfrak{Q}'}$ $(P \in \mathfrak{S}' \setminus \mathfrak{Q}')$ is already generated by those mappings $\tilde{P}|_{\mathfrak{Q}'}$, where $P$ is on the line through $0',\infty'$ or where $P$ is a point of the tangent to $\mathfrak{Q}'$ at $\infty'$. But our computations made above show that these mappings generate the group $PGL(2,K)$. Thus the groups $G$ and $G'$ are isomorphic as permutation groups acting on $\mathfrak{Q}$, $\mathfrak{Q}'$ respectively. The isomorphism from $(G, \mathfrak{Q})$ to $(G', \mathfrak{Q}')$ is given by the pair $(\iota,\gamma)$, where $\iota : G \longrightarrow G'$ is defined by $\sigma \longrightarrow \gamma\sigma\gamma^{-1}$ for all $\sigma \in G$.

(9) The bijection $\gamma : \mathfrak{Q} \longrightarrow \mathfrak{Q}'$ can be extended to a collineation from the projective plane $\mathfrak{S}$ onto the projective plane $\mathfrak{S}'$.

For the proof of (9) we conclude very similarly as

in the proof of (4): For any point $P \in \mathfrak{P} \setminus \mathcal{O}$ the mapping $\tilde{P}|_{\mathcal{O}}$ has order 2 or order 1 . The last alternative is possible by (8) if and only if K has characteristic 2 . The image $\iota(\tilde{P}|_{\mathcal{O}}) = \gamma\tilde{P}|_{\mathcal{O}}\gamma^{-1}$ has the same order. From (8) we deduce, that there is a unique point $P' \in \mathfrak{P}' \setminus \mathcal{O}'$ satisfying $P'|_{\mathcal{O}'} = \iota(\tilde{P}|_{\mathcal{O}})$ . Now we extend the mapping $\gamma : \mathcal{O} \longrightarrow \mathcal{O}'$ to a mapping $\gamma : \mathfrak{P} \longrightarrow \mathfrak{P}'$ by setting $\gamma(P) = P'$ . In the same way as in the proof of (4) it follows, that this extension defines a collineation from $\mathfrak{P}$ to $\mathfrak{P}'$ . Therefore the given pascalian oval $\mathcal{O}$ in the projective plane $\mathfrak{P}$ has been recognized as a conic in a projective plane over a field. This finishes the proof of theorem 2.

## 4. CHARACTERIZATIONS OF CONICS INVOLVING REFLECTIONS

The characterization of a conic by the Pascal property in theorem 2 of section 3 can also be regarded as a characterization involving reflections. Indeed we have shown that a pascalian oval $\mathcal{O}$ is symmetric to any point outside of $\mathcal{O}$ which is not a knot of $\mathcal{O}$. Furthermore it has been shown in (3) that the reflections satisfy a relation, described in (3). The rest of the proof of theorem 2 bases only on these two properties.

In this section we always consider an oval $\mathfrak{O}$ in a *Moufang* or *Desarguesian* projective plane and we assume that there are many reflections of $\mathfrak{O}$. However we do not demand any relation for the reflections. Our main question is the following: *How many reflections of* $\mathfrak{O}$ *are necessary in order to prove that the given oval* $\mathfrak{O}$ *is a conic?*

## 4.1. OVALS SYMMETRIC TO THE POINTS OF A SECANT

THEOREM 1: Let $\mathfrak{O}$ be an oval in a Moufang projective plane $\mathfrak{P}$. Assume that $\mathfrak{O}$ is symmetric to every point of $l \setminus \mathfrak{O}$, where $l$ is a secant of $\mathfrak{O}$. Then $\mathfrak{O}$ is a conic.

PROOF (for Desarguesian planes given by BUEKENHOUT in [4], 4.5)

Assume $l \cap \mathfrak{O}$ consists of the two points A,B and C is the common point of the tangent a to $\mathfrak{O}$ at A and the tangent b at B. The affine plane $\mathfrak{P} \setminus l$ can be coordinatized by an alternative division ring K in such a way that $\mathfrak{P} \setminus l$, $a \setminus \{A\}$, $b \setminus \{B\}$ respectively can be identified with the cartesian products $K^2$, $K \times \{0\}$, $\{0\} \times K$ respectively and such that $(1,1)$ is a point of $\mathfrak{O}$. The reflection $\sigma_m$ at the intersection point of the line $l$ and the line described by $\{(x,-mx) \mid x \in K\}$ $(m \in K^{\times})$ interchanges the points A,B and fixes the point C. Therefore $\{(x,mx) \mid x \in K\}$ is the axis of $\sigma_m$. From this we conclude that the point $(x,y) \in K^2$

is mapped by $\sigma_m$ onto the point $(m^{-1}y, mx)$. The set of all reflections $\sigma_m$, where $m$ runs over $K^\times$, acts transitively on $\mathcal{O}\setminus\{A,B\}$, hence $\mathcal{O}\setminus\{A,B\} = \{\sigma_m(1,1)\,|\,m\in K^\times\} = \{(m^{-1}, m)\,|\,m\in K^\times\}$. Especially $\sigma_m(x,x^{-1}) = (m^{-1}x^{-1}, mx)$ is a point of $\mathcal{O}$ for all $m,x\in K^\times$. Thus we obtain $m^{-1}x^{-1} = (mx)^{-1}$ and from the law $(mx)^{-1} = x^{-1}m^{-1}$, valid in each alternative division ring (see for example [12], p. 162), we conclude that multiplication in $K$ is commutative and $(K,+,\cdot)$ is by [12], p. 162 a commutative field. The above representation for $\mathcal{O}\setminus\{A,B\}$ shows that $\mathcal{O}$ is a conic.

COROLLARY: *Suppose $\mathcal{O}$ is an oval in a Desarguesian projective plane, symmetric to every point not being a knot and outside of $\mathcal{O}$. Then $\mathcal{O}$ is a conic.*

THEOREM 2: *Let $\mathcal{O}$ be an oval in a Desarguesian affine plane over an ordered field $K$. Assume that $\mathcal{O}$ is symmetric to every point $\neq A,B$ of the convex hull of two distinct points $A,B\in\mathcal{O}$. Then $\mathcal{O}$ is a conic.*

PROOF: Let $\mathcal{P}$ be the projective closure of the given affine plane. Then $\mathcal{P}\setminus l$ is an affine plane. We now modify our considerations, made in the proof of theorem 1, in the following way: For each $m<0$ the reflection $\sigma_m$ exists. From $\sigma_m(1,1) = (m^{-1},m)\in\mathcal{O}$ and from $\sigma_m\,\sigma_{-1}(1,1) = (-m^{-1},-m)\in\mathcal{O}$ for all $m<0$

we obtain $\mathcal{O} \setminus \{A,B\} = \{(x^{-1},x) \mid x \in K^{\times}\}$ . In the same way as in the proof of theorem 1 we deduce $m^{-1}x^{-1} = (mx)^{-1} = x^{-1}m^{-1}$ for all $x \in K^{\times}$ and all $m < 0$ . Therefore $K$ is a commutative field and $\mathcal{O} = \{A,B\} \cup \{(x^{-1},x) \mid x \in K^{\times}\}$ is a conic.

4.2 OVALS SYMMETRIC TO THE POINTS OF A TANGENT

THEOREM 3 ( [10] ): *Let $A$ be a point of an oval $\mathcal{O}$ in a projective Moufang plane $\mathcal{P}$ . Suppose that for every point $B \in \mathcal{O} \setminus \{A\}$ there exists a collineation $\sigma_B$ of $\mathcal{P}$ of order 2, such that $\sigma_B(\mathcal{O}) = \mathcal{O}$ and $\mathcal{O} \cap Fix\ \sigma_B = \{A,B\}$ . Then $\mathcal{O}$ is a conic and $\mathcal{P}$ is a pappian plane of characteristic $\neq 2$.*

NOTATION: $Fix\ \sigma$ denotes the set of all fixed points of $\sigma$ .

PROOF: Let $a,b$ be the tangents to $\mathcal{O}$ at $A,B$ resp. The lines $A+B,a,b$ are fixed by $\sigma_B$, but no other line through $A$ or $B$ is left invariant under $\sigma_B$, because otherwise there would be a further fixed point on $\mathcal{O}$ . Thus $C := a \cap b$ is the only fixed point of $\sigma_B$ outside the line $A+B$ and $\sigma_B$ cannot be a BAER-involution. By this we see that $\sigma_B$ is a involutory perspectivity with center $C$ and axis $A+B$ . Furthermore the characteristic of the plane $\mathcal{P}$ and of the alternative division ring $K$ coordinatizing this plane is unequal to 2 . Now let $\mathcal{A}$ be the affine

Moufang plane $\mathfrak{I} \setminus a$ . The pointset of $\mathcal{A}$ can be represented by $K \times K$ in such a way, that the affine lines not passing through $A$ are described by equations $y = m \cdot x + c$ $(m, c \in K)$ . Furthermore we may assume that $(0,0), (1,1)$ are on $\mathcal{O}$ and $K \times \{0\}$ is the tangent to $\mathcal{O}$ at $(0,0)$ . From the incidence properties of an oval we deduce the existence and uniqueness of a map $f : K \longrightarrow K$ with $(x, f(x)) \in \mathcal{O}$ for all $x \in K$ . If $B = (x_o, f(x_o))$ and if $y - f(x_o) = m(x - x_o)$ represents the tangent to $\mathcal{O}$ at $B$ , then $(x,y) \longmapsto (2x_o - x, y - 2m_o x + 2m_o x_o)$ is an involutory perspectivity with the same center and the same axis as $\sigma_B$ . Because an involutory perspectivity is uniquely determined by its center and its axis, we conclude

(i) $\sigma_B(x,y) = (2x_o - x, y - 2m_o x + 2m_o x_o)$ .

We should remark that multiplication with factor 2 is commutative and associative. Setting $x_o = 0$ in (i) we get from $m_o = 0$ in this case $\sigma_{(o,o)}(x,y) = (-x,y)$ and as a consequence of $\sigma_{(o,o)}(\mathcal{O}) = \mathcal{O}$ the equation

(ii) $f(-x) = f(x)$ .

The composition $\sigma_B \sigma_{(o,o)}$ maps $(x,y)$ onto $(x + 2x_o, y + 2m_o x + 2m_o x_o)$ and leaves invariant the oval $\mathcal{O}$ . From this we get the functional equation $f(x + 2x_o) = f(x) + 2m_o x + 2m_o x_o$ . By setting $v = 2x_o$

the coefficient $m_o$ depends on $v$, hence we write $m_o = m(v)$. Taking $0$ for $x$ we see $f(v) = 2m_o x_o = m(v) \cdot v$ and resembling these informations we obtain the equation

(iii) $f(x + v) = f(x) + f(v) + 2\, m(v) \cdot x$ .

Interchanging $x$ and $v$ leads to

(iv) $m(v) \cdot x = m(x) \cdot v$ .

From $(1,1) \in \mathcal{O}$ we derive $m(1) = 1$ and with (iv) also the equality $m(x) = x$ for all $x \in K$. Thus (iv) is the commutative law for the multiplication and $K$ is a commutative field (see [12], p. 162). Hence $f(v) = m(v) \cdot v = v^2$ for all $v \in K$ and $\mathcal{O} \setminus \{A\}$ is a parabola in $\mathcal{A}$. Now it is obvious that $\mathcal{O}$ is a conic in $\mathcal{P}$.

<u>THEOREM 4 ([11]): Let $a$ be a tangent to an oval $\mathcal{O}$ in a projective Moufang plane $\mathcal{P}$. Suppose $A = a \cap \mathcal{O}$ and $\mathcal{O}$ is symmetric to every point of $a \setminus \{A\}$. Then $\mathcal{O}$ is a conic and $\mathcal{P}$ is a pappian plane of characteristic $\neq 2$.</u>

In order to prove this result it remains to show the validity of the assumption of theorem 3. For this purpose let $B$ be a point of $\mathcal{O} \setminus \{A\}$. The tangent $b$ to $\mathcal{O}$ at $B$ intersects $a$ in a point $C$. The reflection $\sigma$ of $\mathcal{O}$ at the point $C$ has the axis

$A \neq B$ . Hence $\mathcal{O} \cap Fix\ \sigma = \{A,B\}$ and theorem 3 finishes the proof.

## 4.3 OVALS SYMMETRIC TO THE POINTS OF A PASSING LINE

A line $l$ disjoint to an oval $\mathcal{O}$ is called a *passing line* of $\mathcal{O}$.

THEOREM 5 ( [11] ): Let $l$ be a passing line of an oval $\mathcal{O}$ in a Desarguesian projective plane $\mathcal{P}$ of characteristic unequal to 2. Suppose $\mathcal{O}$ is symmetric to every point of $l$ . Then $\mathcal{O}$ is a conic in the pappian plane $\mathcal{P}$ .

PROOF: We denote by $\sigma_X$ the reflection of $\mathcal{O}$ at the point $X \in l$ and by $\pi(X)$ the axis of $\sigma_X$. Because $\sigma_X$ is of order 2 and the characteristic is unequal to 2, $\sigma_X$ can not be an elation. Hence $X \notin \pi(X)$ for all $X \in l$ . If $C$ is any point on $l$ and if $D := l \cap \pi(C)$ then $\sigma_C \sigma_D \sigma_C = \sigma_D$ . Hence $\sigma_C, \sigma_D$ commute and it is well known that in this situation their product is a perspectivity of order 2 with center $M := \pi(C) \cap \pi(D)$ and axis $l$ . Thus there exists a reflection $\sigma_M$ of $\mathcal{O}$ at $M$ , in particular - by § 2, lemma 1 - the point $M$ is not on $\mathcal{O}$ . The reflection $\sigma_M$ commutes with all elements in $H := \{\sigma_X | X \in l\}$ and therefore with each element in $G$ , the group generated by $H$ . For points $A,B \in \mathcal{O}$ there is exactly one point $L \in l$ such that $A,B,L$ are collinear. The reflection $\sigma_L$ interchanges the points A,B, thus

(i) $H$ acts sharply transitively on the oval $\mathcal{O}$.

Let $R$ be any point of $\mathcal{O}$. We determine the stabilizer $G_R := \{\alpha \in G \mid \alpha(R) = R\}$ of the point $R$ : If $L$ denotes the common point of the tangent to $\mathcal{O}$ at $R$ and of $l$, then $\sigma_L$ is contained in $G_R$. Conversely if $\alpha \neq 1$ is in $G_R$, then $\alpha$ fixes $R$ and all points of the orbit of $R$ under the group $\Gamma(M,l)$; where $\Gamma(M,l)$ consists of all perspectivities with center $M$ and axis $l$ and centralizes $\alpha$. Because $\mathcal{S}$ is a Desarguesian plane, $\Gamma(M,l)$ is linear transitiv and the orbit of $R$ consists of all points of $(R+M) \setminus (l \cup \{M\})$. Hence $R+M$ is the axis of $\alpha$. Because the tangent $R+L$ and the line $l$ are fixed by the central collineation $\alpha$, $L$ is the center of $\alpha$. This yields $\alpha = \sigma_L$ and

(ii) $G_R = \{\sigma_L, 1\}$ .

Our next step in the proof of theorem 5 will be to show that

(iii) $H^2 \cap H = \emptyset$ and $H^3 \subset H$ .

We have proved above, that $\sigma_M$ is the only element of order 2 in $H^2$, hence $H \cap H^2 = \emptyset$. Now let $R,L$ be points as in (ii). If $\alpha \in G$ then - by (i) - there exists an element $h \in H$ with $h(R) = \alpha(R)$. This means $h \cdot \alpha \in G_R = \{1, \sigma_L\}$ and $G = H \cup H^2$. In particular for $\alpha \in H^2$ we get $h\alpha = \sigma_L$, because the other alternative

$h\alpha = 1$ would contradict $H \cap H^2 = \emptyset$ . We deduce $H^2 = H\sigma_L$ and by inverting both sides also $H^2 = \sigma_L H$ . Consequently $H^4 = H^2 \circ H^2 = H\sigma_L{}^2 H = H^2$ . Since $H^4 \cap H \subset H^2 \cap H = \emptyset$ we obtain $H^3 \cap H^2 = \emptyset$ and with $H^3 \subset G = H \cup H^2$ we finally conclude $H^3 \subset H$ . From the theory of metric planes (see for example [7] ) we know that $\mathcal{A} := \mathcal{P} \setminus l$ is a metric affine plane, which can be described in the following algebraic way: The point-set of $\mathcal{A}$ can be identified with a commutative field E , which is a quadratic extension of the field K of coordinates of the plane $\mathcal{A}$. The point M corresponds to the zero element of E . The lines of $\mathcal{A}$ can be re-presented by the cosets $a + Kb$ $(a,b \in E,\ b \neq 0)$ . The reflection $\sigma_X (X \in L)$ can be represented by a mapping $x \longmapsto s\bar{x}$ from E onto itself, where $-$ denotes the generating element of the Galois group of the extension $K \subset E$ and where $s \cdot \bar{s} = 1$ . If $a \in E \setminus \{0\}$ is a point of $\mathcal{Q}$ then - by (i) - $\mathcal{Q} = \{\bar{a} \cdot s \mid s \in E,\ s \cdot \bar{s} = 1\}$ . Because the norm mapping $q : x \longmapsto x \cdot \bar{x}$ from E into K is a quadratic form on the 2-dimensional K-vec-torspace E, we see that $\mathcal{Q} = \{x \in E \mid q(x) = q(\bar{a})\}$ is a conic in the affine plane $\mathcal{A}$. This finishes the proof of theorem 5.

## 5. CHARACTERIZATIONS OF QUADRICS INVOLVING REFLECTIONS

In this section we show how the results of 4 about

conics can be extended to characterizations of quadrics by means of reflections. In view of BUEKENHOUT's results [4] - indicated in section 1 - we only consider ovoids. Our main question is the following: *How many reflections of an ovoid* $\mathcal{O}$ *in a projective space* $\mathcal{R}$ *of dimension greater or equal* 3 *are necessary in order to prove that the given ovoid is a quadric?*

The answer to this question, given in theorem 1, depends on the results of section 4 and on the following

LEMMA (TITS [14]): Let $\mathcal{O}$ be an ovoid in the projective space $\mathcal{R}$ of dimension greater or equal 3. Assume $\mathcal{O}$ contains a point $P$ with the property: Each plane through $P$, not contained in the tangent hyperplane $\mathcal{O}_P$ to $\mathcal{O}$ at $P$ intersects $\mathcal{O}$ in a conic. Then $\mathcal{O}$ is a quadric.

PROOF: Let $Q$ be a fixed point of $\mathcal{O}\setminus\{P\}$. The affine space $\mathcal{R}\setminus\mathcal{O}_P$ can be described by means of linear algebra in the following way: The set $\mathcal{R}\setminus\mathcal{O}_P$ of affine points can be identified with $K \oplus V$, where $V$ is a vectorspace of dimension $\geq 2$ over the commutative field $K$ of coordinates of the space $\mathcal{R}$. The zero vector $0$ corresponds to the point $Q$ and $\{0\} \oplus V$ is the set of all affine points of the hyperplane $\mathcal{O}_Q$. The cosets $a + U$, $U$ being a 1-dimensional linear subspace of the vectorspace $K \oplus V$ are the affine lines and for $U = K \oplus \{0\}$ these cosets represent the affine lines through the point at infinity $P$. Because any such line intersects $\mathfrak{Q} \setminus \{P\}$ in exactly one point, we obtain a mapping $f : V \longrightarrow K$ with the property $\mathfrak{Q} \setminus \{P\} = \{f(v) + v \mid v \in V\}$. Now the assertion of the lemma, that $\mathfrak{Q}$ is a quadric, is equivalent to $f$ being a quadratic form on the vectorspace $V$. On the other side the prepositions of the lemma are equivalent to the property.

(1) For all distinct $v_1, v_2 \in V$, the function $k \longmapsto f((1-k)v_1 + kv_2)$ from $K$ into $K$ is induced by a polynomial of degree 2.

Since $\{0\} \oplus V$ is the set of affine points of $\mathfrak{Q}_Q$ we see that

(2) $f(kv) = k^2 f(v)$ for all $k \in K$ and all $v \in V$.

For the proof that $f : V \longrightarrow K$ is a quadratic form we first handle the case $\dim V = 2$.

Let $e_1, e_2$ be a basis of $V$ and define $g(x_1e_1+x_2e_2) = f(e_1)x_1^2 + f(e_2)x_2^2 + [f(e_1+e_2) - f(e_1) - f(e_2)]\, x_1x_2$; then $g$ is a quadratic form on $V$ and $d := f - g$ satisfies (2) and the condition

(3) For all distinct $v_1, v_2 \in V$ the function $k \longmapsto d((1-k)v_1 + kv_2)$ from $K$ into itself is induced by a polynomial of degree $\leq 2$.

By the definitions of $g$ and $d$ we see that $S :=$ $Ke_1 \cup Ke_2 \cup K(e_1 + e_2)$ consists of solutions of the equation $d(v) = 0$ . Our aim is to show, that $d$ is the zero function. This is obvious if $|K| = 2$, because $V$ is equal to $S$ . In the case $|K| = 3$ we have $K = \{0,1,-1\}$ . The affine space $K \oplus V$ is covered by the 3 parallel affine planes $1 + V$, $-1 + V$, $V$ . The last plane intersects $\mathcal{Q}$ only in the point $Q$ . Therefore the remaining 8 points of $\mathcal{Q} \setminus \{P,Q\}$ are distributed between the planes $-1 + V$ and $1 + V$ . Because each of these planes contains at most 4 points of $\mathcal{Q}$ , we see that either of the equations $f(v) = 1$, $f(v) = -1$ has exactly 4 solutions. Therefore $\sum_{v \in V} f(v) = 0$ . The same is true for $g$ , hence also for the difference $d$ . With $\sum_{v \in V} d(v) = 0$ , we obtain $\sum_{v \in V \setminus S} d(v) = d(1,-1) + d(-1,1) = 0$ . By (2) we have $d(1,-1) = d(-1,1)$. Therefore $d \equiv 0$ . If $|K| \geq 4$ and if $v \in V \setminus S$ there exists a vector $w \in V$ not contained in $S \cup K \cdot v$ . The affine line $v + K \cdot w$ intersects the 3 lines $Ke_1$, $Ke_2$, $K(e_1 + e_2)$ in 3 distinct points. Therefore the polynomial of degree $\leq 2$ inducing the function $k \longmapsto d(v + k \cdot w)$ from $K$ into $K$ has at least 3 roots. Hence $d|_{v+Kw} = 0$ and $d(v) = 0$ . Thus $d : V \longrightarrow K$ is the zero function.

Now we consider the general case. For proving

$f : V \longrightarrow K$ is a quadratic form - by definition of a quadratic form - it remains to show that the restriction $f|_U$ is a quadratic form. Here $U$ stands for an arbitrary 3-dimensional linear subspace of $V$. Let $e_1, e_2, e_3$ be a basis of $U$ and define $g(\sum_{i=1}^{3} x_i e_i) := \sum_{i=1}^{3} f(e_i)x_i^2 + \sum_{1\leq i\leq j\leq 3} [f(e_i+e_j) - f(e_i) - f(e_j)]x_i x_j$. Then $g$ is a quadratic form on $U$ and the function $d := f|_U - g$ from $U$ into $K$ satisfies (2) and (3). For $1 \leq i < j \leq 3$ the set $K e_i \cup K e_j \cup K(e_i + e_j)$ consists of solutions of the equation $d(u) = 0$. By our preceding considerations about the 2-dimensional case we know that for $S := (K e_1 + K\cdot e_2) \cup (K e_1 + K e_3) \cup (K e_2 + K e_3)$ we have $d|_S = 0$. If $v = \sum_{i=1}^{3} x_i e_i \in U \setminus S$ then $x_1, x_2, x_3$ are different from $0$. We can assume $|K| = \infty$, because for a finite field $K$ the space $\mathcal{R}$ is 3-dimensional and $V$ is a 2-dimensional vectorspace (see for example [6], p.48). Let $k_1, k_2, k_3$ be 3 distinct elements of $K \setminus \{0\}$ and set $w := -\sum_{i=1}^{3} x_i k_i^{-1} e_i$. We see that the line $v + K w$ contains the 3 distinct points $v_i := v + k_i w_i$ $(i = 1,2,3)$ of $S$. With (3) we obtain $d|_{v+Kw} = 0$ and $d(v) = 0$. Hence $d|_U = 0$ and $f|_U = g$ is a quadratic form. This settles the proof of the lemma.

THEOREM 1 [11]: *Let $\mathcal{O}$ be an ovoid and let $\varepsilon$ be a hyperplane in a projective space $\mathcal{R}$ of dimension $\geq 3$*

<u>and of characteristic</u> $\neq 2$. <u>Suppose</u> $\mathcal{Q}$ <u>is symmetric to every point of</u> $\varepsilon \setminus \mathcal{Q}$. <u>Then</u> $\mathcal{Q}$ <u>is a quadric</u>.

<u>PROOF:</u> Let $\gamma$ be any plane of $\mathcal{R}$ with $|\gamma \cap \mathcal{Q}| \geq 2$ and let $l$ be any line in $\gamma \cap \varepsilon$. For every point $P \in l \setminus \mathcal{Q}$ there exists a reflection $\sigma_P$ of $\mathcal{Q}$ at this point. The perspectivity $\sigma_P$ with center $P$ leaves invariant the plane $\gamma$ through $P$ and the ovoid $\mathcal{Q}$. Hence $\sigma_P|_\gamma$ is a perspectivity $\neq 1$ of $\gamma$ which fixes the oval $\mathcal{Q} \cap \gamma$. This oval being symmetric to every point of $l \setminus (\gamma \cap \mathcal{Q})$ must be a conic. This is a consequence of the theorems 1,4 and 5 in section 4. With the lemma above we conclude that $\mathcal{Q}$ is a quadric.

<u>REMARK:</u> It is not possible to replace the assumption "$\varepsilon$ a hyperplane" by the weaker assumption "$\varepsilon$ a projective subspace of corang 2". To demonstrate this, we consider the n-dimensional projective geometry $PG(\mathbb{R}^{n+1})$ $(n \geq 3)$ and define

$$\mathcal{Q} := \{ \mathbb{R}(x_0, \ldots, x_n) \mid \sqrt[2]{x_1^4 + x_2^4} + x_3^2 + \ldots + x_n^2 - x_0^2 = 0 \} .$$

Then $\mathcal{Q}$ is symmetric to every point of $\varepsilon \setminus \mathcal{Q}$, where $\varepsilon$ is the projective subspace of corang 2 defined by the equations $x_1 = 0$, $x_2 = 0$. But $\mathcal{Q}$ is not a quadric.

In theorem 1 there is no finiteness condition on

the dimension of $\mathcal{R}$. Thus the existence of reflections at many points has no effect on the dimension. However if we assume that enough hyperplanes occur as axes of reflections, then the finiteness of the dimension is enforced. To demonstrate this, we prove the following theorem 2, which is a sharper result than that given in [9].

THEOREM 2: Let P be a point of an ovoid $\mathcal{O}$ in the projective space $\mathcal{R}$ of dimension $\geq 3$ and of characteristic $\neq 2$. Suppose for every hyperplane $h \neq \mathcal{O}_P$ through P there exists a perspectivity $\sigma_h \neq 1$ with axis h leaving invariant the ovoid $\mathcal{O}$. Then $\mathcal{O}$ is a quadric and $\mathcal{R}$ has finite dimension.

PROOF: By lemma 1 of section 2 the perspectivity $\sigma_h$ has order 2 and its center C is outside of $\mathcal{O}$. C is not contained in h, since the characteristic is different from 2. Furthermore $\sigma_h$ and the center C are uniquely determined by the hyperplane h, because the existence of a second perspectivity $\sigma_h'$ leaving invariant $\mathcal{O}$ and having axis h, would force the product $\tau := \sigma_h \circ \sigma_h'$ to be an elation of order $\neq 2$ with axis h. Therefore, for any point $X \in \mathcal{O} \setminus h$ we would obtain 3 collinear points $X, \tau(X), \tau^2(X)$ on $\mathcal{O}$, a contradiction. Conversely h is uniquely determined by C, as was proved in Lemma 1 of section 2. Now we choose an arbitrary point $Q \in \mathcal{Q} \setminus \{P\}$. Then $D := \mathcal{Q}_P \cap \mathcal{Q}_Q$ is a projective subspace of $\mathcal{R}$ of corang 2. If $\mathcal{R} = PG(V)$, V being a (left-) vector-space over a not necessarily commutative field K, then $V = P \oplus Q \oplus D$ and D - more precisely $PG_o(D)$ - contains no point of $\mathcal{Q}$. Let $\mathcal{M}$ be the set of all hyperplanes of D. For each $M \in \mathcal{M}$ the sum $h := P \oplus Q \oplus M$ is a hyperplane of $\mathcal{R}$ different from $\mathcal{Q}_P$. Denote by $\pi(M)$ the center of $\sigma_h$. The points P,Q are fixed by $\sigma_h$, hence the hyperplanes $\mathcal{Q}_P$, $\mathcal{Q}_Q$ are invariant under $\sigma_h$ and therefore must contain the center $\pi(M)$. Because h and $M = h \cap D$ are determined by $\pi(M)$, we see that

(1) $\pi$ is an injective mapping from $\mathcal{M}$ into $PG_o(D)$.

Suppose $M' \in \mathcal{M}$ and $\pi(M') \subset M$. If $h' := P \oplus Q \oplus M'$ then $\sigma_h \sigma_{h'} \sigma_h$ is a perspectivity of order 2 with center $\sigma_h(\pi(M')) = \pi(M')$ leaving invariant $\mathcal{Q}$. Hence $\sigma_h \sigma_{h'} \sigma_h = \sigma_{h'}$ and the axis h' of $\sigma_{h'}$ is fixed by $\sigma_h$. We conclude $\pi(M) \subset h \cap D = M'$ and have obtained the result

(2) For $M, M' \in \mathcal{M}$, the conditions $\pi(M) \subset M'$ and $\pi(M') \subset M$ are equivalent.

The next step in the proof of theorem 2 is

(3) If $M_1, M_2$ are distinct elements of $\mathcal{M}$, then the bundle $\{M \in \mathcal{M} \mid M_1 \cap M_2 \subset M\}$ of hyperplanes is

mapped by $\pi$ onto the set of all points on the line $\pi(M_1) + \pi(M_2)$ .
Let $R$ be a point on this line $v := \pi(M_1) + \pi(M_2)$ and choose $M_3 \in \mathfrak{M}$ such that $v \cap M_3 = R$ . There exists a hyperplane $M_4 \in \mathfrak{M}$ with $M_1 \cap M_2 \subset M_4$ and $\pi(M_3) \subset M_4$. If $M$ denotes some hyperplane of $\mathfrak{M}$ containing $v$ , then with (2) we obtain $\pi(M) \subset M_1 \cap M_2 \subset M_4$ . Again with (2) we get $\pi(M_4) \subset M$ . This holds for every $M \in \mathfrak{M}$ containing $v$ . Hence $\pi(M_4) \subset v$ . Finally from $\pi(M_3) \subset M_4$ we deduce $\pi(M_4) \subset M_3$ and $\pi(M_4) \subset M_3 \cap v = R$. This finishes the proof of (3). An immediate consequence of (3) is

(4) $\pi(\mathfrak{M})$ is a subspace $S$ of $D$ .

If $\hat{D}$ denotes the dual vectorspace of $D$ , then $\pi^{-1}$ induces a bijection from $PG_o(S)$ onto $PG_o(\hat{D})$ , mapping every line onto a line. This means that $\pi^{-1}$ is a collineation from $PG(S)$ onto $PG(\hat{D})$. Consequently $\dim \hat{D} = \dim S \leq \dim D$ . It is well known that this implies the finiteness of the dimension of $D$ . Furthermore we get $\dim \hat{D} = \dim D$ and $D = S$ . We note

(5) $\mathfrak{R}$ has finite dimension and $\pi$ is a bijection from $\mathfrak{M}$ onto $PG_o(D)$ .

By induction on the dimension of $D$ we deduce the existence of points $P_1, P_2, \ldots, P_d$ with

$D = P_1 \oplus P_2 \oplus \dots \oplus P_d$ and $P_i \subset \pi^{-1}(P_j)$ for all $i \neq j$. For each $i$ $(1 \leq i \leq d)$ define $h_i :=$ $P + Q + \sum_{j \neq i} P_j$. The perspectivity $\sigma_{h_i}$ is induced by the linear map $\lambda : V \longrightarrow V$ with $\lambda|_{h_i} = id_{h_i}$ and $\lambda|_{P_i} = -id_{P_i}$. Therefore the product $\sigma_Q = \sigma_{h_1} \circ \sigma_{h_2} \circ \dots \circ \sigma_{h_d}$ is induced by the linear map $\mu : V \longrightarrow V$ with $\mu|_{P+Q} = id_{P+Q}$ and $\mu|_D = -id_D$. Thus we have proved the following result:

(6) For every point $Q \in \mathcal{Q} \setminus \{P\}$ there exists a collineation $\sigma_Q$ of $\mathcal{R}$ leaving invariant $\mathcal{Q}$. The points $P, Q$ are the only points on $\mathcal{Q}$ stabilized by $\sigma_Q$. Every plane $\gamma$ through $P$ and $Q$ is fixed by $\sigma_Q$ and $\sigma_Q|_\gamma$ has order 2.

If $\gamma$ is an arbitrary plane through $P$ with $|\gamma \cap \mathcal{Q}| \geq 2$ then $\gamma \cap \mathcal{Q}$ is an oval for which - by (6) - all assumptions of theorem 3 in section 4.2 are fulfilled. Therefore this theorem yields that $\gamma \cap \mathcal{Q}$ is a conic. Together with the preceding lemma we conclude that $\mathcal{Q}$ is a quadric.

REFERENCES

[1] ARTZY, R.: Pascal's theorem on an oval. Amer. Math. Mont.75, 143 - 146 (1968)

[2] BUEKENHOUT, F.: Étude intrinsèque des ovals. Rendiconti di Math. (3 - 4) Vol. 25, 1 - 61 (1966)

[3] BUEKENHOUT, F.: Plans projectifs à ovoides pascaliens. Arch. Math. 17, 89 - 93 (1966)

[4] BUEKENHOUT, F.: Ensembles quadratiques des espaces projectifs. Math. Zeitschr. 110, 306 - 318 (1969)

[5] BUEKENHOUT, F.: Characterizations of semi quadrics. A survey. Alti. Coll. Geom. Comb. Roma. Tomo 1, 393 - 421 (1976)

[6] DEMBOWSKI, P.: Finite geometries. Springer Berlin 1968

[7] KARZEL, H. and KIST, G.: Zur Begründung metrischer affiner Ebenen. Abh. Math. Sem. Hamb. 49, 234 - 236 (1979)

[8] KARZEL, H. and SÖRENSEN, K.: Projektive Ebenen mit einem pascalschen Oval. Abh. Math. Sem. Hamb. 36, 123 - 125 (1971)

[9] MÄURER, H.: Ovoidale Möbiusgeometrien mit Inversionen. Arch. Math. 22, 310 - 318 (1971)

[10] MÄURER, H.: Zu Punktepaaren symmetrische Ovoide. Arch. Math. 24, 434 - 439 (1973)

[11] MÄURER, H.: Ovoide mit Symmetrien an den Punkten einer Hyperebene. Abh. Math. Sem. Hamb. 45, 237 - 244 (1976)

[12] PICKERT, G.: Projektive Ebenen. Springer Berlin 1955

[13] RIGBY, J.F.: Pascal ovals in projective planes. Canad. J. M. 21, 1462 - 1476 (1969)

[14] TITS, J.: Ovoides à translations. Rend. Mat. 21, 37 - 59 (1962)

# SOME NEW RESULTS ON GROUPS OF PROJECTIVITIES

Heinz Lüneburg

FB Mathematik, Universität Kaiserslautern

The first examples of groups of projectivities in non-desarguesian planes were computed by A. Barlotti [1959]. He showed that the group of projectivities of the three known non-desarguesian planes of order 9 is in fact the symmetric group of degree 10 and that the group of projectivities of the Hall plane of order 16 contains the alternating group of degree 17. Later on he showed that this group is actually $A_{17}$. In the sequel A. Herzer, J. Joussen, and A. Longwitz determined the group of projectivities for several infinite classes of projective planes (see the bibliography for bibliographical details) showing that it always contains the alternating group of degree q + 1, where q is the order of the plane under consideration. Almost all their results are included in the results by Th. Grundhöfer which will be presented here for the first time. I would like to thank Th. Grundhöfer very much indeed for allowing me to incorporate his material into this note.

At the other extreme of the spectrum, the group of projectivities of a finite desarguesian plane turns out to be sharply triply transitive. This fact actually characterizes the pappian planes whether they are finite or not. This has been known for quite some time and is being pointed out in Prof. Pickert's con-



*P. Plaumann and K. Strambach (eds.), Geometry – von Staudt's Point of View, 231–248.*

tribution to this volume. Moreover, several results are known to enforce the group to be sharply triply transitive. I shall finish this note by one of these results which was already mentioned in Prof. Pickert's paper, namely that $(P_6)$ implies Pappos' Theorem in a finite plane (Schleiermacher [1971]).

Let us start with the most spectacular result of this section.

THEOREM 1 (Th. Grundhöfer). *Let P be a finite projective plane of order* $q = 2^n$. *If P is non-desarguesian, then* $\Pi_1$ *contains the alternating group* $A_{q+1}$ *of degree* $q + 1$.

Proof. As $\Pi_1$ acts triply transitively on a set of $2^n + 1$ points, we can use D. Holt's result [1977] that either $PSL(2,2^n) \subseteq \Pi_1 \subseteq P\Gamma L(2,2^n)$ or $A_{q+1} \subseteq \Pi_1$. The former case cannot occur according to Yaqub's Theorem (see Pickert, section 1.1 D of these proceedings), as *P* is non-desarguesian, q. e. d.

Since finite translation planes always have prime power order, we have

COROLLARY 1 (Th. Grundhöfer). *If P is the projective closure of a non-desarguesian translation plane of order* $q \equiv 0 \bmod 2$, *then* $\Pi_1 \supseteq A_{q+1}$.

Using Herzer's result that the group of projectivities of a finite André plane of even order does not contain an odd permutation (Herzer [1974]), we obtain the

COROLLARY 2. *If P is a finite non-desarguesian André plane of even order* q, *then* $\Pi_1 = A_{q+1}$.

As there is no result similar to Holt's for triply transitive groups of degree $p^n + 1$ for p an odd prime, we must work harder to get some information about $\Pi_l$ in finite translation planes of odd order. So let us start anew. We describe the translation plane under consideration by means of a quasifield $Q(+,\circ)$, where we assume the distributive law $(x + y)\circ z = x\circ z + y\circ z$ to hold. $V = Q \oplus Q$ is the point set of the plane described by Q. The lines through the point (0,0) are the point sets

$$V(\infty) = \{(0,x) \mid x \in Q\} \quad \text{and} \quad V(m) = \{(x,x\circ m) \mid x \in Q\} \quad \text{for } m \in Q.$$

All other lines are translates of these sets, where the translations are the mappings $(x,y) \to (x + a, y + b)$.

We call products of parallel projections affine projectivities and denote by $\Pi_l^{aff}$ the group of all affine projectivities of the line l onto itself.

LEMMA 1. *The mapping* $(x,0) \to (0,x)$ *is an affine projectivity of* $V(0)$ *onto* $V(\infty)$.

Proof. Project V(0) onto V(1) along $V(\infty)$ and V(1) onto $V(\infty)$ along V(0).

LEMMA 2. *The mapping* $(x,0) \to (x + c,0)$ *is an affine projectivity of* V(0) *onto itself for all* $c \in Q$.

Proof. Project V(0) onto V(0) + (c,c) along V(1) and V(0) + (c,c) onto V(0) along $V(\infty)$. From

$$(V(1) + (x,0)) \cap (V(0) + (c,c)) = \{(x + c,c)\}$$

we obtain the desired result.

LEMMA 3. *The mapping* $(x,0) \to (x\circ a - x\circ b,0)$ *is an affine projectivity belonging to* $\Pi_{V(0)}^{aff}$ *for all* $a, b \in Q$ *with* $a \neq b$.

Proof. Project V(0) onto V(a) along $V(\infty)$. Then (x,0) is mapped onto $(x,x\circ a)$. Project V(a) onto $V(\infty)$ along V(b). This is possible because of $a \neq b$. Intersecting $V(b) + (x,x\circ a)$ with $V(\infty)$ yields elements $y, z \in Q$ with $(y,y\circ b) + (x,x\circ a) = (0,z)$. Hence

$y = -x$ and $z = (-x)\circ b + x\circ a = -x\circ b + x\circ a$. Thus $(x,0)$ is mapped onto $(0, x\circ a - x\circ b)$ by the product of the above two parallel projections. We reach the desired result by applying Lemma 1.

The kernel K of Q is the set

$$\{k \mid k \in Q,\ k\circ(x + y) = k\circ x + k\circ y,\ k\circ(x\circ y) = (k\circ x)\circ y \text{ for all } x, y \in Q\}.$$

This yields that the mappings $x \to x\circ a - x\circ b$ are K-linear.

THEOREM 2 (Grundhöfer, Longwitz). *Let* $\Pi$ *be the group generated by all the mappings* $x \to x + c$ *and* $x \to x\circ a - x\circ b$. *Then* $\Pi = \Pi^{aff}_{V(0)}$.

Proof. $\Pi \subseteq \Pi^{aff}_{V(0)}$ by Lemmas 2 and 3. To prove the converse we first show:

Let $\pi$ be a parallel projection of the line L onto the line M. Then there exist two mappings $\gamma_1$, $\gamma_2$ from Q into itself with $(x,y)^\pi = (x^{\gamma_1}, x^{\gamma_2})$ or $(x,y)^\pi = (y^{\gamma_1}, y^{\gamma_2})$ where at least one of the $\gamma_i$'s belong to $\Pi$. If $\gamma_i \notin \Pi$, then there exists $c \in Q$ with $x^{\gamma_i} = c$ for all $x \in Q$.

Using Lemma 1 and 2, we see that we may assume $(0,0) \in L$. Now let $\pi$ be the parallel projection along the line $V(a)$ with $a \in Q \cup \{\infty\}$.

Case 1: $a = \infty$. There are $l, m, c \in Q$ with $L = \{(x, x\circ l) \mid x \in Q\}$ and $M = \{(x, x\circ m + c) \mid x \in Q\}$. It follows $(x, x\circ l)^\pi = (x, x\circ m + c) = (x^{\gamma_1}, x^{\gamma_2})$ with $\gamma_1 = \mathrm{id} \in \Pi$ and $\gamma_2 \in \Pi$ or $x^{\gamma_2} = c$, if $m = 0$.

Case 2: $a \neq \infty$.

2.1: $L = \{(0,y) \mid y \in Q\}$ and $M = \{(c,y) \mid y \in Q\}$. In this case we have

$$(V(a) + (0,y)) \cap M = \{(x, x\circ a) + (0,y)\} = \{(c,z)\}.$$

It follows $c = x$ and $z = x\circ a + y$. Hence $(0,y)^\pi = (c, y + c\circ a)$.

2.2: $L = \{(0,y) \mid y \in Q\}$ and $M = \{(x, x\circ m + c) \mid x \in Q\}$. As L is not parallel to M, we may also assume $c = 0$. Then we get

$$(V(a) + (0,y)) \cap M = \{(x, x\circ a + y)\} = \{(z, z\circ m)\}.$$

Hence z = x and z∘m = x∘a + y. As V(a) is not parallel to M, we have m ≠ a. Therefore, the mapping γ defined by $\xi^\gamma = \xi\circ m - \xi\circ a$ belongs to Π. It follows $x = y^{\gamma^{-1}}$. Thus $(0,y)^\pi = (y^{\gamma_1}, y^{\gamma_1}\circ m) = (y^{\gamma_1}, y^{\gamma_2})$ with $\gamma_1 \in \Pi$ and $\gamma_2 \in \Pi$ or $y^{\gamma_2} = 0$ for all $y \in Q$.

2.3: $L = \{(x, x\circ 1) | x \in Q\}$ and $M = \{(c,y) | y \in Q\}$. Again we may assume c = 0. Then

$$(V(a) + (x,x\circ 1)) \cap M = \{(z,z\circ a) + (x,x\circ 1)\} = \{(0,y)\}.$$

Hence z = -x and $-x\circ a + x\circ 1 = y$. Therefore $(x,x\circ 1)^\pi = (0,x^{\gamma_2})$ with $\gamma_2 \in \Pi$, as a ≠ 1.

2.4: $L = \{(x,x\circ 1) | x \in Q\}$ and $M = \{(x,x\circ m + c) | x \in Q\}$. Now

$$(V(a) + (x,x\circ 1)) \cap M = \{(y + x, y\circ a + x\circ 1)\} = \{(z,z\circ m + c)\}.$$

Hence z = y + x and $y\circ a + x\circ 1 = z\circ m + c$. Let us consider the case m = 1 first. Then there is exactly one y with $y\circ a = y\circ 1 + c$, as a ≠ 1. Thus $(x,x\circ 1)^\pi = (x + y, x\circ 1 + y\circ 1 + c)$ is of the required form. If m ≠ 1, then we may assume c = 0. It follows that $(z - x)\circ a + x\circ 1 = z\circ m$, whence $z\circ a - z\circ m = x\circ a - x\circ 1$. This yields $z = x^{\gamma_1}$ with $\gamma_1 \in \Pi$. Moreover $z\circ m = x^{\gamma_1}\circ m = x^{\gamma_2}$ with $\gamma_2 \in \Pi$ or $x^{\gamma_2} = 0$ for all $x \in Q$.

Now let $\pi \in \Pi^{aff}_{V(0)}$. Then there are parallel projections $\pi_1,\ldots,\pi_t$ with $\pi = \pi_1\cdots\pi_t$. Hence there are mappings $\gamma_1,\ldots,\gamma_t$ and $\eta_1,\ldots,\eta_t$ with

$$(x,0)^\pi = (x^{\gamma_1\cdots\gamma_t}, x^{\eta_1\cdots\eta_t}) = (x^{\gamma_1\cdots\gamma_t}, 0),$$

where $\gamma_i$ and $\eta_j$ are either in Π or are constant mappings. It follows that all $\gamma_i$ are in Π. Hence $\Pi = \Pi^{aff}_{V(0)}$, q. e. d.

COROLLARY (Schleiermacher). *Let* K *denote the kernel of* Q. *Then* $\Pi^{aff}_{V(0)} \subseteq AGL(Q,K)$. *Moreover* $\Pi^{aff}_{V(0)}$ *is the semidirect product of the normal subgroup* $\{x \to x + c | c \in Q\}$ *and the group generated by the mappings* $x \to x\circ a - b\circ x$, *where* a, b ∈ Q *and* a ≠ b.

Next we prove the

GENERALIZED CARTAN-BRAUER-HUA THEOREM. *Let* A *be an abelian group, let* $\Sigma$ *be a subset and* L *be a subfield of* $\mathrm{End}_{\mathbb{Z}}(A)$ *with* $1 \in \Sigma$ *and* $1 \in L$. *If for all* $\sigma, \tau \in \Sigma$, *we have* $\sigma L = L\sigma$ *and* $(\sigma - \tau)L = L(\sigma - \tau)$, *then either* $\Sigma \subseteq L$ *or* $\Sigma$ *centralizes* L.

Proof. Let $\sigma \in \Sigma - L$ and $l \in L$. As $1 \in \Sigma$, there are $l'$, $l''$ in L such that $\sigma l = l'\sigma$ and $(\sigma - 1)l = l''(\sigma - 1)$. This yields $l'\sigma - l = \sigma l - l = (\sigma - 1)l = l''\sigma - l''$. Hence $(l' - l'')\sigma = l - l''$. Assume $l' \neq l''$. As $1 \in L$, the inverse of an element in L is the inverse of this element in $\mathrm{End}_{\mathbb{Z}}(A)$. Hence $\sigma = (l' - l'')^{-1}(l - l'')$ is in L, a contradiction. Thus $l' = l'' = l$. This proves that each element in $\Sigma - L$ centralizes L.

Assume that $\Sigma$ is not contained in L. Then there exists $\sigma \in \Sigma - L$. Pick $k \in \Sigma \cap L$ and $l \in L$. Then there exists $l' \in L$ with $(\sigma - k)l = l'(\sigma - k)$. It follows

$$l\sigma - kl = \sigma l - kl = (\sigma - k)l = l'(\sigma - k) = l'\sigma - l'k.$$

Hence $(l - l')\sigma = -l'k + kl$. Since $\sigma \notin L$, we find $l = l'$ and $lk = kl$. As $\Sigma = (\Sigma - L) \cup (\Sigma \cap L)$, we see that $\Sigma$ centralizes L, q. e. d.

COROLLARY (Grundhöfer). *Let* Q *be a quasifield. If* L *is a subfield of* $\mathrm{End}_{\mathbb{Z}}(Q,+)$ *which is normalized by the group* $\Pi_0$ *generated by the mappings* $x \to x \circ a - x \circ b$ *with* $a, b \in Q$ *and* $a \neq b$, *then either* $\Pi_0 = L^* = L - \{0\}$ *and the plane coordinatized by* Q *is desarguesian or* L *is centralized by* $\Pi_0$ *and* L *is antiisomorphic to a subfield of the kernel of* Q.

Proof. As $(Q,+)$ is a group, we have by the generalized Cartan-Brauer-Hua Theorem that either $\Pi_0 \subseteq L^*$ or that L is centralized by all mappings of the form $x \to x \circ a$ with $a \in Q$. In the latter case, L is contained in the outer kernel of Q. Therefore L is antiisomorphic to a subfield of the kernel of Q (see e. g. Lüneburg [1980, Theorem 5.4, p. 24]). In the former case, $L^*$ operates sharply transitively on $Q - \{0\}$, whence $\Pi_0 = L^*$ in that case. This implies that Q is isomorphic to L and hence that the plane coordinatized by Q is desarguesian, q. e. d.

THEOREM 3 (Grundhöfer). *Let* Q *be a finite quasifield. Let* q *be the order of its kernel* K *and* $|Q| = q^n$. *If* $\Pi_0$ *contains a Singer cycle of* GL(Q,K), *then* $\Pi^{aff}_{V(0)} = AGL(n,q)$.

Proof. According to Kantor [1980], the group $\Pi_0$ contains a normal subgroup isomorphic to $GL(n/s,q^s)$. Hence $\Pi_0$ normalizes a subfield of $End_{\mathbb{Z}}(Q,+)$ isomorphic to $GF(q^s)$. By the Corollary to the generalized Cartan-Brauer-Hua Theorem, $GF(q^s)$ is isomorphic to a subfield af GF(q). Hence s = 1. Using the Corollary to Theorem 2, we obtain $GL(n,q) \subseteq \Pi_0 \subseteq GL(n,q)$, i. e. $GL(n,q) = \Pi_0$, q. e. d.

LEMMA 4. *Let* U *be a subgroup of* $GF(q)^*$ *with* $|U| \geq \sqrt{q}$. *Then* $\langle 1 - u | u \in U - \{1\}\rangle = GF(q)^*$.

Proof. Let $GF(q)^* = \langle w\rangle$ and $U = \langle w^s\rangle$, where s is a divisor of q - 1. Let furthermore $\langle w^t\rangle$ be the group generated by all the 1 - u, where again t is a divisor of q - 1. We may assume by way of contradiction that t > 1. As $x^s \in U$ for every $x \neq 0$, we find to each such x which also satisfies $x^s \neq 1$ a $y \in GF(q)^*$ with $1 - x^s = y^t$. Now we count the number of pairs (x,y) in GF(q) satisfying $1 - x^s = y^t$. For x = 0 there are t solutions y, since t divides q - 1. Thus we have t solutions (0,y). There are s solutions x of $1 - x^s = 0$ and hence s solutions of the form (x,0). Now there are q - 1 - s elements $x \in GF(q)$ with $x^s \neq 0, 1$ and hence (q - 1 - s)t solutions (x,y) with $x^s \neq 0, 1$. Therefore the total number of solutions is $N = t + s + (q - 1 - s)t = qt - s(t - 1)$. According to a theorem of Davenport-Hasse [1935], we find $|N - q| \leq (s - 1)(t - 1)\sqrt{q}$, i. e. $q - s \leq (s - 1)\sqrt{q}$, as t > 1. This yields $q \leq (s - 1)\sqrt{q} + s$. As $s = (q - 1)/|U| \leq (q - 1)/\sqrt{q} < \sqrt{q}$, we obtain the contradiction $q < (\sqrt{q} - 1)\sqrt{q} + \sqrt{q} = q$.

THEOREM 4 (Grundhöfer). *If* A *is a finite André plane, then*

$\Pi^{aff}_{V(0)} = AGL(n,q)$, *where* $q$ *is the order of the kernel of* A *and* $n$ *is the rank of the coordinatizing* André *system over the kernel.*

Proof. If Q is a coordinatizing André system, then the set of all mappings of the form $x \to x \circ a$ contains a subgroup of order $(q^n - 1)/(q - 1)$ which is part of a Singer cycle in GL(n,q) (see e. g. Lüneburg [1980, Theorem 12.4, 2), p. 56 and p. 48, 49]). This group is just the norm-1-group André used in the construction of his systems. As $(q^n - 1)/(q - 1) \geq q^{n-1} \geq q^{n/2}$ for all $n \geq 2$, we see that $\Pi^{aff}_{V(0)}$ contains a Singer cycle by Lemma 4 and 3. Hence, by Theorem 3, we have that $\Pi^{aff}_{V(0)} = AGL(n,q)$, q. e. d.

LEMMA 5. *Let* $n$ *be a prime and assume that* $\alpha \in GL(n,q)$ *is such that* $o(\alpha)$ *divides* $q^n - 1$ *but does not divide* $q - 1$. *Then* $\alpha$ *is contained in a Singer cycle of* GL(n,q).

Proof. As $o(\alpha)$ divides $q^n - 1$, we have $\gcd(o(\alpha),q) = 1$. Therefore $V = GF(q)^{(n)}$ is a completely reducible $\alpha$-module by Maschke's Theorem. Let $V = \oplus_{i=1}^{t} V_i$ where all the $V_i$ are irreducible $\alpha$-modules. Let $\alpha_i$ be the restriction of $\alpha$ to $V_i$. Then $o(\alpha) = \operatorname{lcm}(o(\alpha_i) | i = 1,\dots,t)$. Now $o(\alpha_i)$ divides $|V_i| - 1 = q^{n(i)} - 1$. Moreover $\gcd(q^{n(i)} - 1, q^n - 1) = q^{\gcd(n(i),n)} - 1$. As $n$ is a prime, $\gcd(n(i),n) = 1$ or $n$. Therefore, there exists an $i$ such that $n(i) = n$, since otherwise $q - 1 \equiv 0 \bmod o(\alpha)$. Hence V itself is an irreducible $\alpha$- module. Using Schur's Lemma, we obtain the desired result, q. e. d.

THEOREM 5. (Grundhöfer). *If* A *is a finite nearfield plane, then* $\Pi^{aff}_{V(0)} = AGL(n,q)$, *where* $q$ *and* $n$ *have the usual meaning.*

Proof. If the nearfield F coordinatizing A is a Dickson nearfield, then the group of all mappings of the form $x \to x \circ a$ with $a \in F^*$ contains a subgroup of order $(q^n - 1)/n$ being part of a Singer cycle (see e. g. Lüneburg [1980, p. 32]). Assume $(q^n - 1)/n < q^{n/2}$. Then $q^{n/2} - q^{-n/2} < n$. Hence $q^n - 2 + q^{-n} < n^2$ yielding $q^n - 1 \leq n^2$. In particular $2^n - 1 \leq n^2$. Now $2^5 - 1 > 5^2$.

If $n \geqq 5$ and $2^n - 1 > n^2$, then

$$2^{n+1} - 1 = 2^{n+1} - 2 + 1 > 2n^2 + 1 \geqq n^2 + 2n + 1 = (n + 1)^2.$$

This shows that $n \leqq 4$. If $n = 2$, then $q^2 - 1 \leqq 4$ and $q = 2$. If $n = 3$, then $q^3 - 1 \leqq 9$ and $q = 2$. If $n = 4$, then $q^4 - 1 \leqq 16$ and $q = 2$. As there does not exist a nearfield of order $2^n$ with kernel $GF(2)$, we have that $(q^n - 1)/n \geqq q^{n/2}$. Hence $\Pi^{aff}_{V(0)}$ contains a Singer cycle by Lemma 4 and 3. Thus $\Pi^{aff}_{V(0)} = AGL(n,q)$ by Theorem 3.

If F is not a finite Dickson nearfield, then $n = 2$ and $q \in \{5, 7, 11, 23, 29, 59\}$. The structure of $F^*$ is as follows:

$$11^2\colon F^* = SL(2,5),$$
$$29^2\colon F^* = SL(2,5) \times \mathbb{Z}_7,$$
$$59^2\colon F^* = SL(2,5) \times \mathbb{Z}_{29}.$$

The first case has to be handled separately. We shall leave it aside here. In the second case, $F^*$ contains a cyclic subgroup of order $5 \cdot 7 > 29$ and in the third case a subgroup of order $5 \cdot 29 > 59$. In these two cases, using Lemma 5, our standard argument will work.

There are furthermore the cases

$$5^2\colon F^* = SL(2,3),$$
$$11^2\colon F^* = SL(2,3) \times \mathbb{Z}_5.$$

In the first case there is a cyclic subgroup of order $6 > 5$ and in the second a cyclic subgroup of order $6 \cdot 5 > 11$. Again the standard argument works.

Finally there are the cases

$$7^2\colon F^* = G,$$
$$23^2\colon F^* = G \times \mathbb{Z}_{11}.$$

In these cases G is an extension of $SL(2,3)$ the Sylow 2-subgroups of which being generalized quaternion groups of order 16. Hence G contains a cyclic subgroup of order 8. Therefore $F^*$ contains a cyclic subgroup of order $8 > 7$ in the first case and a cyclic subgroup of order $8 \cdot 11 > 23$ in the second. Again the standard argument yields the desired result.

THEOREM 6 (Grundhöfer). *Let A be a generalized André plane of order* $q^n$, *where* q *is the order of its kernel. If* $\varphi(n) \geq n/2 + 1$, *where* $\varphi$ *is Euler's totient function, then* $\Pi^{aff}_{V(0)} = AGL(n,q)$.

Proof. A generalized André system always contains a subgroup U of order $\Phi_n(q)$ which is part of a Singer cycle. Here $\Phi_n$ denotes the n-th cyclotomic polynomial. $\varphi(n) \geq n/2 + 1$ then yields $|U| \geq q^{n/2}$, etc.

Using results of Hering [1974] one obtains:

THEOREM 7 (Grundhöfer). *Let A be a translation plane of order* $q^n$ *with kernel* GF(q). *Then the following is true:*

*a) If* $n = 3$, *then* $\Pi^{aff}_{V(0)} \supseteq ASL(n,q)$.

*b) If* $n = 2$ *and* $q \neq 5, 7, 11, 19, 29$, *then* $\Pi^{aff}_{V(0)} \supseteq ASL(n,q)$.

THEOREM 8 (Grundhöfer). *Let A be a translation plane of order* $q^n$ *with kernel* GF(q) *and* $n \geq 2$. *If* $\Pi^{aff}_{V(0)} \supseteq ASL(n,q)$ *and* $n \geq 3$ *or* $\Pi^{aff}_{V(0)} \supseteq AGL(2,q)$ *and* $n = 2$, *then the group of projectivities of the projective closure of A contains* $A_{q^n+1}$.

Proof. If this were not the case, then $\Pi^{proj}_{V(0),\infty}$ would contain a normal subgroup of order $q^n$ by a result of B. Mortimer [1977] in contradiction to the Lüneburg-Yaqub Theorem.

COROLLARY 1. *Let A be a translation plane of order* $q^n$ *with kernel* GF(q), $n \geq 2$. *If* q *is odd and* $\Pi^{aff}_{V(0)} \supseteq AGL(n,q)$, *then the group of projectivities of the projective closure of A is* $S_{q^n+1}$.

Proof. This follows from the fact that a Singer cycle in GL(n,q) is an odd permutation and Theorem 8.

COROLLARY 2. *Let A be a translation plane of order* $q^n$ *with kernel* GF(q), $n \geq 2$. *Furthermore, let* $\Pi$ *denote the group of projectivities of the projective closure of A.*

*a) If* q *is even and if A is an André plane, then* $\Pi = S_{q^n+1}$.

*b) If q is odd and if A is either an André plane or a nearfield plane, then* $\Pi = S_{q^n+1}$.
*c) If q is even and if A is a nearfield plane, then* $\Pi \supseteq A_{q^n+1}$.
*d) If* $n = 3$, *then* $\Pi \supseteq A_{q^3+1}$.

In order to prove Schleiermacher's result mentioned at the beginning of this section, we need some preparation. Let A be an affine plane and let t be a binary operation on the set of points of A. We call t a *ratio* of A, if t satisfies the following conditions:

(R1) It is $P\ t\ P = P$ for all points P of A.
(R2) The points P, Q and $P\ t\ Q$ are collinear.
(R3) Given P and R there are unique points X and Y such that $P\ t\ X = R = Y\ t\ P$.
(R4) If $\rho$ is a parallel projection of the line l onto another line, then $(P\ t\ Q)^\rho = P^\rho\ t\ Q^\rho$ for all points P and Q on l.

THEOREM 9 (Lüneburg). *Let A be an affine plane and let t be a ratio on A. If P is a point of A, then there exists a central collineation* $\rho(P)$ *with centre P and axis* $l_\infty$ *with* $X^{\rho(P)}\ t\ X = P$ *for all points X of A. In particular* $\rho(P) \neq 1$.

Proof. By (R3), there is exactly one point $X^{\rho(P)}$ and exactly one point $X^{\sigma(P)}$ with $X^{\rho(P)}\ t\ X = P = X\ t\ X^{\sigma(P)}$ for all points X of A. It follows $X^{\rho(P)}\ t\ X = X^{\rho(P)}\ t\ X^{\rho(P)\sigma(P)}$, whence $\rho(P)\sigma(P) = 1$. Similarly $\sigma(P)\rho(P) = 1$. Hence $\rho(P)$ is a bijection. Moreover $\rho(P)^{-1} = \sigma(P)$. As $P\ t\ P = P = P^{\rho(P)}\ t\ P$, we have $P^{\rho(P)} = P$. Let l be a line of A. If P is on l, then $X^{\rho(P)}$ is on l for all X on l by (R2). As the same is true for $\sigma(P)$, we get $l^{\rho(P)} = l$. (Here we consider lines as point sets.) Assume $P \notin l$. Pick $X \in l$ and let m be the line parallel to l which is incident with $X^{\rho(P)}$. Let Y be a second point on l. Consider the parallel projection $\pi$ along l which maps the line joining P and X onto the line joining P and Y. Then $X^\pi = Y$ and $P^\pi = P$. Using (R4) we obtain

$$X^{\rho(P)\pi} \, t \, Y = X^{\rho(P)\pi} \, t \, X^{\pi} = (X^{\rho(P)} \, t \, X)^{\pi} = P^{\pi} = P = Y^{\rho(P)} \, t \, Y.$$

Thus $X^{\rho(P)\pi} = Y^{\rho(P)}$, whence $Y^{\rho(P)} \in m$, i. e. $l^{\rho(P)} \subseteq m$. Similarly $m^{\sigma(P)} \subseteq l$. Hence $m = m^{\sigma(P)\rho(P)} \subseteq l^{\rho(P)}$. Therefore $l^{\rho(P)} = m$. This proves that $\rho(P)$ is a central collineation with centre P and axis $l_{\infty}$. If $X^{\rho(P)} = X$, then $X = X \, t \, X = X^{\rho(P)} \, t \, X = P$. Hence $\rho(P) \neq 1$, q. e. d.

THEOREM 10 (Lüneburg). *Let A be an affine plane. Then A admits a ratio* $\tau$, *if and only if A is a translation plane whose kernel contains at least 3 elements.*

Proof. Let A be a translation plane whose kernel K contains more than 2 elements. We can identify the points of A with the vectors of a left vector space V over K. Let $t \in K - \{0, 1\}$ and define $\tau$ by $u \, \tau \, v = (1 - t)u + tv$. Then it is easily seen that $\tau$ is a ratio on A.

Assume conversely that A admits a ratio $\tau$. We have to show that a given point P is mapped onto a given point Q by a translation. Put $R = P \, t \, Q$. By Theorem 9, there are homologies $\rho(P)$ and $\sigma(R)$ with $X^{\rho(P)} \, \tau \, X = P$ and $X \, \tau \, X^{\sigma(R)} = R$ for all points X. Now $P \, \tau \, P^{\rho(P)\sigma(R)} = P \, \tau \, P^{\sigma(R)} = R = P \, \tau \, Q$. Therefore $\rho(P)\sigma(R)$ maps P onto Q. It remains to show that $\rho(P)\sigma(R)$ is a translation. Let l be the line joining P and Q. (We may assume $P \neq Q$.) Let m be a line parallel to and distinct from l and pick $X \in m$. Then $X \notin l$ and hence $X^{\rho(P)} \notin l$, as l is fixed by $\rho(P)$. Therefore $X \neq P$ and $X^{\rho(P)} \neq R$. Consider the parallel projection $\lambda$ from the line XP onto the line $X^{\rho(P)}R$ along l. Then $P^{\lambda} = R$ and $X^{\rho(P)\lambda} = X^{\rho(P)}$, as $XP \cap X^{\rho(P)}R = X^{\rho(P)}$. Moreover

$$X^{\rho(P)} \, \tau \, X^{\lambda} = X^{\rho(P)\lambda} \, \tau \, X^{\lambda} = P^{\lambda} = R = X^{\rho(P)} \, \tau \, X^{\rho(P)\sigma(R)}.$$

Therefore $X^{\rho(P)\sigma(R)} = X^{\lambda} \in m$. Thus $m^{\rho(P)\sigma(R)} = m$ which proves that $\rho(P)\sigma(R)$ is a translation.

Since $\rho(P) \neq 1$ the kernel K of A contains an element distinct from 0 and 1, q. e. d.

THEOREM 11 (Lüneburg). *Let* A *be a translation plane coordinatized by the quasifield* Q *and let* K *be the kernel of* A. *If* $\tau$ *is a ratio on* A, *then there is exactly one* $t \in K - \{0,1\}$ *such that* $P \tau Q = (1 - t)P + tQ$ *for all points* P *and* Q *of* A. *Conversely, given* $t \in K - \{0,1\}$, *then* $P \tau Q = (1 - t)P + tQ$ *defines a ratio* $\tau$ *on* A.

Proof. Put $0 = (0,0)$. Then there exists $\rho(0)$ with $P^{\rho(0)} \tau P = 0$ for all points P of A by Theorem 9. As $\rho(0) \neq 1$, there exists $s \in K - \{0,1\}$ with $P^{\rho(0)} = sP$ for all P. Put $t = s(s - 1)^{-1}$ and define the ratio $\tau'$ by $X \tau' Y = (1 - t)X + tY$. (Here we use, as we did in the proof of Theorem 10, the last statement of the Theorem which is really an easy exercise.) Then

$$P^{\rho(0)} \tau' P = (sP) \tau' P = (s - t(s - 1))P = 0 = P^{\rho(0)} \tau P.$$

Given two distinct points X and Y and given $P \neq 0$, then there is an affine projectivity $\pi$ with $P^{\rho(0)\pi} = X$ and $P^{\pi} = Y$. By (R4), we get

$$X \tau' Y = P^{\rho(0)\pi} \tau' P^{\pi} = (P^{\rho(0)} \tau' P)^{\pi} = (P^{\rho(0)} \tau P)^{\pi} = X \tau Y.$$

Hence $\tau' = \tau$.

Assume that $P \tau Q = (1 - t')P + t'Q$. Then $tQ = 0 \tau Q = t'Q$ for all points Q, whence $t = t'$ proving the uniqueness.

THEOREM 12. *Let* A *be an affine plane and let* $\Pi$ *be the group of all affine projectivities of the line* 1 *onto itself. Furthermore, let* P, Q, R *be three distinct points on* 1. *Then there exists a ratio* $\tau$ *of* A *with* $P \tau Q = R$, *if and only if* $\Pi_{P,Q} = \Pi_{P,Q,R}$.

Proof. Assume that there is such a ratio. If $\pi \in \Pi_{P,Q}$, then $R = P \tau Q = P^{\pi} \tau Q^{\pi} = (P \tau Q)^{\pi} = R^{\pi}$ by (R4). Hence $\Pi_{P,Q} \subseteq \Pi_{P,Q,R}$.

Assume conversely that $\Pi_{P,Q} = \Pi_{P,Q,R}$. We define a ternary relation T on the triples of collinear points by $(X,X,X) \in T$ for all X and, moreover, $(X,Y,Z) \in T$ for all those triples for which there exists an affine projectivity $\pi$ with $P^{\pi} = X$, $Q^{\pi} = Y$, $R^{\pi} = Z$. It follows from $\Pi_{P,Q} = \Pi_{P,Q,R}$ that X and Y determine Z with

$(X,Y,Z) \in T$ uniquely. Hence there is a binary operation $\tau$ on the set of points of $A$ with $(X,Y,X \tau Y) \in T$. Obviously, $\tau$ is a ratio with the required properties, q. e. d.

COROLLARY 1. *Let* $Q$ *be a quasifield with kernel* $K$ *and let* $0 \neq v \in V(0)$. *Then the set of fixed points of* $\Pi^{aff}_{V(0),0,v}$ *is the set* $Kv$.

Proof. This follows from Theorems 12 and 11.

COROLLARY 2. *Let* $A$ *be an affine plane and let* $\Pi$ *be the group of all affine projectivities of the line* $l$ *of* $A$ *onto itself. Then* $A$ *is desarguesian, if and only if* $\Pi$ *is sharply 2-transitive.*

Proof. Let $A$ be coordinatized by $Q$. We may assume that $\Pi = \Pi^{aff}_{V(0)}$. Let $0 \neq v \in V(0)$. Then $\Pi$ is sharply 2-transitive, if and only if $\Pi_{0,v} = \{1\}$. This is the case, by Corollary 1, if and only if $Kv = V(0)$, i. e. if and only if $K = Q$, q. e. d.

COROLLARY 3 (Schleiermacher). *Let* $P$ *be a projective plane and let* $\Pi$ *be the group of projectivities of the line* $l$ *onto itself. Furthermore, let* $P$, $Q$, $R$ *be three distinct points on* $l$. *Then* $P$ *is a Moufang plane, if there exists a fourth point* $S$ *on* $l$ *with* $\Pi_{P,Q,R} = \Pi_{P,Q,R,S}$.

Proof. As $\Pi$ is similar to all groups of projectivities of $P$ and as $\Pi$ is triply transitive on $l$, the fact postulated for $l$ and $P$, $Q$, $R$ holds for alle lines $l$ and all triples $P$, $Q$, $R$ on $l$. Now let $l_\infty$ be any line of $P$ and let $l$ be a line of the affine plane $A$ derived from $P$ by $l_\infty$. Put $R = l \cap l_\infty$. As $\Lambda = \Pi^{aff}_{l} \subseteq \Pi_R$, we see that the group $\Lambda_{P,Q}$ always has a third fixed point on $l - \{R\}$. Thus $A$ is a translation plane by Theorem 12 and 10. As $l_\infty$ was arbitrary, $P$ is a Moufang plane. q. e. d.

Using the technique used in this proof together with Corollary 2 one sees that $(P_3)$ not only implies Pappos' Theorem but also Desargues' Theorem. Of course, Pappos' Theorem implies Desargues' Theorem by Hessenberg's Theorem. For neat proofs of Hessenberg's Theorem see Herzer [1972a] and Lüneburg [1969 and 1980a].

Ratios can also be used to prove that $\Pi = \Pi^{aff}_{V(0)} \subseteq AGL(Q,K)$ provided K, the kernel of the quasifield Q, contains more than two elements: Consider $k \in K$. If $k = 0$ or 1, then $(kv)^\pi = kv^\pi$ for all $v \in V(0)$ and all $\pi \in \Pi_0$. Let k be distinct from 0 and 1. Then $\tau$ defined by $P \tau Q = (1 - k)P + kQ$ is a ratio on A. Moreover, $0 \tau v = k v$ for all $v \in V(0)$. Hence, for $\pi \in \Pi_0$, we get $(kv)^\pi = (0 \tau v)^\pi = 0^\pi \tau v^\pi = 0 \tau v^\pi = kv^\pi$. Moreover, for v and w in V(0) we obtain

$$k(v + w)^\pi = 0 \tau (v + w)^\pi = 0^\pi \tau (v + w)^\pi = (0 \tau (v + w))^\pi.$$

Putting $z = (1 - k)^{-1}kv$, we have $z \tau w = 0 \tau (v + w)$. Hence

$$k(v + w)^\pi = (z \tau w)^\pi = z^\pi \tau w^\pi = (1 - k)(1 - k)^{-1}kv + kw = $$
$$= k(v^\pi + w^\pi).$$

Therefore $(v + w)^\pi = v^\pi + w^\pi$, as $k \neq 0$.

Now we are ready to prove

THEOREM 13 (Schleiermacher). *Let P be a finite projective plane satisfying* $(P_6)$. *Then P is desarguesian.*

Proof. Consider the group $\Pi_1$ of projectivities of the line 1 onto itself.

(1) If for a tripel A, B, C of points on 1 the group $(\Pi_1)_{A,B,C}$ has a fourth fixed point, then P is desarguesian.

This follows from Corollary 3 to Theorem 12 and the fact that finite Moufang planes are desarguesian.

(2) $(\Pi_1)_{A,B,C,D}$ acts intransitively on $1 - \{A,B,C,D\}$ for all quadruples of distinct points.

If $(\Pi_1)_{A,B,C,D}$ acts transitively, then $P$ is non-desarguesian. Hence D is not a fixed point of $(\Pi_1)_{A,B,C}$ by (1). Therefore, $(\Pi_1)_{A,B,C}$ acts 2-transitively on $l - \{A,B,C\}$. This yields that $\Pi_1$ acts 5-transitively on l. Using this and $(P_6)$, we get by a result of Gorenstein and Hughes (see e. g. Lüneburg [1969a, (8.6), p. 51]) that $\Pi_1$ is permutation isomorphic to $S_5$, $S_6$, $S_7$, $A_7$, $A_8$ or $M_{12}$. But $S_5$, $S_6$ and $A_8$ cannot occur, as the planes of order 4, 5, 7 are desarguesian. $S_7$, $A_7$ cannot occur, as there is no plane of order 6. Finally, $M_{12}$ cannot occur, as in this case $P$ satisfies $(P_5)$, and, therefore, is desarguesian.

For the remainder of the proof, we adopt Prof. Pickert's notation. Let us consider the pentagon $E_1$, $E_2$, $E_3$, $E_4$, $E_5$, $d_1 = E_2E_4$, $d_2 = E_3E_5$, $d_3 = E_4E_1$, $d_4 = E_5E_2$, $d_5 = E_1E_3$. Put $l = d_1$, $A = E_2$, $B = d_5 \cap l$, $C = d_2 \cap l$, $D = E_4$. Moreover let

$$\pi = (d_1E_1d_2)(d_2E_2d_3)(d_3E_3d_4)(d_4E_4d_5)(d_5E_5d_1).$$

Finally, let $Y = E_1E_5 \cap l$. Then $\pi \in (\Pi_1)_{A,B,C,D,Y}$. Thus if $(\Pi_1)_{A,B,C,D,Y} = \{1\}$, then $E_1, \ldots, E_5$ is a desarguesian pentagon. Hence we may assume $(\Pi_1)_{A,B,C,D,Y} \neq \{1\}$.

Put $\Gamma = (\Pi_1)_{A,B,C,D}$ and let $\Delta$ be the orbit of Y under $\Gamma$. We show that $\Gamma$ acts regularly on $\Omega - \Delta$, where $\Omega = l - \{A,B,C,D\}$, and that $|\Omega - \Delta| \geq 4$. If $\Gamma = \Gamma_Y$, then $\Gamma$ acts regularly on $\Omega - \Delta = \Omega - \{Y\}$, as $P$ satisfies $(P_6)$. Moreover, if q is the order of $P$, then $q \geq 8$ and hence $|\Omega - \Delta| = q + 1 - 5 \geq 4$. If $\Gamma \neq \Gamma_Y$, then, using $(P_6)$ once again, $\Gamma$ acts as a Frobenius group on $\Delta$. Moreover, since Frobenius groups have only one conjugacy class of Frobenius complements, $\Gamma$ acts regularly on $\Omega - \Delta$. As Frobenius groups are non-abelian and as $\Gamma$ acts intransitively on $\Omega$ by (2), we see that $|\Omega - \Delta| \geq 6$. Hence $|\Omega - \Delta| \geq 4$ and $\Gamma$ acts regularly on $\Omega - \Delta$ in either case. This yields $(\Pi_1)_{A,B,C,D,Y'} = \{1\}$ for all $Y' \in \Omega - \Delta$.

Now let $E_1'$, $E_2$, $E_3'$, $E_4$, $E_5'$ be a pentagon with $B = d_5' \cap l$ and $C = d_2' \cap l$ and $Y' = E_1'E_5' \cap l \in \Omega - \Delta$. Then $E_1'$, $E_2$, $E_3'$, $E_4$, $E_5'$ is desarguesian, as $(\Pi_1)_{A,B,C,D,Y'} = \{1\}$. Using this and $|\Omega - \Delta| \geq 4$

and Lemma 2 of Prof. Pickert's note, one sees that $E_1, \ldots, E_5$ is also desarguesian. Hence *P* is desarguesian, q. e. d.

BIBLIOGRAPHY

A. Barlotti, La determinazione del gruppo delle proiettività di una retta in sè in alcuni particolari piani grafici finiti non-desarguesiani. Boll. Un. Mat. III. Ser. 14, (1959), 543-547.

H. Davenport & H. Hasse, Die Nullstellen der Kongruenzzetafunktion in gewissen zyklischen Fällen. J. r. a. Math. 172, (1935), 151-182.

Ch. Hering, Transitive Linear Groups and Linear Groups which Contain Irreducible Subgroups of Prime Order. Geom. Ded. 2, (1974), 425-460.

A. Herzer, Über die Gruppe der Projektivitäten einer Geraden auf sich in einigen endlichen Fastkörperebenen. Geom. Ded. 1, (1972), 47-64.

- " - , Dualitäten mit zwei Geraden aus absoluten Punkten in projektiven Ebenen. Math. Z. 129, (1972a), 235-257.

- " - , Über die Gruppe der Projektivitäten einer Geraden auf sich in einigen endlichen Translationsebenen. Geom. Ded. 2, (1973), 363-385.

- " - , Die Gruppe $\Pi(g)$ in den endlichen Hall-Ebenen. Geom. Ded. 2, (1973),1-12.

- " -, Die Gruppe $\Pi(g)$ in den endlichen André-Ebenen gerader Ordnung. Geom. Ded. 3, (1974), 241-249.

D. Holt, Triply Transitive Permutation Groups in which an Involution Central in a Sylow 2-Subgroup Fixes a Unique Point. J. Lond. Math. Soc. (2) 15, (1977), 55-65.

J. Joussen, Zum Transitivitätsverhalten der Projektivitätengruppe einer endlichen Fastkörperebene. Abh. Math. Sem. Hamb.35, (1971), 230-241.

A. Longwitz, Die Gruppe der Projektivitäten einer Geraden auf sich in André-Ebenen vom Grad 2. Geom. Ded. 4, (1975), 263-270.

- " - , Die Gruppe der Projektivitäten einer Geraden auf sich in endlichen André-Ebenen. Geom. Ded. 8, (1979), 501-511.

H. Lüneburg, An Axiomatic Treatment of Ratios in an Affine Plane. Arch. Math. 18, (1967), 444-448.

- " - , Lectures on Projective Planes. Chicago 1969.

- " - , Transitive Erweiterungen endlicher Permutationsgruppen. Berlin-Heidelberg-New York 1969a.

- " - , Translation Planes. Berlin-Heidelberg-New York 1980.

- " - , Grundlagen der ebenen Geometrie. 7 Studienbriefe. Hagen 1980a.

B. Mortimer, Permutation Groups Containing Affine Groups of the Same Degree. J. Lond. Math. Soc. (2) 15, (1977), 445-455.

A. Schleiermacher, Über projektive Ebenen, in denen jede Projektivität mit sechs Fixpunkten die Identität ist. Math. Z. 123, (1971), 325-339.

# THEOREMS ABOUT REIDEMEISTER CONDITIONS

Ascher Wagner

University of Birmingham, England

CONTENTS



## 0. INTRODUCTION

The organizers of this Meeting have asked me to talk about the Kegel-Lüneburg Theorem. To understand this theorem one needs to know some very beautiful older results in geometry and also quite a lot of fairly sophisticated finite group theory. In view of the limited time at my disposal I have had to make a selection of the topics covered. Naturally, since this is a meeting devoted to geometry, I have put the emphasis on explaining the geometry.

As far as the finite group theory is concerned, I have generally contented myself with giving appropriate references. I have, however, made an exception with the proof of Theorem 5.4. Here I have not been able to resist the temptation to include the simplifications made by Prof. Kurzweil. I have endeavoured to sketch his unpublished proof in the Appendix.



*P. Plaumann and K. Strambach (eds.), Geometry – von Staudt's Point of View, 249–273.*

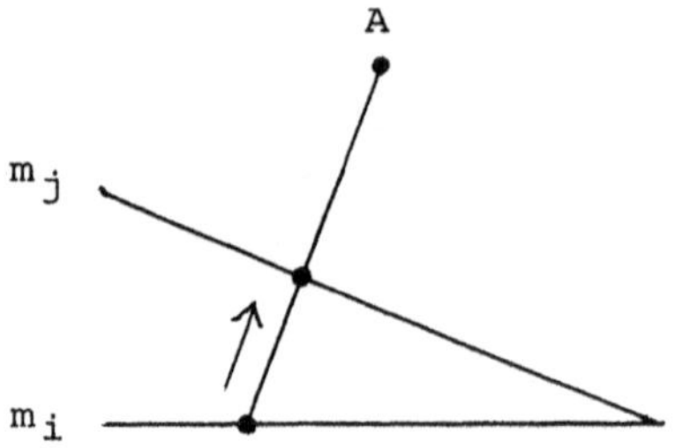


fig.1

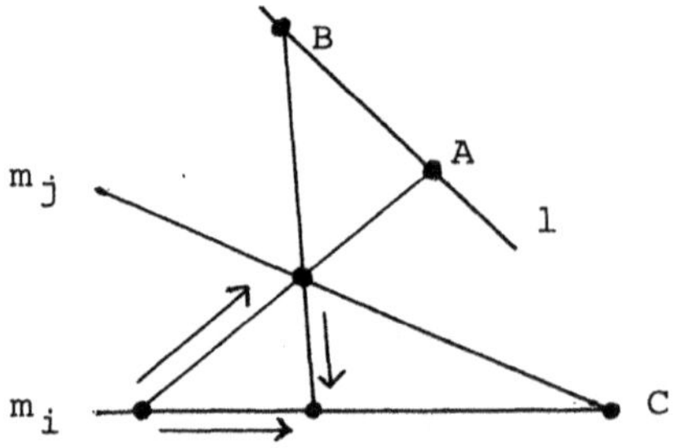


fig.2

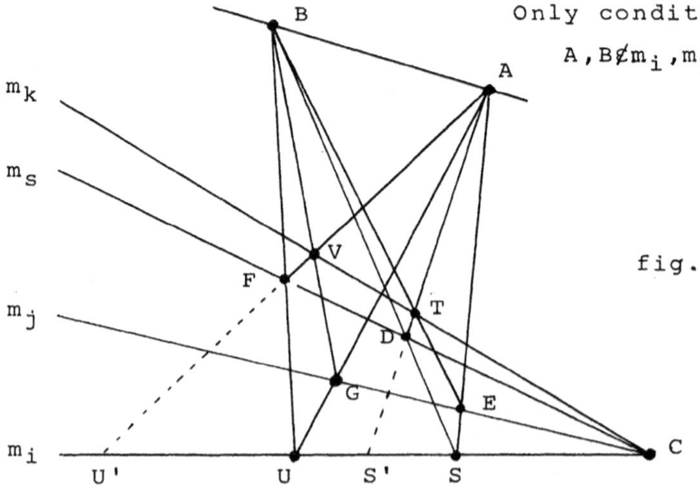


Only condition is $A, B \notin m_i, m_j, m_k$.

fig.3

As far as the geometry is concerned, I have restricted myself to trying to convey the essential points of the arguments. Often, I hope, the diagrams will speak for themselves. For a detailed treatment one must read the original papers.

## 1. NOTATION AND BASIC CONCEPTS

Throughout we shall use P to denote a projective plane. No assumptions are made about P other than those stated from time to time. In particular P is not assumed to be finite. The points of P shall be denoted by upper case letters, the lines of P by lower case letters.

Though we shall be concerned only with projective planes, many of the figures can be drawn more neatly if we draw an affine picture. Thus we shall sometimes pick out a suitable line in P, call it $l_\infty$ and then draw the figure in the affine plane $P-l_\infty$.

Let $m_i$ and $m_j$ be any two lines in P, and A any point not on either of these lines. Then the <u>perspectivity</u> from A of the points of $m_i$ onto the points of $m_j$, see fig.1, shall be denoted by (i,j,A).

Take A,B,C to be points in P with C≠A,B. Let $m_x$ be lines through C not incident with A or B, where x belongs to some index set. Denote the line AB by l. Then we shall write L(i,A,B,C) to denote the set of (i,j,A)(j,i,B) for all $m_j$, see fig.2. In the special and for us very important case that A,B,C are collinear, it is clearly sufficient to write L(i,A,B) instead of L(i,A,B,C).

The notation introduced in the last two paragraphs will be adhered to throughout these lectures.

It is clear that L(i,A,B,C) is a set of permutations on the points of $m_i$. Looking at fig.2 it is further obvious that:

<u>Result 1.1</u>. L(i,A,B,C) fixes the points C and $m_i \cap l$ and is sharply transitive on the remaining points of $m_i$.

The question now arises as to when L(i,A,B,C) is a group. Reidemeister [10] showed that this set is a group precisely when a certain configuration holds. We make the following definition: <u>P satisfies the Reidemeister</u>

condition if every partial plane in P isomorphic to the partial plane described by fig.3 has the points corresponding to C,D,F collinear. To define P satisfies the little Reidemeister condition, we make the additional assumption that in fig.3 the points A,B,C are collinear; otherwise the definition is as before.

Result 1.2. (i) L(i,A,B,C) is a group for all choices of i,A,B,C if, and only if, P satisfies the Reidemeister condition.
(ii) L(i,A,B) is a group for all choices of i,A,B if, and only if, P satisfies the little Reidemeister condition.

Proof. L(i,A,B,C) is a group if, and only if, there exists an $m_s$ such that

$$(i,j,A)(j,i,B)\{(i,k,A)(k,i,B)\}^{-1}=(i,s,A)(s,i,B) \qquad (1)$$

Now $\{(i,k,A)(k,i,B)\}^{-1}=(i,k,B)(k,i,A)$.

Looking at fig.3, it is clear that the left-hand side of (1) maps S onto S' and U onto U'. The result is now obvious.

Result 1.3. (i) If P is desarguesian, then P satisfies the Reidemeister condition.
(ii) If P is Moufang, then P satisfies the little Reidemeister condition.

Proof. In fig.3 draw the lines UV and ST. The rest is obvious.

The subject matter of these lectures may now be stated: is the converse of Result 1.3 true?

## 2. A THEOREM OF KLINGENBERG

The status of the Reidemeister condition was for a long time unclear. So it came as some surprise when Klingenberg [6] showed that the converse of Result 1.3(i) holds. (After this lecture was given Prof. Klingenberg, who was present, told me that Ruth Moufang's initial reaction was one of incredulity.)

Theorem 2.1. If P satisfies the Reidemeister condition then P is desarguesian.

In fact the proof is exceedingly simple - it depends on drawing the right picture. (Note: Desargues Theorem is self-dual.)

Let us first dualize the Reidemeister condition (i.e. fig.3) and make one of the lines containing four points $l_\infty$. We obtain fig.4 where the conclusion is that $ST \parallel QR$.

<u>Proof of 2.1</u>. Let the triangles $A_1A_2A_3$ and $B_1B_2B_3$ be in perspective from the point O. We shall assume that the corresponding sides of the triangles do not meet in collinear points. Then without loss of generality we may assume that the join of $A_1A_2 \cap B_1B_2$ and $A_2A_3 \cap B_2B_3$ does not pass through O. We may now draw an affine picture as in fig.5 with $A_1A_3 \parallel B_1B_3$ and $A_2A_3 \parallel B_2B_3$. Let these lines meet in S. Draw $ST \parallel OB_1$, $SU \parallel B_1B_3$ and $TV \parallel OB_2$, see fig.5. Clearly $V \neq W$. Now identify the lines h and g of fig.4 with the lines $A_1A_2$ and $OA_3$ respectively. It follows that $A_2A_3 \parallel UV$. Similarly, identifying the lines h and g with the lines $B_1B_2$ and $OA_3$ respectively shows that $B_2B_3 \parallel UW$. Hence $UV \parallel UW$ and it follows that $V=W$, a contradiction.

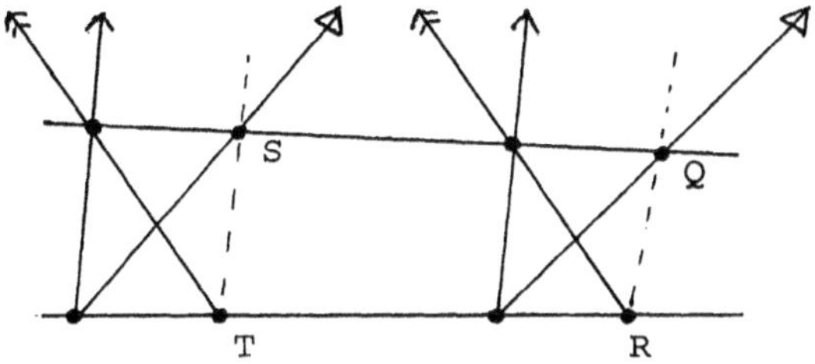


fig.4

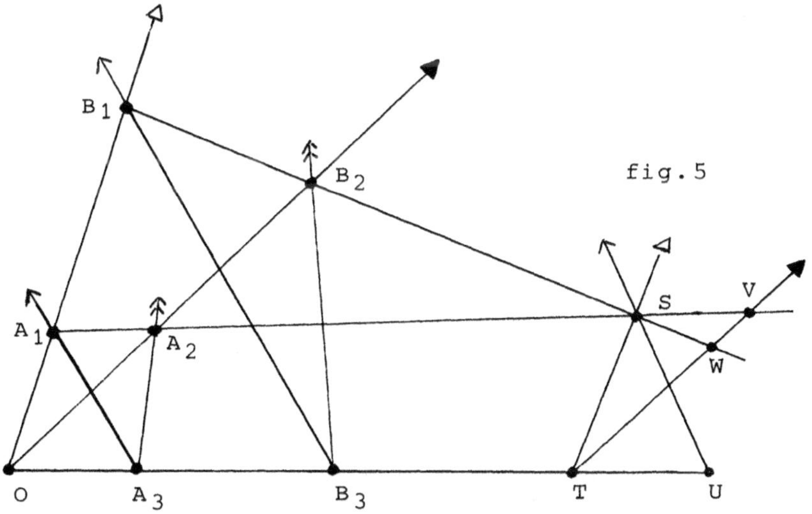


fig.5

[Remark: Reference [6] is generally given as the source for Theorem 2.1. It is a curious fact, however, that in [5] Klingenberg had already shown that the condition indicated by fig.4 implies Desargues Theorem and for this he used the proof I have essentially reproduced here. In the earlier paper Klingenberg was studying affine specializations of various conditions and it would seem that at first he missed the fact that projectively fig.4 is the dual of fig.3.]

It remains to consider whether the converse of Result 1.3(ii) is true. This appears to be a much more difficult matter. Despite a considerable amount of work by a number of people, no-one has so far succeeded either in proving the converse in general or in constructing a counter-example. I think it is probably fair to say that most people who have considered this problem are inclined to believe that the converse is in general false.

The converse is, however, true if the plane is finite - this is the Kegel-Lüneburg Theorem. The proof of this theorem makes use of some results of Moufang and Gleason which originally appeared in a somewhat different context. I will now turn to discussing these results.

## 3. A THEOREM OF MOUFANG

In [8] Moufang considers configurations of rank 8. Among the many results proved in this long paper is a theorem which we shall need concerning the little hexagonality condition. The little hexagonality condition is a specialization of the little Reidemeister condition and is obtained by letting the points D and G of fig.3 coincide. The resulting picture is drawn in fig.6. We shall say that P satisfies the little hexagonality condition if every partial plane in P isomorphic to the partial plane described by fig.6 has the points corresponding to C,D,F collinear.

It will be convenient to have two affine pictures of the little hexagonality condition. If we make a line of fig.6 containing three points (e.g. AB) $l_\infty$, we obtain fig.7 where the conclusion is that D,E,F are collinear. If, however, we make a line of fig.6 containing four points (e.g. CF) $l_\infty$, we obtain fig.8 where the conclusion is that AV and BU are parallel.

We also make the following definitions. A non-degenerate quadrangle is a set of four points no three of which are collinear. Further, if Q is a non-degenerate quadrangle in P we shall say that Q generates the subplane $P' \subsetneq P$ if P' is the intersection of all subplanes in P containing Q. We note that in any Moufang plane (hence also in every desarguesian plane) every non-degenerate quadrangle generates a pappian subplane.

The theorem of Moufang which we shall need may now be stated.

Theorem 3.1. Every non-degenerate quadrangle in P generates a pappian subplane if, and only if, P satisfies the little hexagonality condition.

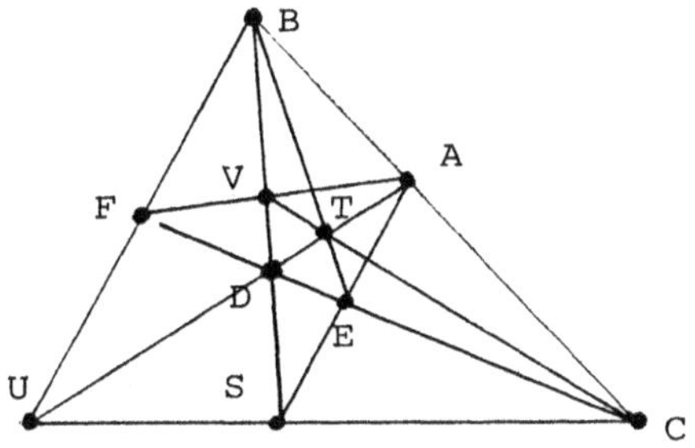


fig.6

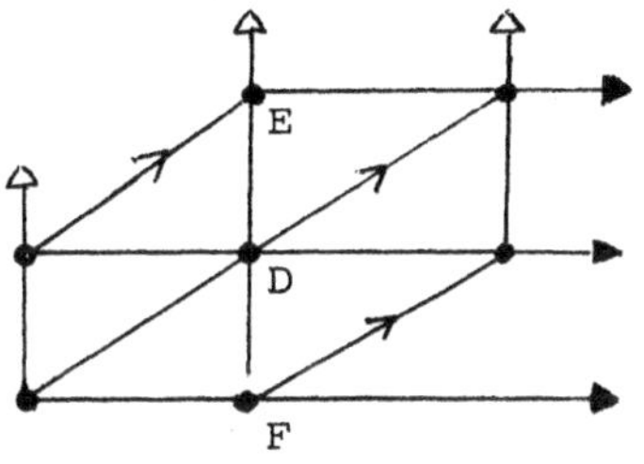


fig.7

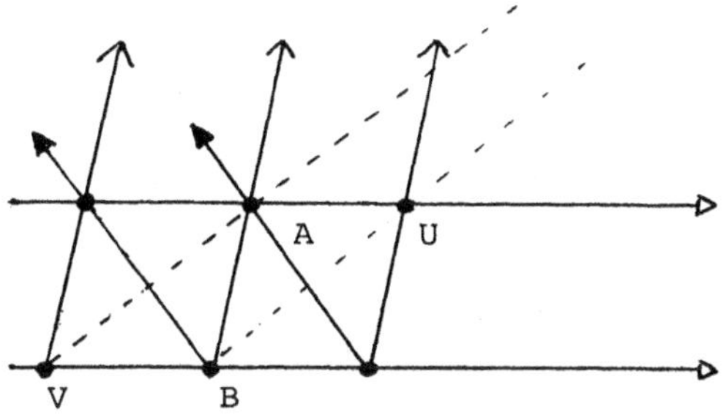


fig.8

The "only if" part of the proof is easy. The "if" part of the proof is rather long if there are infinite subplanes in P generated by quadrangles. If, however, all such subplanes are finite - and this is certainly the case when P is finite - then the proof is considerably easier and shorter. In the Kegel-Lüneburg Theorem P is finite, so I shall only give the proof for the "if" part under the assumption that P is finite and indicate what is involved in proving the general case. A complete proof is given in Pickert [9], and also of course in [8].

Proof of 3.1. (i) Suppose every subplane of P generated by a non-degenerate quadrangle is pappian. Now the partial plane represented by fig.6 is generated by a non-degenerate quadrangle, e.g. DEST, hence is contained in a pappian plane. It follows that C,D,F are collinear, i.e. P satisfies the little hexagonality condition.
(ii) We suppose that P satisfies the little hexagonality condition.

We shall need to introduce coordinates. We shall do this in the manner of Marshall Hall, see for example [3. p. 354]. Take a point in the plane and call it O. Draw three lines through O and call them the x-axis, the y-axis and f respectively. Some line not through O we shall call $l_\infty$. We now give coordinates to all the affine points, i.e. with each affine point we associate an ordered pair of elements from a set with the same cardinality as the number of points on an affine line. O is given coordinates (a,a). Some point E of f is given coordinates (e,e) and all points of f have both coordinates equal. Further, all the points on a line parallel to the y-axis have the same first coordinate and all points on a line parallel to the x-axis have the same second coordinate. Further, any point on a line joining (a,a) to (e,c) has coordinates of the form (x,xc) and all points on a line through (a,d) parallel to f have coordinates of the form (x,x+d), see fig.9. If n is a natural number we shall write ne to denote (n-1)e+e, with the convention that Oe=O.

Let P' be the subplane generated by the quadrangle Q: (a,e),(e,e),(e,2e),(a,2e). Clearly the point X, where the x-axis meets $l_\infty$, belongs to P'. Similarly the point Y, where the y-axis meets $l_\infty$, belongs to P'. Hence $l_\infty \varepsilon P'$, hence also $f \cap l_\infty = F \varepsilon P'$. Now construct from Q in succession the points (2e,2e),(2e,3e),(e,3e). Identifying this partial plane with fig.7 shows that the join of (a,2e) to (e,3e) is parallel to f, see fig.10. If w

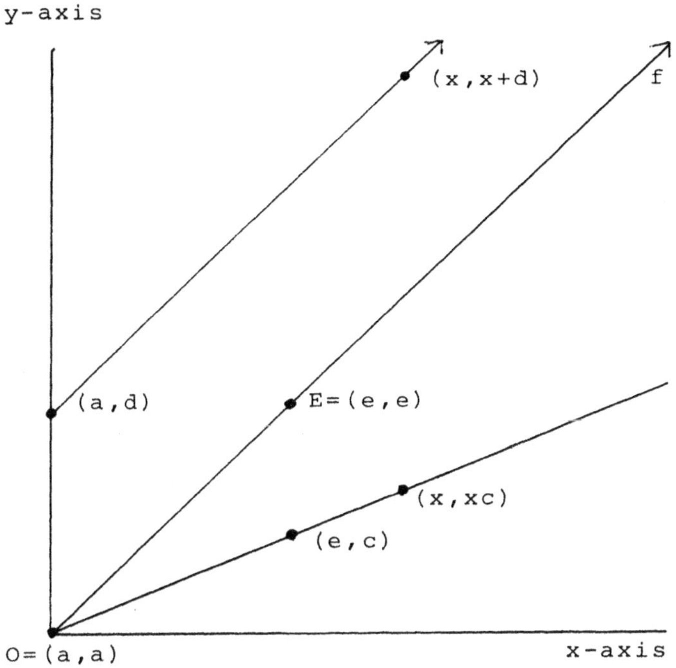


fig.9

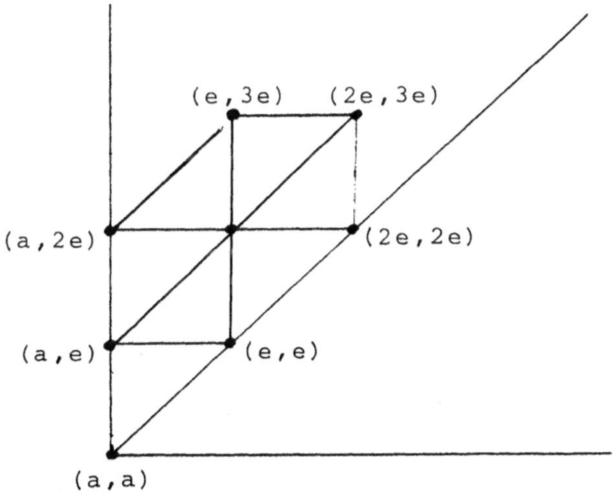


fig.10

let the quadrangle (e,2e), (2e,2e), (2e,3e), (e,3e) play the role of Q and repeat the process we may obtain as many "squares" as we please having a vertex on f, see fig.11. We may also repeat the process on the "adjacent diagonal" by starting with the quadrangle (a,2e), (e,2e), (e,3e), (a,3e). Now we may repeat the process on the "next adjacent diagonal". Repeating this process as often as we please we obtain the situation described by fig.12.

If we now look at fig.12 and recall the rule giving the coordinates of a point on a line parallel to f we see that

$$(m+n)e=me+ne \qquad (1)$$

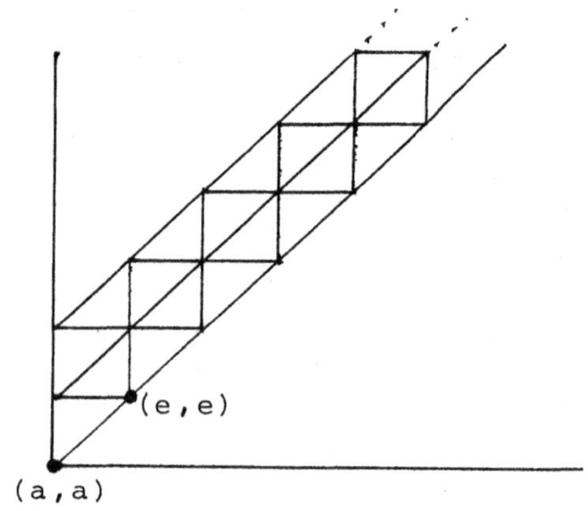


fig.11

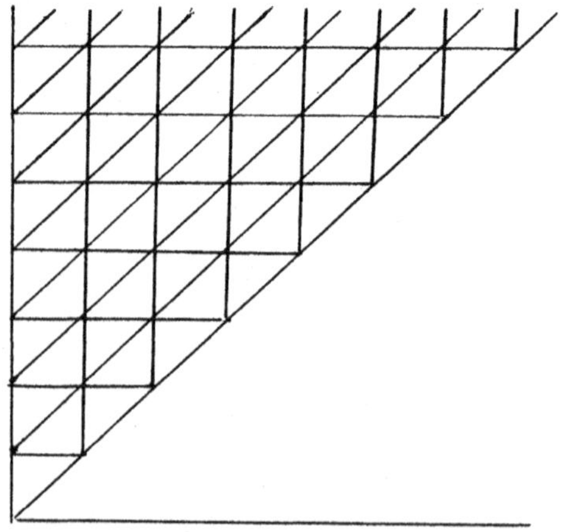

fig.12

Let us say that the line joining (re,se) to ((r+1)e,(s+m)e), where r,s,m are natural numbers or zero, has <u>slope</u> m. We wish to show that all lines with the same slope are parallel. This is certainly true when m=0 and also, as we have just shown, when m=1. Let us assume that $m \geq 2$ and that this result is true for all lines of slope less than m. Consider fig.13 and compare it to fig.8. It is clear that ZS is parallel to YR and that these lines have slope m. Consequently all lines of slope m in the same "vertical column" of fig.12 are parallel. Now consider fig.14 and compare it to fig.8. It is clear that XS is parallel to YT and

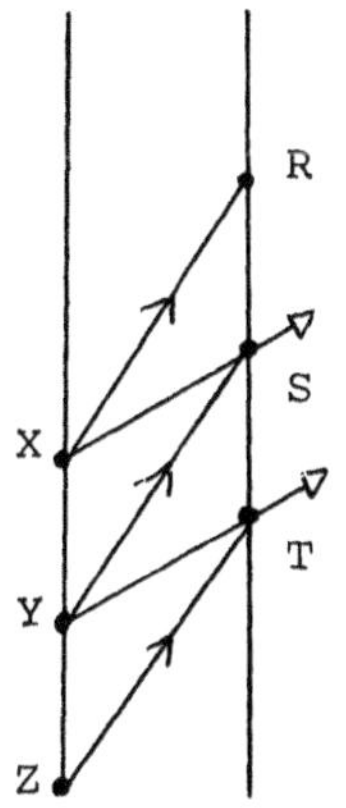


X=(re,(s+2)e)

Y=(re,(s+1)e)

Z=(re,se)

R=((r+1)e,(s+m+1)e)

S=((r+1)e,(s+m)e)

T=((r+1)e,(s+m-1)e)

fig.13

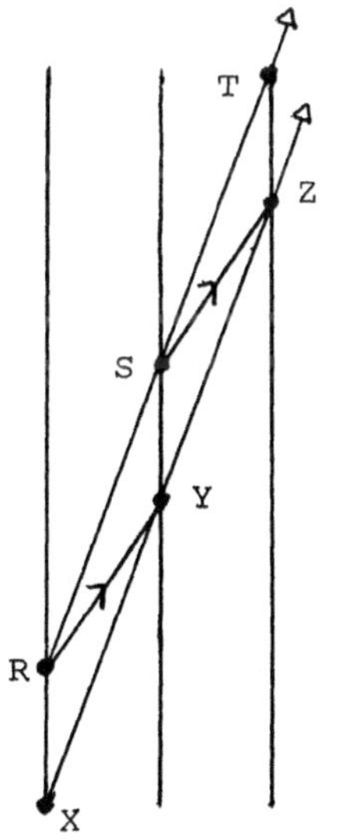


X=(re,se)

Y=((r+1)e,(s+m-1)e)

Z=((r+2)e,(s+m-2)e)

R=(re,(s+1)e)

S=((r+1)e,(s+m)e)

T=((r+2)e,(s+m-1)e)

fig.14

these lines have slope m. Consequently all lines of slope m in adjacent "vertical columns" are parallel. It follows by induction that all lines with the same slope are parallel. It is now easily seen that the line joining (a,a) to (e,ne) contains the point (me,(mn)e). Recalling the rule giving the coordinates of a point on a line through O, we see that

$$(me)(ne)=(mn)e. \qquad (2)$$

Using the fact that a line through O meeting the x-axis in two distinct points must be the x-axis, we see that

$$(me)(ne)=a \text{ implies that } me=a \text{ or } ne=a. \qquad (3)$$

We now wish to show that when P' is finite the set S={a,ne}, where n is a natural number, is, with respect to the addition and multiplication we have introduced, a finite field of prime order. It is clear from the definitions (or see [3 p. 355]) that a and e are the zero and unit respectively. Also, it immediately follows from (1) and (2) that addition and multiplication are associative and commutative and that the distributive law holds. From the finiteness of P' it follows that for some m and n, with m > n, me=ne. Consequently ne+(m-n)e=ne, i.e. a line parallel to f through (a,(m-n)e) contains the point (ne,ne). But this implies that the line meets f, which is possible only if (m-n)e=a. It follows that every element of S has an additive inverse. Also from (3) it is clear that the least natural number p, such that pe=a, is a prime. It follows that every element of S, different from a, has a multiplicative inverse. Consequently S is a finite field of prime order. That a plane coordinatized by a field is pappian is well-known.

To prove this theorem when P' is infinite, one must also introduce points whose coordinates are ((-m)e,(-n)e). This is easily done by extending fig.12 to the left and downwards. Then one must show that all lines with the same rational slope are parallel, then introduce points with rational coordinates, and finally show that joining these points gives no further lines. The method is clear, the details are tedious. In this case, of course, the coordinatizing field is the rationals.

## 4. GLEASON'S PAPER ON FANO PLANES

A <u>Fano plane</u> is a plane where every non-degenerate quadrangle has collinear diagonal points.

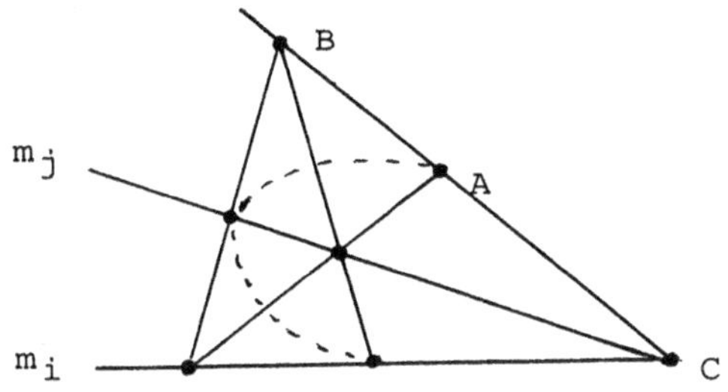


fig.15

Ostensibly the main theorem of Gleason [2] is

<u>Theorem 4.1</u>. Every finite Fano plane is desarguesian.

This is a very beautiful and striking result. To prove this theorem Gleason first proved a number of preliminary results, many of which are in fact of great independent interest and utility. All the arguments used in this paper are quite elementary - a truly amazing paper!

The next lemma shows the connection between Fano planes and the little Reidemeister condition. We shall use the notation of Section 1.

<u>Lemma 4.2</u>. P is a Fano plane. Then
(i) Every element of L(i,A,B) has order 2.
(ii) The product of any two elements of L(i,A,B) has order 2.
(iii) L(i,A,B) is an elementary abelian group.

<u>Proof</u>. (i) This is obvious from fig.15. Note that (i) implies that (i,j,A)(j,i,B)=(i,j,B)(j,i,A).

(ii) (i,j,A)(j,i,B)(i,k,A)(k,i,B)(i,j,A)(j,i,B)(i,k,A)(k,i,B)
=(i,j,B)(j,i,A)(i,k,A)(k,i,B)(i,j,B)(j,i,A)(i,k,A)(k,i,B)
=(i,j,B)(j,k,A)(k,j,B)(j,k,A)(k,i,B)
=(i,j,B)(j,k,B)(k,j,A)(j,k,A)(k,i,B)
=(i,k,B)(k,i,B)=1

(iii) By (ii) all elements of L(i,A,B) commute. Let M denote the group generated by L(i,A,B). Since L(i,A,B) is transitive on $m_i - C$, see Result 1.1, it follows that M is transitive on $m_i - C$. Since M is abelian, M is sharply transitive on $m_i - C$. However, L(i,A,B) also has this property. Hence M=L(i,A,B).

[Remark: if one tries to prove directly, i.e. without first proving (ii), that L(i,A,B) is a group, one obtains a completely unmanageable figure. I think the proof of 4.2 is a good example of the ingenuity of Gleason's arguments.]

The analogue of Theorem 4.1 for infinite planes would be: every Fano plane is a Moufang plane. Despite a great deal of work by many people, this is still an open problem. Of course a counter-example would, in view of 4.2 (iii), also solve the problem raised at the end of Section 2.

The remaining results of this Section are quite general and do not assume that the plane is Fano.

Lemma 4.3. (i) Let $\alpha$ be a (C,l)-elation. Then the restriction of $\alpha$ to $m_i$ commutes with every element of L(i,A,B).
(ii) Suppose $\beta$ is a permutation on $m_i$ which commutes with every element of L(i,A,B) and every element of L(i,X,A), when $X \in l$, $X \neq C$. Then $\beta$ is the restriction to $m_i$ of a (C,l)-elation.

Proof. (i) Let $\lambda = (i,j,A)(j,i,B)$. Consider fig.16. Then $S\alpha^{-1}\lambda\alpha = S'$, also $S\lambda = S'$. This is true for all $S \in m_i$, hence $\alpha^{-1}\lambda\alpha = \lambda$.

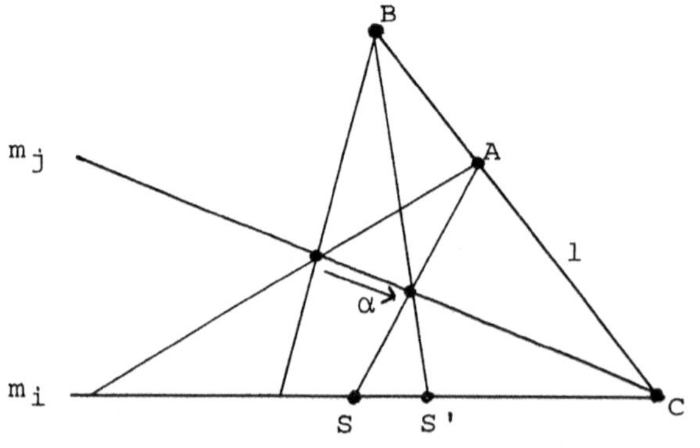


fig.16

(ii) Define a map $\alpha$ of points of P onto points of P by:

$Y\alpha = Y$ if $Y \in l$.

$Y\alpha = Y(j,i,A)\beta(i,j,A)$ if $Y \in m_j$

so in particular $Y\alpha = Y\beta$ if $Y \in m_i$.
Let $Y \in m_j$, $Z \in m_k$ and let $YZ \cap l = X$, see fig.17. To show that $\alpha$ is a collineation it is sufficient to show that $Y\alpha, Z\alpha, X$ are collinear. If $j=k$ this is obvious. Now assume $j \neq k$. Hence $X \neq C$. Now:

$$
\begin{aligned}
(Y\alpha)(j,k,X) &= Y(j,i,A)\beta(i,j,A)(j,k,X) \\
&= Y(j,i,A)\beta(i,j,A)(j,i,X)(i,k,X)(k,i,A)(i,k,A) \\
&= Y(j,i,A)(i,j,A)(j,i,X)(i,k,X)(k,i,A)\beta(i,k,A) \\
&= Y(j,k,X)(k,i,A)\beta(i,k,A) = Z\alpha, \text{ see fig.17.}
\end{aligned}
$$

Hence $Y\alpha, Z\alpha, X$ are collinear.

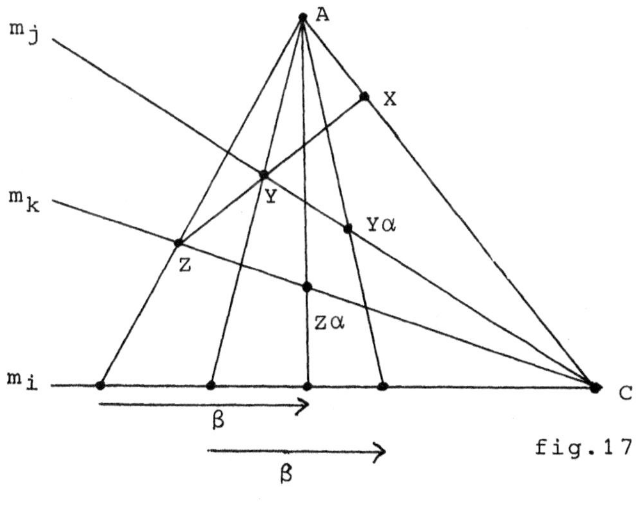


fig.17

Lemma 4.4. Suppose $L(i,A,B)$, $L(i,A,X)$, $L(i,B,X)$ are groups, where $X \in l$, $X \notin m_i$. Then:

(i) $L(i,A,B)=L(i,B,A)$

(ii) $L(i,A,B)L(i,A,X)=L(i,A,X)L(i,A,B) \supsetneq L(i,B,X)$

Proof. (i) Elements of $L(i,B,A)$ are inverses of elements of $L(i,A,B)$ and vice versa.

(ii) $(i,j,B)(j,i,X)=(i,j,B)(j,i,A)(i,j,A)(j,i,X)$

Therefore

$$L(i,B,X) \subsetneq L(i,B,A)L(i,A,X)=L(i,A,B)L(i,A,X).$$

Also

$$(i,j,B)(j,i,X)(i,k,B)(k,i,X)=(i,s,B)(s,i,X)$$

for some $m_s$ since $L(i,B,X)$ is a group. Hence

$$(i,j,A)(j,i,X)(i,k,B)(k,i,A)$$

$$=(i,j,A)(j,i,B)(i,s,B)(s,i,A)(i,s,A)(s,i,X)(i,k,X)(k,i,A).$$

Consequently $L(i,A,X)L(i,B,A) \subsetneq L(i,A,B)L(i,A,X)$. Taking inverses reverses the inclusion. The equality in (ii) now follows.

Theorem 4.5. Let P be finite and suppose that there exists a $(C,l)$-elation $\neq 1$ for every incident point-line pair $(C,l)$ in P. Then P is desarguesian.

For a proof see [2]. This is probably the most frequently used result in [2] and a number of stronger versions of this theorem have subsequently been proved. Since the large subject of elation groups is really outside the scope of this meeting, I have reluctantly decided to omit the moderately lengthy proof of Theorem 4.5.

Theorem 4.6. If P is finite of prime power order and satisfies the little Reidemeister condition, then P is desarguesian.

Proof. By Result 1.1 all the groups $L(i,A,X)$ are p-groups. Let M denote the product of all the $L(i,A,X)$, where $X \in l$, $X \notin m_i$. By Lemma 4.4 M is also a p-group. Let $\beta \neq 1$ be an element in the centre of M. By Lemma 4.3 there exists a $(C,l)$-elation $\neq 1$. The result now follows from Theorem 4.5.

Theorem 4.1 now follows immediately from Lemma 4.2 (iii) and Theorem 4.6.

[Remark: Dembowski [1. p.163] points out that Lemmas 4.3, 4.4 and 5.2 are also true if $C \notin l$ (of course in Lemma 4.3 "elation" must then be changed to "homology"). I have not stated these lemmas in their most general form, chiefly because the notation used then becomes a little longer. Another reason is that a slight complication arises in Lemma 4.3 (ii) when $C \notin l$, namely $\alpha$ cannot be directly defined for the points of CA. In this case one defines $\alpha$ for the points of P-CA and shows that it is a collineation on this affine plane. It is then clear how $\alpha$ may be extended to the points of CA.]

## 5. THE KEGEL-LÜNEBURG THEOREM

Theorem 5.1. (Kegel-Lüneburg [4]). If P is finite and satisfies the little Reidemeister condition then P is desarguesian.

Before we can begin the proof of this theorem we shall need some preliminary results from [4] and [7]. Again, we shall use the notation introduced in Section 1.

Lemma 5.2. P is (C,l)-desarguesian if, and only if, $L(i,A,B)=L(i,A,X)$ for all lines $m_i \neq l$ through C and all points A,B,X on l but different from C.

Proof. Let $D_1D_2D_3$ and $E_1E_2E_3$ be two triangles in perspective from the point C and let $D_1D_2 \cap E_1E_2 = A \in l$, $D_1D_3 \cap E_1E_3 = B \in l$ and $D_2D_3 \cap l = X$. The result is now easily verified by considering fig.18.

Lemma 5.3. Suppose P is finite, has order n and satisfies the little Reidemeister condition. Then:

(i) $L(i,A,B)$ has order n.

(ii) Every element of $L(i,A,B)$ has order p, where p is a prime.

(iii) If p is a prime and $p|n$ then there exist points $D,E \in m_i$ such that D,E,A,B generate a plane of order p.

Proof. This result is an immediate consequence of 1.1, 1.2 and 3.1.

Theorem 5.4. Let G be a finite group and suppose every element of G has prime order. Then either

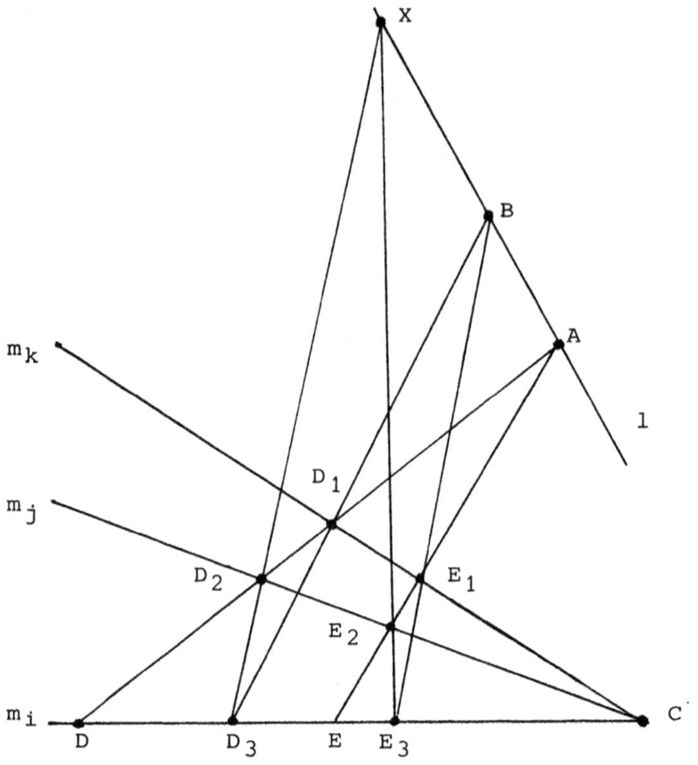


fig.18

(i) $|G|=p^r$, where p is a prime. or
(ii) $|G|=p^r q$, where p and q are distinct primes and G is a Frobenius group with kernel of order $p^r$. or
(iii) $G \cong A_5$.

In [7] this result is deduced from some recent and difficult papers on groups with a partition. We shall discuss the proof of 5.4 more fully in the Appendix.

Theorem 5.5. Let G be a group. Suppose $U, V \subset G$, $G=UV=VU$ and that $U \cong V \cong A_5$. Then $G \cong A_6$ or $G \cong U \times V$.

For the rather difficult proof of this theorem see [4].

Lemma 5.6. Let $U=L(i,A,B)$, $V=L(i,A,X)$, $W=L(i,B,X)$ and suppose U,V,W are groups. Then:
(i) $VW=UW=UV$.
(ii) If $U \neq V$ then U,V,W belong to different conjugacy classes in UV.

Proof. This is an immediate consequence of Lemma 4.4. We may now prove the main result of these lectures.

Proof of 5.1. We shall denote the order of P by n. It follows from 5.3 and 5.4 that $n=p^r$, $p^r q$ or 60.

If $n=p^r$ then 5.1 follows from 4.6.

We now suppose $n=p^r q$. Let S(i,A,X) denote the Sylow p-subgroup of L(i,A,X). Since the L(i,A,X), with $X \in l$ and $X \neq C$, commute pairwise, the S(i,A,X) also commute pairwise see [11]. Denote by T the product of all the S(i,A,X), with $X \in l$ and $X \neq C$. T is a p-group. The orbits of T on the points of $m_i - C$ have length $\geq p^r$, since each S(i,A,X) is regular on the points of $m_i - C$. Now $p^r$ is the highest power of p dividing n, hence some orbit has length $p^r$. Denote one such orbit by $\Theta$. Then $\Theta$ is also an orbit for each S(i,A,X). Let D and E be any two points of $\Theta$. For each choice of X the plane generated by D,E,A,X has order p. Let $\beta$ be the element of L(l,D,E) mapping A onto X. (Here L(l,D,E) has the obvious meaning; l plays the role of $m_i$ and vice versa.) Clearly $\beta$ has order p. Hence S(l,D,E) is transitive on the points of l-C. Consequently n is a power of p, which is a contradiction.

It remains to consider the case that n=60. Using the notation of 5.6 it follows from 5.2 that we may choose $m_i$ and l so that $U \neq V$. Now $U \cong V \cong W \cong A_5$, and by 5.5 either $UV \cong A_6$ or $UV \cong U \times V$.

Suppose $UV \cong A_6$. $A_6$ contains only two conjugacy classes of groups isomorphic to $A_5$, which is incompatible with 5.6 (ii).

Now suppose $UV \cong U \times V$. Denote UV by G. There exist $\bar{U}, \bar{V} \subset G$ such that $\bar{U} \cong \bar{V} \cong A_5$ and $G = \bar{U} \times \bar{V}$. Any subgroup of G isomorphic to $A_5$ is $\bar{U}$, $\bar{V}$ or a "diagonal", i.e. a group of the form $\{uu^\sigma\}$, where $u \in \bar{U}$ and $\sigma$ is an isomorphism of $\bar{U}$ onto $\bar{V}$. The automorphism group of $A_5$ is $S_5$ and it is readily verified that every automorphism of $A_5$ fixes an element $\neq 1$. It is easily seen that this implies that every two "diagonals" of G have a non-trivial intersection. Since $|G|=3600$ it is clear that

$$V \cap W = U \cap W = U \cap V = 1.$$

Consequently only one of U,V,W is a "diagonal". Without loss of generality we may assume $U=\bar{U}$, $V=\bar{V}$ and W is a "diagonal". Let D be a point of $m_i$ different from C. If $G_D$ denotes the stabilizer of D in G, then $|G_D|=60$.

Since U and V are transitive on $m_i - C$, it is clear that when $G_D$ is written as products of an element of U with an element of V, every element of U will occur. Consequently $G_D \cong A_5$. But $G_D \neq U,V$ since these have no fixed points on $m_i - C$. Consequently $G_D$ is a "diagonal", hence $G_D \cap W \neq 1$. This implies that an element $\neq 1$ in W fixes D, which is a contradiction. This completes the proof of 5.1.

[I am indebted to Prof. Kegel for showing me the argument used above to eliminate the case $UV = A_6$; this argument is much easier than the original one used in [4].]

## 6. GROUP THEORETIC APPENDIX

I am much indebted to Professor Kurzweil for showing me Lemmas 6.5 and 6.6. (I have however slightly departed from his proof of Lemma 6.6.) I am also much indebted to Professor Bender for valuable discussions and in particular for introducing me to the little trick involved in 6.1.

We shall say that the subgroups $A_1, A_2, \dots, A_m$ of G are a <u>partition</u> of G if G is the set theoretic union of these subgroups and $A_i \cap A_j = 1$ for $i \neq j$.

<u>Lemma 6.1</u>. Let A and B be subgroups of the finite group G and suppose

(i) A is abelian, $B \subset N_G(A)$ and $A \cap C_G(b) = 1$ for all $1 \neq b \in B$

(ii) B is a partition of s subgroups.

Then A has exponent a divisor of $s-1$.

<u>Proof</u>. Let the partition of B be $B_1, B_2, \dots, B_s$. Let

$$f = \prod_{x \in B} a^x \quad \text{and} \quad f_i = \prod_{x \in B_i} a^x \qquad \text{where } 1 \neq a \in A.$$

Since A is abelian all elements of B commute with f, hence $f=1$. Similarly $f_i = 1$. Since the $B_i$ are a partition of B

$$a^{s-1} f = \prod_{i=1}^{s} f_i \ , \text{ hence } a^{s-1} = 1.$$

<u>Corollary 6.2</u>. Let A,B satisfy 6.1(i). If B contains the direct product of two cyclic groups of prime order q, then q divides the order of A.

<u>Proof</u>. W.l.o.g we may assume that B is the direct product of two cyclic groups of order q. Then B is partitioned by the q+1 subgroups of order q.

Corollary 6.3. Let A,B satisfy 6.1(i). If B is a Frobenius group with kernel K, then $(|K|,|A|)\neq 1$.

Proof. B is partitioned by the kernel and the $|K|$ Frobenius complements.

Lemma 6.4. Let G be a soluble group and suppose every element of G has prime order. Then either

(i) $|G|=p^r$, where p is a prime, or

(ii) $|G|=p^r q$, where p and q are distinct primes and G is a Frobenius group with kernel of order $p^r$.

Proof. Note that if $S,T\subset G$, S and T have relatively prime order and T normalizes S, then $N_G(T)\cap S=1$; otherwise two elements $\neq 1$ in S and T commute and their product has composite order.

The proof will be by induction on the order of G. Let M be a minimal normal subgroup of G. M is elementary abelian. Let $|M|=p^n$. Write $H=G/M$. By the induction hypothesis H is described by (i) or (ii) above.

(a) Suppose $|H|=s^r$, s a prime. Assume $s\neq p$ since otherwise there is nothing to prove. Let A be a Sylow s-subgroup of G. Apply 6.2 to M and A. Then A cannot contain the direct product of two cyclic groups of order s, hence A has order s. It follows that G is descibed by (ii).

(b) Suppose $|H|=s^r t$, where s and t are distinct primes. Let K be the kernel of the Frobenius group H and let L be the pre-image of K in G. It is clear that if s=p then G is described by (ii). Henceforth we assume $s\neq p$. It follows that L is described by (ii) and hence r=1. Let A be a Sylow s-subgroup of G and let $N=N_G(A)$. Then $N\cap M=1$, $N\cong H$, N is a Frobenius group with kernel A and $|N|=st$.

If $t\neq p$ we may apply 6.3 to M and N to give a contradiction.

Now assume t=p. Denote the Frobenius complements of A in N by $B_1,\ldots,B_s$. Let $x\in B_i$ and $1\neq m\in M$. Then

$$x^{-1}m\in G-L \quad \text{hence} \quad (x^{-1}m)^p=1=x^{-1}mx\ldots x^{-p}mx^p.$$

Let $f_i=\prod_{x\in B_i} m^x$, $f=\prod_{x\in N} m^x$, $f'=\prod_{x\in A} m^x$. Clearly $f_i=1$.

Also, since N and A centralize f and f' respectively, f=f'=1. N is partitioned by A and the $B_i$, hence

$$m^s f=f'\prod_i f_i=1, \quad \text{which is a contradiction.}$$

[Note: Of course the situations of 6.2 and 6.3 cannot arise. From what we have shown it follows that there exists a p-group P in AB with $A_\cap P \neq 1 \neq B_\cap P$. Since $A_\cap P \trianglelefteq P$ it follows that $Z(P) \leq A$, hence the last part of 6.1(i) is violated.]

<u>Definition</u>. The group G is said to satisfy condition E(p) if every element of G whose order is divisible by the prime p, has order p.

<u>Lemma 6.5</u>. Let G be a finite group of even order and suppose G satisfies E(2). Denote by S a Sylow 2-subgroup of G and let $|S|=q$. Then either:

(i) $|G|=2r$, where r is odd, and G is a Frobenius group with kernel of order r, or

(ii) $S \triangleleft G$, or

(iii) $|G|=(q+1)q(q-1)$, and G acts sharply 3-transitively on the Sylow 2-subgroups of G.

<u>Proof</u>. It follows readily from E(2) that

(1) S is elementary abelian

(2) for all $t \in S-1$, $C_G(t)=S$

(3) any two distinct Sylow 2-subgroups of G have trivial intersection.

If q=2 then S is its own normalizer in G and (i) follows. Denote $N_G(S)$ by H. If H=G then (ii) follows. Henceforth we shall assume that $H \subset G$ and that $q>2$.

Denote by J the set of all involutions in G. Clearly the product of two non-commuting involutions has odd order and it follows that

(4) all elements of J are conjugate in G.

Using (3) it follows that

(5) H/S is sharply transitive on S-1 and $|H|=q(q-1)$.

Let $x \in G-H$ and let $u,v \in J_\cap Sx$. Write $u=t_1x$, $v=t_2x$, where $t_1,t_2 \in S$ and assume $t_1 \neq t_2$. From $u^2=v^2$ follows $t_2t_1x=xt_2t_1$ and using (2) then shows that $x \in S$, a contradiction. Hence

(6) $|J_\cap Sx| \leq 1$.

Now $|\{Sx :- x \notin H\}| = |G:S| - |H:S|$

Also $|J-(J_\cap H)| = |G:S| - |\{S-1\}| = |G:S| - |H:S|$. Now using (6) gives

(7) $|J_\cap Sx| = 1$.

Let $g\in G-H$. From (7) it follows that $Hg$ contains $q-1$ involutions. Denote these by $u=u_1,u_2,\ldots,u_{q-1}$. Clearly

(8) $Hg=Hu$, hence also $H^g=H^u$.

Denote $H\cap H^u$ by $K$. Since $S\lhd H$ it follows that $K\cap S\unlhd K$ and that $K\cap S$ is a Sylow 2-subgroup of $K$. Hence $K\cap S$ is characteristic in $K$. Therefore $(K\cap S)^u=K\cap S$. Also $(K\cap S)^u=K\cap S^u=1$, hence

(9) $K\cap S=1$.

Now $\{1=uu_1,uu_2,\ldots,uu_{q-1}\}\subseteq K$, hence $|K|\geq q-1$. However $H\geq SK$, therefore

(10) $H=SK$ and $K=\{1=uu_1,uu_2,\ldots,uu_{q-1}\}$.

It follows that

(11) $k^u=k^{-1}$, for all $k$ in $K$, hence $K$ is abelian.

Let $t\in S-1$. Suppose $t(uu_i)t=uu_j$. Then $t(uu_itu_iu)=u_iu_j$ and it follows from (9) and (10) that $u_iu_j=1$. It now follows from (2) that $uu_i=1$. Consequently $K\cap K^t=1$ for all $t\in S-1$. Hence

(12) $H$ is a Frobenius group with kernel $S$ and $K$ is a Frobenius complement.

Apply 6.2 to $H$ and $K$. Using the fact that $K$ is abelian, it follows that

(13) $K$ is cyclic.

Let $n\in N_G(K)-H$. By (7) $n=tv$, where $t\in S$, $v\in J$. Now $K^n=K=H\cap H^n=H\cap H^v=K^v$. Therefore $K^t=K^v=K$, hence by (12) $t=1$ and $v$ is one of $u=u_1,u_2,\ldots u_{q-1}$. It follows that

(14) $N_G(K)=K\cdot\langle u\rangle$ and $N_G(K)$ is dihedral.

It now follows that

(15) if $H,Hu,Hw$ are distinct, then so are $H\cap H^u,H\cap H^w$.

Hence

(16) $|G:H|=q+1$ and $|G|=q(q+1)(q-1)$.

It is now easily seen that $G$ acts sharply triply transitively on the conjugates of $H$, hence also on the Sylow 2-subgroups of $G$.

<u>Lemma 6.6</u>. Let $G$ be a finite group of even order satisfying $E(2)$, $E(3)$ and be as described by 6.5(iii). Then $G\cong A_5$.

Proof. Suppose 3 divides $q-1$. Then from (13) it follows that $q-1=3$, hence $G\cong A_5$.

Henceforth we shall assume that 3 divides $q+1$.

Suppose $t_1u$ and $t_2u$, where $t_1,t_2\in S-1$ and $u\in J$, both have order 3. Then

$$(ut_2t_1u)t_1(ut_1t_2u)=t_2$$

hence $ut_1t_2u\in H$, which is only possible if $t_1t_2=1$. Consequently

(17) $Su$ contains at most one element of order 3.

Let $Y$ be the set of elements of order 3 in $G$.

(18) $|Y|\leqslant q(q-1)$

If $y\in Y$, then clearly $C_G(y)$ has order a power of 3, hence has order a divisor of $q+1$. It follows that $y$ lies in an orbit of length at least $q(q-1)$, hence

(19) $|Y|=q(q-1)$, $|C_G(y)|=q+1=$ a power of 3.

We now regard $G$ as a permutation group on the Sylow 2-subgroups. It is clear that elements of $Y$ fix no points. Let $x\in G$ have no fixed points. Let $r$ be the length of the shortest orbit of $x$. It is easily seen that all orbits of $x$ have length $r$, hence $r$ divides $q+1$. Consequently $x\in Y$.

Let $g_i$ be the number of elements of $G$ fixing exactly $i$ points, $i=0,1,2$. Then

$$g_0+g_1+g_2=q(q+1)(q-1)-1$$

$$g_0=q(q-1)$$

$$g_1=(q+1)(q-1)$$

$$g_2=(1/2)q(q+1)(q-2).$$

These give a cubic equation in $q$ with roots $0,1,2$. This contradicts the assumption that $q>2$.

To complete the proof of Theorem 5.4 we need the fact that finite non-soluble groups have even order. Rather than use the Odd Order Theorem, we may use the rather easier Feit-Hall-Thompson Theorem that finite non-soluble groups in which the centralizer of every element $\neq 1$ is nilpotent, have even order. The groups in Theorem 5.4 have this property.

REFERENCES

[1] Dembowski, P.: Finite Geometries, Springer Verlag 1968.

[2] Gleason, A. M.: "Finite Fano planes", 1956, Amer. J. Math. 78, pp. 797-807.

[3] Hall, M.: The theory of groups, Macmillan 1959.

[4] Kegel, O. H. and Lüneburg, H.: "Über die kleine Reidemeisterbedingung."II, 1963, Arch. Math. 14, pp. 7-10.

[5] Klingenberg, W.: "Beziehungen zwischen einigen affinen Schließungssätzen." 1952, Abh. Hamburg 18, pp. 120-143.

[6] Klingenberg, W.: "Beweis des Desarguesschen Satzes aus der Reidemeisterfigur und verwandte Sätze." 1955, Abh. Hamburg 19, pp. 158-175.

[7] Lüneburg, H.: "Über die kleine Reidemeister-bedingung." 1961, Arch. Math. 12, pp. 382-384.

[8] Moufang, R.: "Zur Struktur der projektiven Geometrie der Ebene." 1931, Math. Ann. 105, pp. 536-601.

[9] Pickert, G.: Projektive Ebenen, Springer Verlag 1955.

[10] Reidemeister, K.: "Topologische Fragen der Differentialgeometrie. V. Gewebe und Gruppen." 1929, Math. Z. 29, pp. 427-435.

[11] Wielandt, H.: "Über das Produkt paarweise vertauschbarer nilpotenter Gruppen." 1951, Math. Z. 55, pp. 1-7.

# PERMUTATION GROUPS WITH FEW FIXED POINTS

E. E. Shult

Kansas State University

The author is surveying here permutation groups carrying hypotheses likely to be encountered in the context of the sort of geometric problems this conference has studied. Accordingly I have omitted a vast and very beautiful literature on doubly transitive groups carrying strong arithmetic hypotheses on their degrees and on character degrees, and prompt apologies are extended for references omitted for this reason, similar reasons or for those omitted simply by accident.

My warm thanks are extended to Jennifer Key who helped me with references, and to Westfield College, University of London, who extended hospitality to me during the writing of this manuscript. Finally, of course, I extend my gratitude to Professors Plaummer and Strambach for the warm hospitality they showed as hosts of this conference.

## 1. INTRODUCTION.

It often strikes one, that a great deal of the mathematics which today we regard as "significant" has resulted from steering a course between two extremes. There are on the one hand, axiom systems which are so tight, so over-specified, that they leave one with a completely determined system about which no interesting questions may be asked. Consider, for example, the following axioms: We have two sets of objects, "points" and "arcs", connected by an incidence relation. Each point is assumed to be incident with two arcs, each arc is assumed to be incident with two points. Such an object, of course, is a join of polygons and little else can be said of it. On the other hand, it is not



*P. Plaumann and K. Strambach (eds.), Geometry – von Staudt's Point of View, 275–311.*

difficult to find axiom systems which are so wildly loose, that any hope of ever classifying the objects in it must be abanded. Steiner triple systems have long been recognized as an example of this type: for any set of reasonable parameters, at least one Steiner system always exists (although almost certainly the number of systems grows exponentially with $v$ or $k$). So we have on the one hand systems which are too tight, on the other, systems which are too free.

The interesting mathematics, of course, lies in between these two extremes. In fact, ideally, one really hopes to have the best of both worlds: One wishes to obtain a complete classification of the objects within the world created by the axioms (as happens in a "tight" axiom system), but at the same time one wishes to have the axioms so seemingly loose and modest, that a surprising "logical distance" separates the axioms from the conclusion, -- even to the point that in the face of the axioms, the conclusion seems nearly miraculous. Indeed, the sustaining fascination of mathematics seems to draw from the many very satisfying instances in which this ideal has been acheived.

Now comes the bad news. In the nature of things, one must also expect to find axiom systems which appear very tight and impose a great deal of structure, but fall just short of allowing a complete classification of the basic objects of the system. In fact, it could be possible that for any reasonable conjecture about the system, there <u>are</u> indeed finite counterexamples existing, but the axioms are so tight that the counterexamples are not permitted to occur in cases small enough to be humanly constructible -- at least for a long time. The question whether there are finite projective planes of non-prime power order may well fall into this category. Even if the celebrated issue of the existence of the plane of order 10 were decided in favor of its extinction, orders 12 and 18 must be approached next, and perhaps somewhere far up the line the "criminal" plane sits, just out of grasp of the most advanced mathematical explorer, but by its very existence obstructing any hoped for classification theory. It is certainly a fact that such mathematical "traps" abound -- indeed (as any number theorist may attest) they have spawned some of the deepest problems of all of mathematics.

Consider for a moment, the case of finite geometry. If we restrict our attention to a nice, but rather broad class of geometries, say, those admitting a "Buekenhout diagram" (in this way I hope to excuse myself from the morass of arbitrary graphs), one may observe that the very difficult and deep problems involving possible "criminal" obstructions to a classification theory of the sort described in the above paragraph, seem to occur mainly in the rank 2 geometries. One hopes to weed out these unknown (possibly unknowable) criminal geometries by assuming that the geometries emitted from the axiom system already posses a fairly rich group of symmetries. Thus one considers "Moufang" conditions on the Buekenhout geometries. What this means for finite projective planes has long ago become classical pedagogy. This great achievement for projective planes has been equalled in recent years by the analogous classification of Moufang generalized quadrangles by Professor Tits. (Ideas involving Moufang conditions on general incidence graphs can be found in a series of interesting papers by Richard Weiss [66, 67].)

The interposition of group theory into other branches of mathematics has a venerable and respectable history. Its first major success, the Galois Theory, nearly coincided with the discovery of groups themselves and is today learned by every serious student of mathematics in the Western World. In the examples of rank 2 geometries cited above, the presence of "Moufang groups" certainly serves to distinguish the nice "classical" examples from the (possible) freak criminal geometries which cause so much trouble. By assuming symmetries in this way, one sees that the mathematician's ideal of a difficult, deep, but complete classification theory (in the sense of the second paragraph of this section) can actually be regained.

This leaves of course those objects with few automorphisms still lying around. One idea has been to find other ways to introduce groups into the object. For example, one may discuss partial isomorphisms which are not (or are not necessarily) automorphisms. A classic example are the perspectivities of projective planes. Or one may introduce other group-producing formalisms (e.g. homotopy and homology groups of simplices, monodromy groups of equations, etc.). The hope, of course, is to shed some light on an obscure subject by transfering the problem to groups.

In a strange sense, I believe this effort has been misguided. Often, it has served only to transform a fact in one field to a fact about groups without really revealing much further. Even the fundamental theorem of Galois Theory performs a transformation of this type and we consider it as the first of two examples.

And it is true, the group-theoretic description of solvability is quite simple. Given a group (equivalently: a reasonable description of a group) it is not usually diffult to decide whether it is solvable. But if the group is described <u>only</u> as the Galois group of some equation of degree 43, we see that in terms of the context of the original subject (splitting fields of equations), deciding solvability is in fact <u>not</u> really easy at all. It is merely the existence of non-solvable groups, which divulges the most significant fact about Galois theory -- that is, the groups serve to tell us how really difficult and hopeless a general theory of the solvability of equations really is.

We now consider a second example. One of the ideas of von Staudt was to consider the group generated by those permutations induced on a line which are products of perspectivities, and this looked promising, for the group so produced is certainly triply transitive on the points of the line. One may recall that as recently as two decades ago, it was generally believed that if one had a doubly or triply transitive group on hand, one probably knew a lot about the group. This view was implicitly maintained by many geometers: It was somehow difficult to be a doubly transitive group. The history of science takes strange turns, and here the irony of it is (1) that geometers had no right to the assumption, and (2) they were right.

But the situation is not easy to describe. Groups and geometry have been closely related since the days of Felix Klein. The belief was that by introducing groups, (1) the subject was now amenable to a better understood theory, namely the theory of groups, and (2) that the group theory could then be easily translated back to give results in the original subject helpful in obtaining a classification theory. The main point of this essay is to survey the historical irony underlying these assumptions:

1. At the time von Staudt and others found ways to realize doubly transitive groups here and there among the projective planes, one had no basis to expect a classification of doubly transitive groups, and hence no hope (in lieu of further assumptions) of added information from that quarter.

2. Optimism concerning the ultimate classification of finite doubly transitive groups grew to a peak much later -- perhaps around 1970, in the wake of a series of sweeping theorems -- only to rescind in frustration and failure, largely due to the lack of the use of induction.

3. The effort in 2. finally gave way to one of the greatest international scientific efforts of all time: the ulitmate classification of all finite simple groups by induction against the known list--a truly courageous enterprise (The proof is of course <u>unteachable</u> as it stands, and may not even be known to our grandchildren.). The final classification of all finite simple groups contains within it the long-awaited classification of the finite 2-transitive groups, at least up to a small gap in the case that the group has a regular normal subgroup. (The size of this gap can be estimated from the theorems of Cameron and Kantor [15] and Kantor [35].)

4. Nonetheless, the classification of the finite simple

groups totally depends on a "ridge" of theorems of Suzuki, Bender, Brauer, Feit, Aschbacher and others involving low rank groups which emerged from theorems attempting to classify doubly transitive groups.

My objective is to survey these points.

## 2. THE GEOMETRIC CONTEXTS

Let $\pi$ be a finite projective plane. As remarked earlier, the assumption of a relatively rich group of automorphisms tends to force the plane to be isomorphic to the plane $\pi_F \simeq PG(2,F)$ for some field F. One need only recall the famous Ostrom-Wagner theorem [47], that if the collineation group of $\pi$ is doubly transitive on the points of $\pi$, then $\pi$ is $\pi_F$.

But even if our plane does not possess many collineations at all, there are still groups which can be made to act "richly" on only smaller portions of the plane, and among these we find, for example, the famous <u>group of projectivities of a line</u>.

For definiteness, fix two lines $L_1$ and $L_2$ of the plane $\pi$, and let p be a point lying on neither line. We form a mapping $L_1 \to L_2$, which we denote rather clumsily as $P(p:L_1 \to L_2) = f$. For each point q in $L_1$, f(q) is the unique point of $L_2$ lying on the line through q and p. Clearly the mapping f is 1 - 1. We call such a mapping $P(p:L_1 \to L_2)$ a <u>perspectivity</u> and by a <u>projectivity</u> we understand any composition of perspectivities, so allïgned that the range of each perspectivity is the domain of the next in the order in which the composition is taken. Since these perspectivities are all <u>partial</u> isomorphisms of $\pi$, they do not always extend to collineations of $\pi$. Thus if we have a chain of perspectivities,

$$L_1 \xrightarrow{p_1} L_2 \xrightarrow{p_2} L_3 \xrightarrow{p_3} \cdots \quad L_n \xrightarrow{p_n} L_{n+1} ,$$

the composition is a projectivity and is a 1-1 mapping $L_1 \to L_{n+1}$, again a partial isomorphism of $\pi$.

An immediate observation concerning projectivities is the following

Lemma

Let $p, q, r$ and $p', q', r'$ be two ordered triplets of collinear points. Then there is a projectivity which takes one triplet to the other in the prescribed order.

Now taking $L_1$ and $L_2$ as before, we may obtain a permutation of the points of $L_1$ by composing two perspectivities $P(q;L_1 \to L_2)$ and $P(r;L_2 \to L_1)$ where both $q$ and $r$ lie off $L_1 \cup L_2$. This composition we shall denote $\alpha_{qr}$. If $s$ is also a point not lying on $L_1$ or $L_2$, then clearly

$$\alpha_{qr} \circ \alpha_{rs} = \alpha_{qs} . \qquad (2.1)$$

Of course, each $\alpha_{qr}$ must fix the point $m$ on $L_1 \cap L_2$, but it is quite clear that the set $K = \{\alpha_{xy} | \{x,y\} \quad \pi-(L_1 \cup L_2)\}$ is a doubly transitive set on the $n$ points of $L_1 - \{m\}$ (where $n$ is the order of $\pi$). First, it is clear that $K$ is transitive. Fixing points $x$, $y$ and $y'$ on $L_1 - \{m\}$, one can find $\alpha_{uv}$ fixing $x$ and moving $y$ to $y'$ in the following way: Choose $z$ arbitrarily on $L_2 - (m)$ and form the line $M$ through $x$ and $z$. Choose $u$ arbitrarily on $M - (L_1 \cup L_2)$, form the line $y + u$ (through $y$ and $u$) which meets $L_2$ at a point $w$. Then setting $v = (y' + w) \cap M$ we see $\alpha_{uv}$ performs the desired task (see figure 1.). Indeed we have shown the $n-1$ elements of $K_{u,M} = \{\alpha_{uv} | v \in M - (L_1 \cup L_2)\}$ form a sharply transitive set of permutations of the $n-1$ elements of $L_1 - \{z,m\}$.

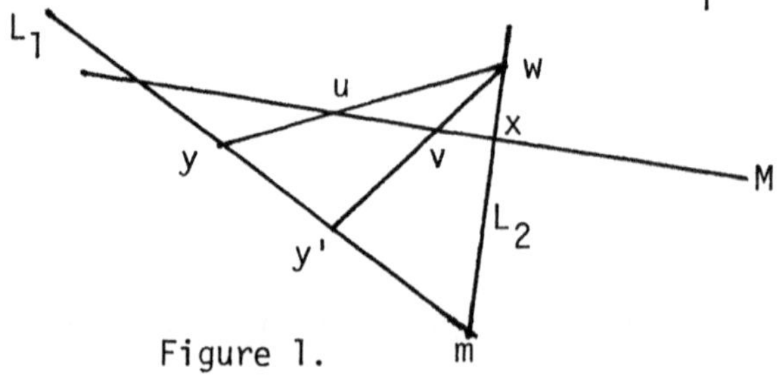


Figure 1.

We obtain, then, a doubly transitive group $G = G(L_1, L_2)$ of permutation of $L_1 - m$, generated by the $n^2 - n$ elements of the 2-transitive set K. Does this group G tell us anything about $\pi$? If it did, and if, in turn, there were only a very limited number of doubly transitive groups, then we would have some striking information about an arbitrary plane $\pi$, and I believe this was one of the philosophical motivations for this approach.

The second part, seems to be correct at least: there are a limited number of possibilities for a doubly transitive group G. In fact, if one allows the line $L_2$ to vary, the groups $G(L_1, L_2)$ generate a triply transitive group $G(L_1)$ on $L_1$ and there are an even more limited number of these 3-transitive groups around. But we only know this, as will be explained later, because of the full-scale classification of all finite simple groups -- not because of a self-contained classification of the 2- or 3-transitive groups themselves.

On the other hand, there is good reason to doubt the first part of the strategy; that the group G will tell us much about $\pi$. Undoubtedly there are very many sorts of planes $\pi$ and lines L such that the 3-transitive group G(L) described above is the alternating group on n+1 letters, $A_{n+1}$. In fact the trouble with the alternating groups is that they are too easy to generate. By the same token, of course, if $G(L_1)$ is not $A_{n+1}$ or, equivalently, if $G(L_1, L_2)$ is not $A_n$, then something very tight might well be being asserted about $\pi$. As a guide, we remark that in the plane $\pi_F \simeq PG(2,F)$, $G(L_1)$ is the 3-transitive group PGL(2,F) and K is itself a group.

One should be interested, then in relatively simple criteria that force a 3-transitive group to be GL(2,F). Indeed, many such criteria are well-established within the confines of the theory of doubly transitive groups itself, and these theorems are outlined in the succeeding sections. In fact it is one of the achievements of the theory presented in those sections that one can get by with very little: It is usually enough to know that the stabilizer of a point in $G(L_1, L_2)$ contains a normal solvable subgroup at all.

One may now ask: Is it possible for a plane to contain a

line L such that G(L) is one of the "strange" 3-transitive groups, $M_{11}$, $M_{12}$, $M_{23}$ or $M_{24}$ (or even $M_{11}$ on 12 letters)?

If not, one is left with the two extremes, for G(L):PGL(2,q) or $A_n$ (or $S_n$, the symmetric group on n Letters).

So far, we have been considering projective planes. It does not seem to be well known, but there is another rank 2 geometry that also admits a permutation group of partial isomorphisms in a way very analagous to the groups of projectivities G(L) which we have just met. The geometries in question are generalized quadrangles, and their classification seems to be confronted with the same deep problems which confound the classification of the finite projective planes.

As far as I know, the fact that these permutations are even present in generalized quadrangles, and the possibly limited possibilities for the groups which they can generate, have never been really exploited in the theory of generalized quandrangles. The discovery of these permutations may have been informally repeated many times, but they seem to be first recorded is a relatively obscure paper by S. Dixmier and F. Zora [18]. The sets of fixed points of these permutations have been subjected to the sort of quadratic mean-variance analysis originally introduced by Bose [10] and that is the use Dixmier and Zora make of them. Nonetheless, the latter showed these rather elaborate permutations were present, representing some sort of geometrical invariant.

We call $(\mathcal{P},\mathcal{L})$ a linear incidence system, if $\mathcal{P}$ is a set of "points" with special subsets $\mathcal{L}$, of $\mathcal{P}$, called "lines", such that every pair of distinct points lies on at most one line. Two points are said to be collinear if they in fact do lie on a line. A linear incidence system $(\mathcal{P},\mathcal{L})$ is called a generalized quadrangle if for every non-incident point-line pair, (p,L), p is collinear with exactly one point of L, and if, in addition, there exists no point collinear with all remaining points of $\mathcal{P}$. Two examples are (1); the $n \times m$ grid; that is, any finite regulus, and (2) its dual, the graph $K_{n,m}$ in which edges represent lines.

Aside from these two examples, for each finite generalized quadrangle, Q, there exist two positive integers s and t such that every line in Q has 1+s points and every point of Q lies on 1+t lines; Q then contains (1+s)(1+st) points and (1+t)(1+st) lines, and is said to have order (s,t).

Just as there is a canonical "classical" example for the finite projective planes, namely PG(2,q), there are exactly five "classical" examples of generalized quadranges. These have inherited a rather strange notation over recent years; the symbols are Q (for "quadratic"), H (for "Hermitian" -- or is it for the "Dickson-ese" term "hyperabelian"?) and W (for "symplectic" -- I can offer no explanation for that one.). Out of ignorance, I prefer to name them according to their associated simple groups. Under this nomenclature, the classical quadrangles are, $O^+(4,q)$ (a(q+1)-by-(q+1) regulus); O(5,q) and its dual Sp(4,q); $O^-(6,q)$ and its dual U(4,q); and finally U(5,q) and its dual, which seems not to have another nice name. These have respective orders

(q,1) (q,q) ((q,q) for its dual), $(q,q^2)$ $((q^2,q)$ for its dual) and $(q^2q^3)$ (and $(q^3,q^2)$ for its unnamed dual). There are also generalized quandragles which look "classical" from the point of view of a single point -- the constructions of Tit's from ovoids [64], and I suppose these find their best analogy in projective planes, among their "cousins", the dual translation planes.

But just as there are strange finite planes (ranging from the well known Hall planes to the nearly bizarre "one-time" constructions of Hering and Walker), there are also "strange" generalized quadrangles. The first notable examples were discovered by Marshall Hall and Sekeres [26], and have order

$(2^n-1,2^n+1)$. More recently, Payne (See [48]) has given a simpler general construction of generalized quadrangles of order (q-1,q+1), which include the Hall-Sekeres examples (q being any prime power). If these were all that had appeared, one might resume some degree of optimism for a classification theory for generalized quadrangles, but Kantor [36] has recently discovered a class of generalized quadrangles of order $(q^2,q)$ where $q \equiv 2 \bmod 3$, which can be constructed from a generalized hexagon of order (q,q). So life in Generalized Quandrangle Land promises to be as bad as it is in Projective Plane Land.

Let us turn now to the Dixmier-Zora permutations. Consider first a generalized quadrangle Q of order (s,t) (recall from the previous that nothing is being thrown away by this assumption for if the quadrangle does not even possess an order, its isomorphism type is determined). Fix for all time two points x and y which are not collinear. The trace of x and y consists of the set of 1+t vertices collinear with both x and y and is written tr(x,y). Since by hypothesis, each line through x bears a unique point collinear with y, and vice versa, tr(x,y) can be used to index the 1+t lines through x or the 1+t lines through y. The permutations introduced, are to be permutations of tr(x,y), or equivalently, permutations of the sets $\mathcal{L}_x$ or $\mathcal{L}_y$ (all lines on x or all lines on y) which it indexes. There will be one such permutation $P_z$ for each of the s(st-t-1) points z not collinear with either x or y. We shall describe these as permutations of tr(x,y). Throughout, if u is collinear with v, we write u+v for the line through u and v. Now choose z not collinear with either x or y. For each vertex w on tr(x,y), z is collinear with some point v on the line x+w. In turn, y must be collinear with a unique point m on w+v. Then, in turn, x is collinear with a point q on y+m. This mumbo-jumbo shows that given z, the choice of w in tr(x,y) completely determines the point q to which it is to be mapped under the mapping $P_z$. If one looks sharply, this is illustrated in figure 2.

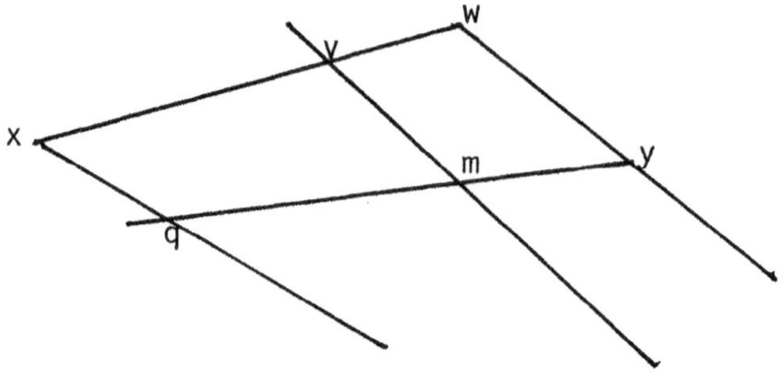


Figure 2.

So what is one to make of these permutations? Do they carry a deep secret about the geometry of the quadrangle? More specifically, after recognizing that the fixed points sets of the $P_z$ is amenable to the statistical analysis of the sort Bose introduced; does it make sense to ask if there is some form of Polya counting of the cycle lengths which would lend even further geometric information about Q?

In any event, if $s > 1$, one can see that the set $K = \{P_z \mid z$ not collinear with either x or y$\}$ is doubly transitive on tr(x,y). First we show transitivity of K. Fix a, b $\varepsilon$ tr(x,y) and choose c on the line x+a different from both x and a. Then by hypothesis, c is collinear with some point d on the line b+y and this point cannot be either b or y (see figure 3). For any point z on c+d not equal to c or d, $P_z$ transforms a to b.

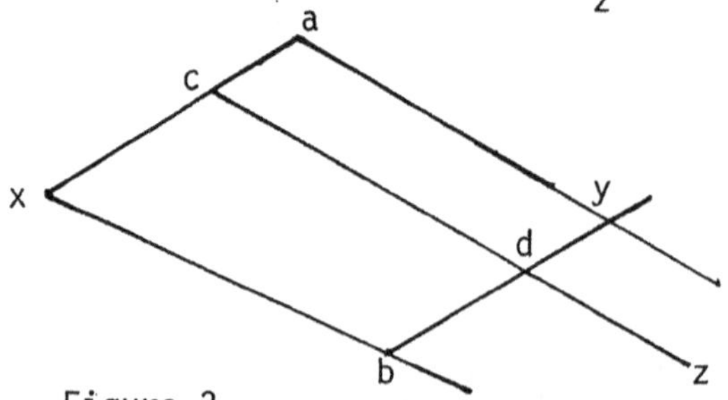


Figure 3.

Next we fix a in tr(x,y) and select two further points b and c in tr(x,y). Our object is to find a point z such that $P_z$ fixes a, but moves b to c. Choose u on line x+b distinct from x and b. Then u is collinear with a point v on the line y+c, and this point v is distinct from c and y. Now the point a is not collinear with u or v, since otherwise a triangle would be formed. Thus a is collinear with a point z on u+v distinct from both u and v (see figure 4). $P_z$ now fixes a, but moves b to c. Thus K is a 2-transitive set of permutations of K.

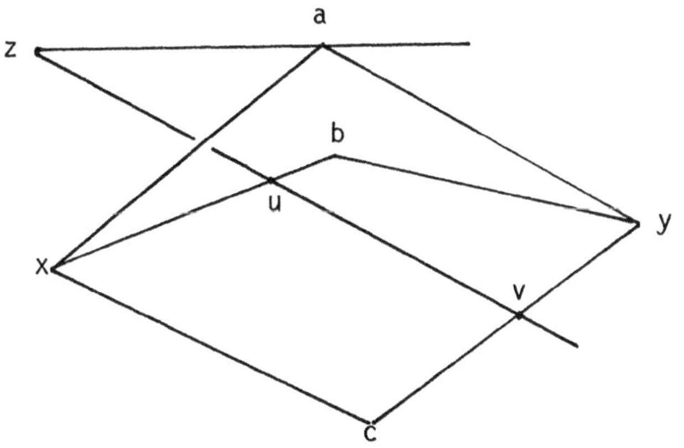


Figure 4.

Aside from multiple transitivity of the sets of permutations, there is still another point of similarity between the set K of projectivities of a line onto itself in a projective plane, and the set K of permutations $P_z$ on tr(x,y) in a generalized quadrangle. That is that in the "classical" cases, the permutations lift to automorphisms of the geometry. In the Desaurguesian plane $\pi_F$, all perspectivities $L_1 \to L_2$, lift (not necessarily uniquely) to collineations of the plane. As a "classical" generalized quadrangle consider Sp(4,q), whose points are the points of PG(3,q) and whose lines are the isotropic lines of a non-degenerate symplectic polarity. Let $V = V(4,q)$ be the underlying vector space. Then x+y is a non-isotropic line and we have an orthogonal decomposition $V = \langle x,y\rangle \perp P$ where P is a non-degenerate 2-subspace of V whose 1-spaces comprize tr(x,y). If z is a 1-space in $\langle x,y\rangle$, $P_z$ induces the identity on tr(x,y). Otherwise $z = \langle u\rangle$ where the vector $u = u_1 + u_2$, $u_2 \in \langle x,y\rangle$, $u_2 \in P$, and $u_2 \neq 0$ while $u_1$ does not lie in $\langle x\rangle \cup \langle y\rangle$. Then $P_z$ induces on $tr\langle x,y\rangle$ the same action as a transvection on P with direction $\langle u_2\rangle = z^{\perp} \cap P$, and so $P_z$ lifts.

The quadrangles U(4,q) and U(5,q) yield similar "liftings" for $P_z$, but the situation for their duals and for 9(5,q), the dual of Sp(4,q), seems somewhat more obscure but may be computed in Sp(4,q) and the U(n,q)'s by showing that a dually defined permutation $P_L$ of a trace of lines is induced by a collineation of the quadrangle.

It is worth noting here, that in the "classical" quadrangle, $U(5,q^2)$, K generates the 2-transitive group $PSU(3,q^2)$ on $1 + q^3$ letters, and so characterizations of this group might be of interest. These characterizations exist.

## 3. THE DOUBLY TRANSITIVE GROUPS.

The known doubly transitive groups are listed below in Table I. If G is a doubly transitive group of permutations of the set $\Omega$, and N is an abelian normal subgroup of G, then N is regular on $\Omega$ and so $G = G_\alpha N$ where $N \cap G_\alpha = 1$. Maximality of $G_\alpha$ in G forces N to be elementary abelian of order $p^n$, for some prime p, and so $G_\alpha$ is a subgroup of $GL(n,p)$ which is transitive on the set $V^\#$ of non-zero vectors of the underlying vector space $V = V(n,p)$. Conversely, if H is any subgroup of $GL(n,p)$ transitive on the non-identity vectors of V, then the semidirect product $G = VH$, of the additive group of V and H, acts doubly transitively on the set $\Omega$ of cosets of H in G, with V acting as a regular normal subgroup. In this case, we say G is type RNS (for "regular normal subgroup"), and as just observed, G is "known" up to a choice of subgroup H of $GL(n,p)$ transitive on $V^\#$.

How well, then, are groups of type RNS really known? Two things can be said:

(1) If H contains a subgroup inducing a Singer cycle on the set $PG(n-1,p)$ of 1-spaces of V, then H contains $SL(n/k,p^k)$ for some k dividing n (Kantor, [35]).

(2) If G is triply transitive of type RNS, then $p = 2$, and either $H = GL(n,2)$ or $H \simeq Alt(7)$ while $n = 4$. (This determination began as a problem of Wagner [65], and was not finally resolved until a decade later in the impressive theorem of Cameron and Kantor [15].)

Next suppose G does not have a normal abelian subgroup. Then Socle(G), the join of all minimal normal subgroups of G, is the direct product of non-abelian simple groups. Since G is primitive, it is not difficult to see that Socle(G) is itself a simple group X and that only the identity element of G centralizes X. In this case

$$X \leq G \leq Aut(X),$$

the full automorphism group of X. In this case, we say G is

type X, in Table I. In all known cases but one, the subgroup X of G is also doubly transitive, but there seems to be no direct proof that this should normally be the case. The one exception is $X \sim SL(2,8)$ which is a rank 4 permutation group of degree 28, where $G = S\Gamma L(2,8)$, SL(2,8) extended by the automorphism of order 3 induced by a field automorphism. In this case, and all remaining cases, no proper subgroup of G remains doubly transitive, and (if one excuses $M_{22}$) no proper normal extension of G can act on $\Omega$, so there is no interval of group containments, from which to choose G. In these cases, referred in the table as "exceptional cases", the type of G, is simply its isomorphism type.

As usual, $A_n$ denotes the alternating group on n, letters, and PSL(n,q), the familiar projective special linear groups. The 3-dimensional projective special unitary groups $PSU(3,q^2)$ and the Ree groups $^2G_2(q)$ both act on a block design with $\lambda = 1$ and $k = 1 + q$, called a unital. The Suzuki groups Sz(q) act on the points of an inversive plane. The two doubly transitive representations of the symplectic groups Sp(2n,2) are on cosets of its orthogonal subgroups $O^+(2n,2)$ and $O^-(2n,2)$. From the isomorphism $A_8 \sim SL(4,2)$ one observes that $A_7$ is a subgroup of SL(4,2) = PSL(4,2) remaining doubly transitive on the points of PG(3,2). The Mathieu groups have a venerable history known to most readers. PSL(2,11) acts on a biplane of 11 points, and admits a transitive extension to $M_{11}$ acting on the 12 points and 22 blocks of an extended Hadamard design. The Higman-Sims group HS and the Conway group (.3) both act on equiangular line systems in Euclidean space, in dimensions 22 and 23, respectively. The symmetric groups, as well as the more complicated (and perhaps to geometers, less familiar examples numbered 7, 8, 18 and 19, all have a relatively simple-minded constructions either (a) in terms of the graph extension theorem [56] or (b) in terms of 2-graphs (using the fact that the automorphism group of any graph in a switching class of graphs is a subgroup of the full automorphism group of the 2-graph defined by the class. See J. J. Seidel [52] and Seidel and Taylor [53] for more complete details.)

The recent completion of the classification of the finite simple groups makes it appear that the groups listed in Table I do indeed cover all finite doubly transitive groups -- at least up to the ambiguities in the RNS case mentioned above. Assuming, for example, the classification complete, Socle(G) = X is a known simple group. If X is not doubly transitive then X is a rank k group with exactly one subdegree of length 1 and the remaining of length m, so $n = 1 + (k-1)m$. In addition, the permutation character of G is $1 + \chi$ and $\chi$ restricted to X is (k-1) irreducible characters of X all conjugate in G. The presence of such G-conjugate characters of X such that 1

| Type | | $\Omega$ |
|---|---|---|
| **Case A: Non-trivial solvable radical** | | |
| 1. | RNS | $\lvert N\rvert = p^s$ where $p$ is a prime and $N$ is the unique minimal normal subgroup of $G$. |
| **Case B: Trivial solvable radical. $X \le G \le \mathrm{Aut}(X)$** | | |
| 2. | $A_n$ | $n-2$ or $n$-fold transitive on $n$ letters |
| 3. | $PSL(n,q)$ | $1 + q + q^2 + \cdots + q^{n-1}$, $(n,q) \neq (2,2)$ or $(2,3)$, $n > 1$, $q$ a prime power |
| 4. | $PSU(3,q^2)$ | $1 + q^3$ |
| 5. | $Sz(q)$ | $1 + q^2$, $q = 2^{2m+1}$, $m \ge 1$ |
| 6. | ${}^2G_2(q)$ | $1 + q^3$, $q = 3^{2m+1}$, $m \ge 1$ |
| 7. | $Sp(2n,2)$ | $2^{n-1}(2^n - 1)$ $n \ge 3$ |
| 8. | $Sp(2n,2)$ | $2^{n-1}(2^n + 1)$ $n \ge 3$ |
| **Case C: Trivial solvable radical. Exceptional Cases.** | | |
| 9. | $A_7$ | 15 conjugates within either of the two classes of subgroups isomorphic to $SL(3,2)$. |
| 10. | $S\Gamma L(2,8)$ | 28 3-Sylow subgroups |
| 11. | $M_{11}$ | on 11 letters |
| 12. | $M_{12}$ | on 12 letters |
| 13. | $M_{22}$ or $\mathrm{Aut}(M_{22})$ | on 22 letters |
| 14. | $M_{23}$ | on 23 letters |
| 15. | $M_{24}$ | on 24 letters |
| 16. | $PSL(2,11)$ | on 11 conjugates of $A_5$ in either of two classes |
| 17. | $M_{11}$ | triply transitive on 12 letters |
| 18. | HS | 176 letters |
| 19. | (.3) | 276 letters |

Table I. The known doubly transitive groups.

plus their sum possesses all the qualities of a permutation character can be checked from the character table for $X$ when $X$

is sporadic, and can be checked in general by the methods of Curtis, Kantor and Seitz [16] when X is Chevalley or alternating. Otherwise X is itself doubly transitive. When X is Chevalley, X appears in Table I by the paper of Curtis, Kantor and Seitz just cited. The possible doubly transitive representations of the 26 sporadic groups can in principle be decided by examining when $1 + \chi$ satisfies Gluck's criterion [24] to be a generalized permutation character, as $\chi$ ranges over the rational irreducible characters of X.

Considering the power, and failry early arrival of some of the more sweeping characterization theorems for doubly transitive groups to be described in the subsequent sections, it seems a shame that the classification of the doubly transitive groups (i.e., the theorem that every doubly transitive group is one of those of Table I) should have to await the ultimate lengthy classification of all finite simple groups. Considering that the latter is a far more universal result, this is of course, a mixed complaint.

## 4. FROBENIUS AND ZASSENHAUS GROUPS.

We consider first those permutation groups for which all non-identity elements possess few fixed points. Theorems identifying such groups are useful, for example, in telling us when the group of projectivities of a line of a projective plane is PGL(2,F) as it ought to be for the "classical" planes $\pi_F$ (where F is a field). In succeeding sections we will survey the theory when only involutions in the permutation group possess few fixed points.

One of the earliest theorems on permutation groups, apparently known to Frobenius, represents one of the earliest triumphs of the theory of group characters.

Frobenius' Theorem:

Suppose G is a transitive group of permutations of the set $\Omega$ with the property that only the identity element of G fixes two or more letters. Then the sub set K consisting of the identity element, together with every permutation of G fixing no letters, forms a normal regular subgroup.

It is remarkable that after seven decades, there is still no known proof of this theorem which does not require the theory of group characters. There is a version of this theorem due to Wielandt [68] in which $G_\alpha$ contains a normal subgroup $N_\alpha$ with the property that any element of $G_\alpha$ fixing at least two letters, lies in $N_\alpha$, and in the conclusion K is replaced by the set of all elements of $G$ not conjugate to an element of $G_\alpha - N_\alpha$ ; the proof is a variation on the character-theoretic proof of Frobenius' Theorem.

Let $G$ and $K$ be as in Frobenius' Theorem. We can produce a labeled graph on $\Omega$ as follows: Fix two distinct letters $\alpha$ and $\beta$. Let $k_{\alpha\beta}$ be the number of elements of $K$ sending $\alpha$ to $\beta$. The inverses of these elements will comprise all $k_{\beta\alpha}$ elements of $K$ taking $\beta$ to $\alpha$ so $k_{\beta\alpha} = k_{\alpha\beta}$. If $k_{\alpha\beta} \neq 0$ we draw an edge between $\alpha$ and $\beta$ labeled "$k_{\alpha\beta}$" and from the previous sentence, there is no apparent reason to attach a direction to this edge. In this way $\Omega$ becomes an undirected graph with its edges labeled. Now $K$ is a group (and hence the normal subgroup of the conclusion of Frobenius' Theorem) if and only if each $k_{\alpha\beta} = 1$ and $\Omega$ is the complete graph. In the case $G$ is doubly transitive, $\Omega$ is forced to be the complete graph and we are done. So there is an easy character-free proof of the Frobenius Theorem in the case that $G$ is doubly transitive.

In the general case, an elementary proof seems difficult and it is not clear whether the graph $\Omega$ can be effectively exploited in the proof. The sum of the labels on all edges leaving a point is $n - 1$ where $n = |\Omega| = |K|$. For primes $p$ dividing $n$, p-Sylow subgroups of $G$ lie in $K$ and form cliques in $\Omega$. It would seem that the number of p-Sylow subgroups of $G$ should be proved to divide $n$ also, but this point seems obscure.

It was also quite a number of years after Frobenius Theorem was well known that J. Thompson proved that the "Frobenius kernel" $K$ of Frobenius' Theorem, must be a nilpotent group [63]. The proof is "elementary" in the sense that no character theory is required for this, but at the time represented an entirely novel way of forcing a group-theoretic proof, the Thompson-ordering of local subgroups. This device was soon to appear again and figures now in many places among the papers of Achbacher, Gorenstein and Lyons in what has become known among group theorists as "local analysis".

The next stage up the line, after Frobenius groups, was occupied by Zassenhaus groups, named after Professor Zassenhaus who instigated a study of them in 1936. By a <u>Zassenhaus</u> group we understand a transitive permutation group in which any permutation fixing at least three letters is the identity.

If $G$ is a Zassenhaus group on $\Omega$, then $G_\alpha$ is a Frobenius group on $\Omega - (\alpha)$ and so is the semi-direct product $G_\alpha = KG_{\alpha\beta}$ where $K$ is the Frobenius kernal and is regular on $\Omega - (\alpha)$.

Zassenhaus himself considered the case that $G$ was triply

transitive [69]. In that case $G_\alpha$ is a doubly transitive Frobenius group and so K is an elementary abelian p-group. G is then classified in Zassenhaus' Theorem [69]. If G is a triply transitive Zassenhaus group, then G is a linear fractional group over a near field. If G is finite, G is type PSL(2,q) (note that this class includes the sharply triply transitive group $M_{10}$).

The ultimate acheivement here was the eventual classification of all Zassenhaus groups in a series of deep papers by Feit, Ito and Suzuki. Each of these papers in turn introduced new methodology to the theory of finite permutation groups. In a paper, which for the first time brought into being the notion of coherence of exceptional characters Feit [19] proved that K, the Frobenius kernel of $G_\alpha$, is a p-group and that $[K:[K,K]] \leq 4|G_{\alpha\beta}|^2+$ or K is abelian. Moreover, he proved that if K was abelian, then either G = PSL(2,q) or is one of the groups of Zassenhaus' theorem, in any event type PSL(2,q). At this point the problem divided into two case, even degree and odd degree.

In 1964, both cases were resolved.

First, in the even-degree case, Ito [34] showed that K must be abelian, so that the G is known by the result of Feit. Ito's proof is ingenious and makes use of block theory, in particular the upper bound $p^{2d-2}$ of the number of complex irreducible characters of G lying in a p-block of defect d. (Five years later George Glauberman [23] found a proof of Ito's result free of the theory of modular characters. It does, however make use of the Frobenius-Schur theorem on the Schur index of real representations.)

Second suppose G had odd degree. Then either G is known or K is a non-abelian 2-group. Suzuki observed that the 2-point stabilizer $G_{\alpha\beta}$ is cyclic (not surprising as $KG_{\alpha\beta} = G_\alpha$ is a Frobenius group), is inverted by an involution t transposing $\alpha$ and $\beta$, and acts transitively on the involutions of K. This, by a result of G. Higman [31], determined the isomorphism type of K. At this point Suzuki [59] brought to the fore, a magnificent extension of the methods of Zassenhaus. We know $G = G_\alpha + G_\alpha tK$ and so each element of $G - G_\alpha$ must have a unique expression $k_1tk_2$ where $k_i \in K$ and $r \in G_{\alpha\beta}$. Now for every non-identity element y of K (we denote this set $K^\#$), tyt lies in $G - G_\alpha$ and so admits the unique factorization $y = k_1rtk_2$. This produces maps $f: y \to k_2$, $g: y \to r$, and $h: y \to k_1$, all with range $K^\#$ and domains K, $G_{\alpha\beta}$ and K, respectively. With the structure of K to guide one, many properties of the mappings f, g and k are worked out by Suzuki, until finally, they uniquely determine the multiplication table for G -- namely that of the Suzuki groups

Sz(q). This method of "multiplication functions" (or "structure equations" as they are often called) which Suzuki developed, has served as the framework for the long-awaited solution of the "Ree group problem" -- which has recently been solved by Bombieri. Combining all these results, we have:

THEOREM
(Zassenhaus [69], Feit [19], Ito [34], Suzuki [59]). A doubly transitive group in which no non-identity element fixes as many as three letters is of type RNS, PSL(2,q) or Sz(q).

Before closing this section one ought to reflect what the remarkable series of papers of Feit, Ito and Suzuki (confined to the years 1962 and 1964) really meant for group theory in general. The methodological advances which they represented have already been remarked on. The Brauer theory had already broken many previously impassable frontiers of group theory and still forms an indispensable part of the low 2-rank analysis of finite groups, perhaps best epitomized in the famous theorem of Alperin, Brauer and Gorenstein [1]. Coherence had became a refined science. Dade's paper [17] on lifting characters refined Feit's further ideas on coherence in the "odd order" paper (Feit and Thompson [20]) and appears rather conveniently (and appropriately from the point of view of the discussion of this section) following a remarkable paper of Suzuki in the Annals of Mathematics vol. 79 , for the latter paper itself represents a tour de force of the "multiplication function" method.

The classification of Zassenhaus groups itself, although greatly generalized in many directions, as subsequent sections of this survey will demonstrate, still is essential. As has happened throughout finite group theory, nearly every paper generalizing theorem A also utilizes Theorem A in its proof and so cannot divest itself of its logical dependence on it. It is thus not surprising to observe that many of the major finite group theory theorems beginning in 1970 no longer seemed to require character theory. It was because by this time there was a sufficient wealth of difficult theorems (whose proof did require character theory!) to facilitate subsequent proofs on the basis of their conclusions alone. Similarily all of the generalizations of the Zassenhaus-group classification which appear in the next section, all of the main characterizations of the rest of this paper -- and indeed, most of all subsequent finite group theory, depend on the basic Feit-Ito-Suzuki theorem. In a logical sense, they can't live without this theorem. For its age, it was a remarkable theorem which today is due a certain homage from the perspective of the present classification of all finite simple groups.

## 5. PERMUTATION GROUPS WHOSE INVOLUTIONS FIX FEW POINTS.

One way of generalizing the classification of Zassenhaus groups would be to relax the requirement that all non-identity elements of a permutation group possess few fixed points to one in which it is required only of involutions, the elements of order 2 in a group. It had long been recognized that involutions play an unusual and somewhat determinative role in finite group theory, so it was not unreasonable to hope for generalizations of this type. The present section is intended to survey this line of development.

The first result, due to Marshall, Hall in 1954, completely preceded the classification of the Zassenhaus groups; and could do this because it was stated for highly multiply transitive groups. Specifically Hall's Theorem [25] showed that any 4-fold transitive group in which on involution fixes as many as five letters must be $S_4$, $A_6$, $A_7$ or $M_{11}$. Some 14 years later, Nagao [40] extended this theorem by requiring (for a 4-fold transitive group) that involutions fix at most five letters, the conclusion being that $S_6$, $S_7$, $A_8$, $A_9$ and $M_{12}$ must be added to Marshall Hall's list. In the series of theorems which appear in the balance of this section it will be seen that the condition of 4-transitivity can eventually be relaxed all the way to simple transitivity in the case of Hall's theorem (in theorems of Noda [41] and Buekenhout [12]) and in most cases of Nagao's theorem (results of Rowlinson [49,50,51] and Buekenhout and Rowlinson [13]).

At the level of 2-transitivity, the first two bona fide generalizations of the Zassenhaus-group classification appeared in 1968. These were the theorems of H. Bender and C. Hering [8,28]. Bender's theorem, perhaps his very first paper, showed that every doubly transitive group in which involutions fix no letters, is of type PSL(2,q) where $q \equiv 3 \bmod 4$. Strictly speaking, this was a generalization of Frobenius' theorem, since the hypothesis of being a Zassenhaus group does not preclude having involutions (in the even degree case) which fix exactly two points. The very difficulty of this first theorem of Bender was that the proof of Frobenius' theorem was not patently extendable in this direction; its neat character-theoretic proof had to be abandoned altogether. If one imagined himself today doing this theorem while fully armed with every known theorem of group theory which is not actually a direct logical descendant of it, it is still not an easy theorem.

This meant that in Hering's theorem, one could actually assume $n \equiv 2 \bmod (4)$ (for otherwise a normal subgroup of index 2 satisfies the conditions of Bender's theorem, so $G$ is known) and that some involution $t$ fixing two letters, transposes two letters, $\alpha$ and $\beta$, and normalizes a 2-Sylow subgroup $S$ of $G_{\alpha\beta}$. By hypothesis, $S$ is semi-regular on $\Omega - \{\alpha,\beta\}$ and so $|C_S(t)| = 2$. A lemma of Suzuki then tells us the structure of S, and so we have a start.

These two results are subhumed in

Theorem (Bender [8] and Hering [28]). If $G$ is a doubly transitive group of permutations of $\Omega$ in which each involution fixes 0 or 2 letters, then $G$ is of type $PSL(2,q)$ (on $q + 1$ letters) or $A_6$ (on 6 letters).

The next stage, the case of odd degree, turned out to be one of the most useful of all finite group theory.

The story begins with Suzuki's classification of Zassenhaus groups of odd degree discussed in the last section. Suzuki called these groups ZT groups. Since $G_\alpha$ would be an even-degree Frobenius group, in such a group, there would be a normal nilpotent even-degree sub-group $Q$ of $G_\alpha$ which was regular on $\Omega - (\alpha)$. About one-third of the way through Suzuki's proof it emerges that $Q$ is a 2-group, rather than merely being a nilpotent group of composite but even order. Since $G_{\alpha\beta}$ was seen to be cyclic in the course of the same proof, this motivated the notion of "generalized ZT-group", which made its appearance in [60]. A <u>generalized</u> ZT-<u>group</u> was a doubly transitive group $G$ of permutations of $\Omega$ such that if $\alpha \in \Omega$, $G_\alpha$ contains a normal 2-subgroup $Q$, regular on $\Omega - (\alpha)$, and $G_{\alpha\beta}$ is cyclic of odd order. This was not a non-trivial extension, for an entirely new class of groups were now permitted to enter into the conclusion, those of type $U(3,2^a)$. In what I have already referred to as a <u>tour de force</u> of the "multiplication function method" Suzuki showed in a later paper [61], that the list of conclusion groups is not increased if in fact $G_{\alpha\beta}$ is assumed only to be solvable and of odd order (one will note here that at the time this paper was submitted, the Feit-Thompson odd-order theorem had not yet made that last phrase a demonstrable redundancy).

Now in both of these theorems, since $G_{\alpha\beta}$ has odd order, no involution is allowed to fix three letters. Looking back, these two theorems of Suzuki represent three points of departure for subsequent work on doubly transitive groups: theorems generalizing the feature that (1) $G_{\alpha\beta}$ is cyclic, (2) that $G_\alpha$ has a normal subgroup regular on the remaining letters, and (3) that involutions fix only one letter. Only (3) lies within the scope of this section, but the lines of development which followed features (1) and (2) are detailed in the next sections and represent a sort of "high point" in the theory of doubly transitive groups which came to blossom around 1969.

Suzuki's two theorems were generalized in a sweeping way by Bender, first for the case that G is doubly transitive in a preprint in 1969, but in published form for simple transitivity.

Bender's Strongly Embedded Theorem [9]. Let $(G,\Omega)$ be a transitive group of permutations of a set $\Omega$, in which each involution of G fixes exactly one letter. Then either (1) all involutions lie in a normal subgroup N of G containing a subgroup $N_1$ of odd order of index 2 in N, or (2) G is doubly transitive on $\Omega$ of type PSL(2,q), Sz(q) or U(3,q) where q is a power of 2.

The structure appearing in part (1) of the conclusion emerges from Glauberman's famous Z*-theorem [22], which requires only that there is an involution possessing no further G-conjugates in some 2-Sylow subgroup of G containing it. The contrary case must eventually lead to a spot in which Suzuki's theorems may be utilized. Bender accomplished this by finding a method of utilizing induction in a non-trivial way, by choosing p-subgroups maximal with respect to fixing certain sets of letters.

The immense utility of Bender's strongly embedded theorem to the theory of finite groups can hardly be overestimated. Although stated here as a theorem on permutation groups, its usefulness to abstract groups in general stems from the various interpretations which can be placed on $\Omega$. Two examples might illustrate this. Suppose H is a proper subgroup of the finite groups G, H has even order is its own normalizer in G, and contains the centralizer in G of each of its involutions. We call such a subgroup H a <u>strongly embedded subgroup of</u> G. (In Bender's strongly embedded theorem stated above, the subgroup $G_\alpha$ is strongly embedded in G, when $\Omega$ contains at least two letters.) Let $\Omega = H^G$. Then G acts on $\Omega$; that is, there is a homomorphism $G \to Sym(\Omega)$ with image $\overline{G}$, and so $(\overline{G},\Omega)$ satisfy the hypotheses on $(G,\Omega)$ in Benders' strongly embedded theorem. As a second example, let G be a finite group and let $S$ be the collection of all 2-Sylow subgroups of G. Suppose "$\sim$" denotes an equivalence relation on $S$, which is preserved by G-conjugation -- that is, if A and B are elements of $S$ and g is any element of G, then $A \sim B$ if and only if $g^{-1}Ag \sim g^{-1}Bg$. Let $S_1,\ldots,S_n$ be the equivalence classes "$\sim$" induces on $S$. Then G transitively permutes the set $\Omega = \{S_i,\ i = 1,\ldots,n\}$ under conjugation and if $\overline{G}$ is the image of G under this action, $G \to Sym(\Omega)$, then again $(\overline{G},\Omega)$ satisfy the hypotheses of Bender's theorem. We shall meet in the next section still further generalizations of this theorem.

The next series of results all extend the two theorems presented so far: The Bender-Hering theorem, and Bender's strongly embedded theorem.

We shall say $G$ is a $(\mu,k)$-group if $(G,\Omega)$ is k-fold transitive and $\mu$ is the maximal number of points fixed by an involution in $G$. For example, the theorem of Marshall Hall cited at the beginning of this section is a theorem on (4,4)-groups. A part of this theorem, for example, asserts there are no (4,0) or (4,1) groups. Note that the degree of a $(\mu,k)$-group has the same parity as $\mu$.

Now if the $(\mu,k)$-groups are known, transitive extensions of these become $(\mu+1,k+1)$-groups. Jennifer King [39] classified the (1,3)-groups, and these are transitive extensions of the groups in Bender's 1968 theorem on 2-transitive groups in which involutions fix no letters. The only ones which extend are of type RNS and are the triply transitive groups $SL(n,2^a)$ and their normal extensions. But in the same paper, and using work in an earlier paper, King [38,39] broke new ground by classifying the (3,2)-groups. These are transitive extensions of simply transitive groups in which some involution fixes exactly two letters -- a class not yet covered by any theorem cited so far. Doubly transitive groups of type PSL(3,2), $S_5$, $A_7$, and three exceptional groups, $A_7$ on 15 letters, PSL(2,11) on 11 letters, and $M_{11}$ on 11 letters, make their entrance here. In the same two papers, transitive extensions of the last class were investigated by King, yield the classification of (4,3)-groups: $M_{12}$, $A_8$ and $S_6$. This last result was finally generalized in 1971 and 1972 by R. Noda and F. Buekenhout, respectively [41,11]. Their combined work completed the classification of (4,2)-groups, Noda handling the case 8 does not divide the degree, Buekenhout the remaining case. The doubly transitive groups appearing here are of these types; RNS degrees 8 and 16, PSL(2,q), PSU(3,3), $A_8$, $M_{11}$ on 12 letters, and $M_{12}$. The groups here are transitive extensions of (3,1)-groups which up to this point were unclassified, but in a remarkable paper dating from the same period Buekenhout [11] classified these groups as well. Almost immediately, partial results began to push the frontier still further. For example P. Rowlinson [49,50] classified $(\mu,1)$-groups for $4 \leq \mu \leq 7$ under the assumption that $G$ contains one class of involutions. Two years later, Rowlinson [51] and Buekenhout and Rowlinson [13] obtained further strong results on (4,1)-groups, under the hypothesis that either (a) a 2-Sylow center of $G$ is semiregular on $\Omega$ or (b) $G$ has no normal subgroup of index 2 and some involution in a 2-Sylow center attains the maximal number of fixed points. The conclusion is that either the fixed point sets of a class of central involutions partition $\Omega$ as a system of imprimitivity, or if $S$ is a 2-Sylow subgroup of $G$, either $|S| \leq 2^8$ or $S$ has a cyclic subgroup of index at most 4. In view of the very powerful group-theoretic characterizations in terms of the structure of 2-Sylow subgroups of $G$ which had become available in the 1972-74 period, the second alternative in the Buekenhout-Rowlinson conclusion was fairly determinative. In a final paper of the series, (part III of [51] in 1976, still further characterization theorems for finite groups by centralizer of involution and 2-Sylow rank were sufficient to classify all primitive or simple (1,4)-groups without further hypotheses.

Efforts to extend the results on $(\mu,k)$-groups could only go so far, and the strain in pushing further in this direction was beginning to be felt. In addition, a large body of results characterizing doubly transitive groups independent of numerical invariants such as $\mu$ or $k$ had been marching forward since 1969 and had come to command center stage for a while.

## 6. OTHER CHARACTERIZATION THEOREMS

The main thrust of the theorems of this section is the characterization of doubly transitive groups on the basis of (1) the structure of the two-point stabilizer $G_{\alpha\beta}$ (or more generally, on the basis of the involution structure of $G_{\alpha\beta}$) or (2) the normal structure of $G_\alpha$.

We consider (1) first. From the previous section we see that a doubly transitive group for which $G_{\alpha\beta}$ has odd order is known--it is a $(\mu,k)$-group, where $\mu < 2$ and $k \geq 2$. It seems natural, then, to focus attention on the involutions of $G_{\alpha\beta}$.

We begin with a set of three theorems which stand somewhere between the theorems of the last section (concerning sizes of fixed point sets of involutions) and the main theorems on the global and 2-structure of $G_{\alpha\beta}$. All three theorems of these involve the relationship (equality or containment) between fixed point sets of involutions that are intimately related in some sense. My interest in recording these three theorems is that, because of their hypotheses, they stand alone, representing an avenue that was never really followed up to any extent.

Theorem 6.1 (Harada, 1968 [27]). Let $G$ be a doubly transitive group of permutations. Assume $G$ contains an involution $t$ such that all non-identity elements of $C_G(t)$, the centralizer in $G$ of $t$, have the same fixed point set as $t$. Then either $G$ is a doubly transitive Frobenius group of even degree, $G$ is type $PSL(2,q)$ or $G$ is type $Sz(q)$.

One can see that the condition that all elements of $C_G(t)^{\#}$ have the same fixed point set as $t$ serves to exclude the groups of type $PSU(3,2^a)$. But if that condition were weakened to assert that any 2-element of $C_G(t)$ or only that any element of $t^G$ $C_G(t)$ had the same fixed point set as t, then the $PSU(3,2^a)$'s would have to reappear in the conclusion. In Harada's proof, the case that the fixed point set F(t) of t is empty must be handled. But as t normalizes $G_{\alpha\beta}$ for some $\alpha$ and $\beta$, the latter subgroup must have odd order. Thus G is a (0,2)-group and results of the last section apply. When F(t) consists of a single letter, Harada shows G is a Zassenhaus group and the results of section 4 apply. When F(t) contains two or more points he shows G possesses one class of involutions and the structure of $C_G(t)$ is determined, allowing the characterization of G.

Theorem 6.2. (Aschbacher, 1972 [2]). Assume G is a doubly transitive group of degree $n \equiv 2 \bmod 4$. Assume G contains an involution t in $G_{\alpha\beta}$, and that the fixed point set of t contains the fixed point set F(s) of each involution s lying in some 2-Sylow subgroup S of $G_{\alpha\beta}$, containing t. Then G is type $A_6$, PSL(2,q) or $PSU(3,q^2)$.

A <u>triangular set of involutions in a group</u> G, is a set of involutions invariant under conjugation by elements from G, and such that the set contains the product of any two of its members which happen to commute. In many contexts it seems plausible to expect theorems to be relativized to some triangular set T; specifically one wishes to replace any reference to "all involutions in a 2-Sylow subgroup S" in the hypothesis, by "elements of $T \cap S$", with the assertions about G in the conclusion replaced by the corresponding assertion about $\langle T \rangle$, the subgroup generated by T. We shall meet later, just such a generalization of Bender's strongly embedded theorem by Aschbacher. The next theorem of this series, although unpublished, was proved in 1974, and to come extent marks such a generalization of theorem 6.2 above.

Theorem 6.3 ([57]). Let G be a doubly transitive group of even degree and let T be a G-invariant set of involutions such that $X = T \cap G_\alpha$ is a triangular set of involutions of $G_\alpha$. Assume that whenever two involutions of X lie in a common 2-Sylow subgroup of $G_{\alpha\beta}$, they have a common fixed point set. Then either (1) $X \subseteq Z^*(G_\alpha)$, (2) G is type RNS or (3) G is type $A_6$.

The conclusion (1) above means that each 2-Sylow subgroup of $G_\alpha$ contains exactly one member of X and by the Glauberman Z*-theorem cited earlier [22], X and $G_\alpha$ have the normal structure described in conclusion (1) of Bender's strongly embedded theorem, of the last chapter.

Comparing Theorems 6.2 and 6.3, we see that the phrase "each involution s lying in some 2-Sylow subgroup S of $G_{\alpha\beta}$ containing t" is being replaced with "each involution s in $X \cap S$"; but not for just one involution t as in Aschbacher's theorem. Rather the hypothesis on t is replaced by one on any element t of X. Also, 6.3 fails to generalize 6.2 in the respect that Aschbacher's requirement of containment of fixed point sets has been strengthened to equality. On the other hand, the condition that $n \equiv 2 \bmod 4$ has been unloaded.

Indeed, if one looks them over, none of these three theorems generalize any of the others; yet they stand together because of a strong similarity in the type of hypothesis involved: sibling theorems seeking a common parent. This raises the following question: In Theorem 6.2, fixing G 2-transitive of even degree, $T = T^G$, $X = T \cap G_\alpha$ a triangular set in $G_\alpha$, can the rest of the hypothesis be replaced with the assertion that there exists some involution t in X such that whenever t and s in X lie in a common 2-Sylow subgroup $G_{\alpha\beta}$, F(t) contains F(s)?

It should be remarked that the groups satisfying conclusion (1) of Theorem 6.3 were never classified by any of the characterization theorems of this section. But in the case that X contains all involutions of $G_\alpha$, we know that $G_{\alpha\beta}$ has generalized quaternion 2-Sylow subgroups and these are classified by the theorems concerning the 2-structure of $G_{\alpha\beta}$ to be considered next.

We conclude the discussion of these three theorems by mentioning a corrollary to 6.3, which marks another way to limit the fixed points sets of involutions in a characterization theorem, not by size, but by its geometric structure.

Corollary 6.4 [57]. Let $(\Omega, L)$ be a linear space (block design with $\lambda = 1$) with an even number of points, and a doubly transitive group of automorphisms G possessing a G-invariant set of involutions T such that $T \cap G_\alpha$ is a triangular set of involutions of $G_\alpha$. Assume no involution in X fixes a non-collinear triplet of points. Then either (1) $X \subseteq Z^*(G_\alpha)$, (2) $(\Omega, L)$ is an affine plane $V(2,2^a)$ over $GF(2^a)$ and $\langle T\rangle$ is the group of substitutions $v \to t(v) + a$, $a \in V(2,2^a)$ $t \in SL(2,2^a)$, or (3) $(\Omega, L)$ is the trivial design with $v = 6$, $k - 2$, $\langle T\rangle = A_6$ .

We turn now to theorems characterizing doubly transitive groups by the global or 2-structure of $G_{\alpha\beta}$ .

The reader will recall that in the first of Suzuki's two papers in the Annals of Mathematics [59,61], $G_{\alpha\beta}$ was cyclic. But in addition many other conditions were present, $G_{\alpha\beta}$ had odd order, and $G_\alpha$ contained a normal subgroup regular on $\Omega - (\alpha)$, etc. Could the 2-transitive groups be classified knowing only that $G_{\alpha\beta}$ was cyclic. This indeed was accomplished in 1970 by Kantor, O'Nan and Seitz [37]. Some comparison with Suzuki's original theorem [59] can be seen in terms of the groups which enter the conclusion. In Suzuki's theorem, they are type RNS, $SL(2,2^a)$, or

Sz(q). But in the Kantor-O'Nan-Seitz theorem, the proof must lead one also to groups of type PSL(2,q), PSU(3,$q^2$), or R(q). Until very recently the uniqueness of the groups of Ree type was not known. What one sees is a collection of conditions which are all equivalent; for example, (i) having a centralizer of involution t of the form $Z_2 \times PSL(2,q)$, $q > 5$, (ii) having a certain character table, or (iii) being a doubly transitive group of degree $1 + q^3$ with $G_{\alpha\beta} \simeq Z_{q-1}$ and other conditions. Any such group has q equal to an odd power of three and its 3-Sylow normalizer uniquely described up to a certain automorphism. Any such group is called a group of Ree type and until the very recent proof that such groups are twisted Chevalley groups, one had always to be content with the phrase "Ree type" in the conclusion of any characterization theorem on doubly transitive groups. The Kantor-O'Nan-Seitz theorem would also have been impossible without two rather deep theorems of Suzuki and O'Nan [62, 43] characterizing the groups PSU(3,q), q odd, from a set of equivalent hypotheses similar to (i)-(iii) above. Indeed without these theorems we may have had to go around saying "PSU(3,q) type" instead of "PSU(3,q)" all these years. The Kantor-O'Nan-Seitz theorem was not published until 1972. At almost the same time Aschbacher [3] classified all doubly transitive groups in which $G_{\alpha\beta}$ was abelian, the proof being independant of the Kantor-O'Nan-Seitz theorem.

These results entail knowledge of the entire global structure of $G_{\alpha\beta}$. Could similar results be obtain be relativizing these hypotheses to the 2-Sylow subgroup of $G_{\alpha\beta}$? Thus instead of asking that $G_{\alpha\beta}$ be cyclic, one requires only that a 2-Sylow subgroup S of $G_{\alpha\beta}$ be cyclic; instead of $G_{\alpha\beta}$ being abelian, only that S be abelian. Doubly transitive groups for which $G_{\alpha\beta}$ has a cyclic 2-Sylow subgroup were first classified in an unpublished Ph.D. Thesis by Karl Keppler in 1973, but this result was almost immediately subsumed by the following theorem of Aschbacher [6], which appeared in 1975:

Theorem 6.5. Let G be a doubly transitive group and suppose the 2-Sylow subgroup S of $G_{\alpha\beta}$ is either cyclic, dihedral or generalized quaternion. Then G is type RNS, PSL(2,q), Sz(q), PSU(3,q), R(q), $A_6$, $A_8$, PSL(2,11) on 11 letters, or $M_{11}$.

The proof of this theorem of Aschbacher is intricate, and, because of the unavailability of smaller doubly transitive sections of the group to which induction could be applied, actually uses characterizations of certain rank-3 permutation groups which appear because there are at most three classes of involutions in $G_{\alpha\beta}$. This theorem and Aschbacher's theorem with $G_{\alpha\beta}$ abelian, mark the furthest advances along the line of characterizing doubly transitive groups by the structure of $G_{\alpha\beta}$. As far as I know, no attempt to classify 2-transitive groups with $G_{\alpha\beta}$ having abelian

2-Sylow subgroups of arbitrary rank ever met with success.

We turn now to the final collection of characterization theorems, those concerning the normal structure of $G_\alpha$.

The first three of this series of very powerful theorems were discovered during the winter of 1969-70. We consider first O'Nan's very beautiful theorem [42]:

Theorem 6.6. Let G be a doubly transitive group. Suppose $G_\alpha$ contains a normal abelian subgroup A which does not act semi-regularly on $\Omega - (\alpha)$, that is A $G_{\alpha\beta} \neq 1$ for some letter $\beta$. Then G is type PSL(n,q) for some n and q.

In a subsequent paper [44], O'Nan found an elegant version of this theorem in the context of designs:

Theorem 6.7. (O'Nan [44]) Let $(\Omega,L)$ be a (finite) linear space admitting a doubly transitive group G of automorphisms. Suppose G contains an element g fixing letter $\alpha$ and stabilizing all lines of $L$ incident with $\alpha$. If g fixes an additional point distinct from $\alpha$, then $(\Omega,L)$ is a projective space over a finite field.

I think any geometer will immediately appreciate the great beauty of this theorem which so eloquently intertwines groups and geometries in such a modest hypothesis.

The next theorem is the celebrated Hering-Kantor-Seitz theorem [30]:

Theorem 6.8. Let G be a doubly transitive group of even degree. Suppose $G_\alpha$ contains a normal subgroup N which is regular on $\Omega - (\alpha)$. Then G is type RNS, PSL(2,q) PSU(3,q) or R(q).

At the same time, in a paper [54] which never eventually found its way to publication, one had

Theorem 6.9 ([54]). Let $(G,\Omega)$ be transitive, and let $1 \neq Q$ be a normal 2-subgroup of $G_\alpha$ semiregular on $\Omega - \alpha$. Then either (i) $N_1 < N < G$, $N/N_1 \cong Z_2$, where $N_1$ has odd order and $N - N_1$ contains the set I of all involutions of Q, or (ii), $\langle I^G \rangle$ is 2-transitive on $\Omega$ of type PSL(2,q), Sz(q) or PSU(3,q), q a power of 2.

A consequence of Theorem 6.9, proved in [55], gave an odd degree analogue of the Hering-Kantor-Seitz theorem:

Theorem 6.10. Let G be a doubly transitive group and suppose $G_\alpha$ contains a normal subgroup Q of even order which acts semi regularly on $\Omega - (\alpha)$. Then G is type RNS, PSL(2,q), Sz(q) or PSU(3,q) with $q = 2^a$.

Theorems 6.8 and 6.10 together comprise the classification of split (B,N)-pairs of rank 1, namely, all doubly transitive groups G for which $G_\alpha$ contains a normal subgroup N, regular on $\Omega - (\alpha)$. This result was needed in the classification by Fong and Seitz [21], of all split (BN)-pairs of rank 2. Since J. Tits had classified all (B,N)-pairs of rank $n \geq 3$, the use of (B,N)-pairs to identify finite groups was now universally useable methodology for finite group theory.

These three theorems, Theorems 6.6, 6.8 and 6.10 which came to light in 1969-70, (together with the Kantor-O'Nan-Seitz Theorem known by the spring of 1970) instilled the field of doubly transitive groups with a great wave of optimism. Extensions of these theorems began to take two paths, one centered most closely around theorems 6.6 and 6.8 concerned the absolute normal structure of $G_\alpha$ , the latter extending theorems 6.9 and 6.10, still concerned involutions in certain ways.

First suppose G is a 2-transitive group. What happens if $G_\alpha$ contains a normal abelian subgroup $Q \neq 1$? We know the answer if Q is not semiregular on $\Omega - (\alpha)$ from 6.6. So assume we Q is semiregular. If Q has even order, we again know G from Theorem 6.10. So we may assume Q has odd order. Again, if Q is actually regular on $\Omega - (\alpha)$, G is known Theorem 6.8. Thus we are left with a group Q of odd order, semiregular but not regular on $\Omega - (\alpha)$. That such a group G should still possess a split (B,N)-pair of rank 1 has come to be known as the Hering conjecture (see Aschbacher's $\mathcal{F}$-sets paper [5]). In fact, in that paper, Aschbacher shows that if $(G,\Omega)$ is a least counterexample to the Hering conjecture, then also $C_G(Q)$ is semi-regular on $\Omega - (\alpha)$. In such a group, we may thus assume $C_G(Q)$ also has odd order.

The Hering conjecture was never proved and indeed manifested itself as one of the most frustrating problems in doubly transitive groups from 1970 to 1976. In its full generality it had to await the complete classification of the finite simple groups. It is clear, though, that if choose an involution t in $G_\alpha$ so $C_Q(t)$ has largest possible rank as an abelian group, then $C_G(t)$ possesses a standard component X -- but unfortunately, of an unknown isomorphism type. The Hering conjecture was solved, however, for the case that G had odd degree, in a magnificent series of theorems of O'Nan [46]. First he proved the following.

Theorem 6.11. Let G be a transitive group of permutations of odd degree. Let P be a normal p-subgroup of $G_\alpha$, semiregular on $\Omega - (\alpha)$, $P \neq 1$. Then either P is cyclic or G is $M_{11}$ acting on 55 letters.

The Hering conjecture for odd degree follows readily, for one replaces Q with a p-Sylow subgroup P, and using the $\mathcal{F}$-sets result of Aschbacher, obtains $C_G(P) \cap G_{\alpha\beta} = 1$ for all $\beta \in \Omega - (\alpha)$. Then $G_{\alpha\beta}$ acts faithfully on P. By 6.11, P is cyclic, so $G_{\alpha\beta}$ is abelian, and so the result follows from Aschbacher's $G_{\alpha\beta}$-abelian theorem [3], and the fact the degree is odd.

Considering the way the Hering conjecture entered the analysis of Q given above, we have

Theorem 6.12 (O'Nan, [46]) Suppose G is a doubly transitive group of odd degree. If $G_\alpha$ contains a non-trivial normal abelian subgroup Q, then G is type RNS, PSL(n,q), Sz(q) or PSU(3,$2^a$).

This means that if G has odd degree and is an "unknown" 2-transitive group, then any minimal normal subgroup N of $G_\alpha$ must be the direct product of simple groups. O'Nan considered later, in 1974, the general case that Soc($G_\alpha$), the join of all minimal normal subgroups of $G_\alpha$, was semisimple. If $N_1$ was a proper subgroup of Soc($G_\alpha$) normal in $G_\alpha$, so $N_1 \times N_2 = \mathrm{Soc}(G_\alpha)$, then either $N_2$ is semiregular (in which case Theorem 6.10 applies) or $N_2$ is not semiregular, and a block design satisfying the hypothesis of Theorem 6.7 can be introduced. In either case G is known, so one can assume Soc($G_\alpha$) is itself minimal normal in $G_\alpha$. In two very deep papers in the transactions of the A.M.S., O'Nan goes on to prove that Soc($G_\alpha$) contains exactly one simple component [44,45]. Thus if G is a heretofore unknown doubly transitive group, then either G is an even-degree candidate for a counter-example of the Hering conjecture or else $G_\alpha$ is in the automorphism group of a simple group X, and contains X. This is about as far as the absolute normal structure of $G_\alpha$ came to be exploited, and very little progress in this direction can be recorded after 1975, except possibly to report the very nice result of D. Holt [32], classifying all doubly transitive groups for which $G_\alpha$ is a solvable group.

We consider next several theorems involving both the normal structure of $G_\alpha$ and the presence of involutions in certain normal subgroups of $G_\alpha$. These theorems began as extensions of Theorem 6.10, but soon ran far beyond the immediate context of doubly transitive groups to yield theorems which were of methodological importance in finite group theory itself. Recall that in Theorem 6.10 G is a 2-transitive group on $\Omega$, and that $G_\alpha$ contains a normal subgroup N of even order, semi-regular on $\Omega - (\alpha)$. The first generalization was due to Hering [29]. One could replace the phrase "N has even order, semi-regular on $\Omega - (\alpha)$" with the hypothesis "N has even order, and a 2-Sylow subgroup of N is semi-regular on $\Omega - (\alpha)$." Now in this entire chain of theorems, beginning with Suzuki's ZT-theorem [59], his two generalizations of this paper in the Annals of Mathematics [60, 61], Bender's strongly embedded theorem [9], Theorems 6.9 and 6.10, and the generalization of Hering just mentioned -- we have an ever expanding characterization of strongly embedded subgroups. Finally, in a somewhat technical theorem appearing in his paper on groups with a proper 2-generated core, Aschbacher essentially gave a characterization of strongly embedded subgroups by any property $P$ with the right subgroup inheritance properties (Theorem 2, of [4]). Suppose $P$ is a group-theoretic property involving a triple $(z,H,G)$ where z is an involution lying in a subgroup H properly contained in G. We require $P$ to contain the assertions that $C_G(z) \leq H$ and if $g^{-1}zg$ lies in H, then g lies in H. The last two assertions imply that H is self-normalizing of odd index in G; moreover if $z \in L \leq G$, then the two assertions required of $P$ hold with G replaced by L and H replaced by $H \cap L$. We wish to prove that $P$ implies $H \cap \langle z^G \rangle$ is strongly embedded in $\langle z^G \rangle$, and for this purpose we assume we have reduced to the case that H contains no non-trivial normal subgroup of G; that is G acts faithfully on $\Omega = H^G$. Now if $C_G(s) \leq H$ for every involution in $\langle z^G \rangle \cap H$ we are done. Thus we have involutions s in H such that $L = G_G(s) \not\leq H$. We then assume $P$ has the inductive property that it inherits to the triple $(z, L \cap H, L)$. This is possible since $L \cap H$ is proper in L and $C_L(z) \leq L \cap H$ and $z^g \in L \cap H$ implies $g \in L \cap H$ for each element $g \in L$. Since $L \neq G$, one may imply induction to obtain that $\langle z^L \rangle \cap H$ is strongly embedded in L. But this is precisely the hypothesis of Aschbacher's theorem 2 of [4]: namely: $H < G$, G acts faithfully on $H^G$, z is an involution in H such that $C_G(z) \leq H$ and $z^g \in H$ implies $g \in H$, and for each involution s in H for which $L = C_G(s) \not\leq H$ we have $\langle z^L \rangle \cap H$ strongly embedded in $\langle L \rangle$. His conclusion: $\langle z^G \rangle \cap H$ is strongly embedded in $\langle z^G \rangle$. Thus Aschbacher's "Theorem 2" contains within it, infinitely many characterizations of strongly embedded subgroups. One of the most important ones is the following.

Theorem 6.13 (Aschbacher). Let T be a triangular set of involutions in a group G. Convert T into a graph $T$ with vertex set T and edge set all commuting pairs of involutions in T. Suppose $T$ is not connected, and let H be the stabilizer in G of a connected component $T$. Then $H \cap \langle T \rangle$ is strongly embedded in $\langle T \rangle$.

A nearly immediate consequence is

Corollary 6.14 ([57]). Let T be a triangular set of involutions in a transitive group of permutations G. If each member of T fixes a single letter, then $G_\alpha \cap \langle T \rangle$ is strongly embedded in $\langle T \rangle$.

Looking back from the vantage point of this corollary, one sees how easy it is to read off "proofs" of the previous theorems. Consider for example Hering's generalization of Theorem 6.10. Here G is 2-transitive on $\Omega$, and $G_\alpha$ contains a normal subgroup N of even order, whose 2-Sylow subgroup S is semi-regular on $\Omega - (\alpha)$. Set I = all involutions in N and T the set of all G-conjugates of elements of I. Each element of T then fixes exactly one letter, from the semi-regularity of S. If two elements s and t of T commute, they fix a common letter, say $\alpha$. Then $t = x^g$ for some involution x in N. Since $F(x^g) = F(x) = \{\alpha\}$ we see $g \in G_\alpha$. Since $N \triangleleft G_\alpha$, $t = x^g \in N$. Similarly $s \in N$ and so st is an involution in N and hence an element of T. Thus T is a triangular set and so Corollary 6.14 applies, to yield the fact that G is type RNS $PSL(2,2^a)$, $Sz(q)$ or $PSU(3,2^a)$.

Of course, this sort of "reproof" of earlier theorems is somewhat illusory, for all of the more recent generalizations -- such as Aschbacher's key "Theorem 2" -- depend at least on Bender's strongly embedded theorem.

An even further generalization of Corollary 6.14 was introduced in Aschbacher's "Standard component" paper [7]. What is being relaxed here, is the requirement that T be a triangular set. If we start with a single involution z and ask for the triangular set which it generates, we must, form the conjugacy class $z^G$, and as a first stage, throw in products of commuting members of $z^G$, as a second stage throw in products of commuting members of the first stage, and so on, until iteration of this procedure of throwing in products of commuting predecessors terminates. If each member of the resultant triangular set fixes exactly one letter, we would be in the situation of Corollary 6.14. Aschbacher relaxes this requirement, insisting that it apply to $z^G$ and its offspring appearing in the first stage only. More precisely

Theorem 6.15. (Aschbacher [7] Let G act on $\Omega$ and let z be an involution fixing exactly one letter. Suppose that whenever $z^g$ is a conjugate of z, distinct from, but commuting with z, then $zz^g$ also fixes exactly one letter. Then $\langle z^G \rangle = \langle u \rangle \times L$ where $L \cap G_\alpha$ is strongly embedded in L, the entire class $z^G$ lies in the coset uL and u has order 1 or 2.

It can be shown that the hypotheses of Theorem 6.15 actually imply that z lies in a 2-Sylow center of G. Aschbacher, in a private conversation with the author as early as 1975, conjectured that if the involution z is 2-central and fixes a unique letter, that $\langle z^G \rangle$ should act as a product of the natural permutation representations of groups acting on cosets of strongly embedded subgroups and alternating or symmetric groups on n letters.

Three years later, such a result was proved, independently by Derek Holt [33] and Fred Smith [58]. The theorem completely generalizes all of the theorems of the last portion of this section, and simply classifies all permutation groups having an involution in the center of some 2-Sylow subgroup which fixes only one letter. For convenience, however, we state this theorem for primitive groups.

Theorem 6.16 (D Holt [33], F. Smith [58]). Let G be a primitive group of permutations on a set $\Omega$. Suppose z is an involution lying in the center of some 2-Sylow subgroup of G and that z fixes just one letter. Then either (1) z lies in $Z^*(G)$ and G contains a regular normal elementary p-subgroup, or (2) G is 2-transitive of type $SL(2,2^a)$, $Sz(q)$ or $PSU(3,2^a)$, or type $A_n$.

There is an immediate corollary of this theorem which figured in Professor Luneberg's lecture;

Corollary 6.17 (Holt-Smith). Let G be a doubly transitive group of degree $n = 1 + 2^k$. Then either G is type RNS, $SL(2,2^a)$ $Sz(q)$, $PSU(3,2^a)$ or $A_n$ .

The theorem of Holt and Smith is a high water mark for theorems involving fixed points sets of involutions and one can see that the normal structure of $G_\alpha$, which appeared in the theorems (for example 6.9 and 6.10) spawning this line of attack, has disappeared altogether.

But the theorem of Holt and Smith was one of the very few structural theorems on doubly transitive groups to make its appearance after 1975. By now the stalemate had set in. The wave of optimism concerning a possible classification of the doubly transitive groups came in a flood of theorems in the years 1969-1972; O'Nan's characterization of PSL(n,q), the rank 1 split (BN)-pair classification, the classification of doubly transitive groups with $G_{\alpha\beta}$ abelian -- all fell in this period. The intractibility of the Hering conjecture signalled a period with a few hard-won results, 1975 to present. The classification of the doubly transitive groups had finally to await the complete classification of all finite simple groups. But the latter is a "proof" pieced together from over 8,000 published pages and perhaps 10,000 pages of preprints scattered everywhere. The old geometers were right. One does know a great deal about a group, knowing only, first, that it is doubly transitive. They were right, but one can't help wishing it was for a better reason.

## REFERENCES

[1] J. L. Alperin, R. Brauer and D. Gorenstein, Finite groups with quasi-dihedral and wreathed Sylow 2-subgroups, Trans. Amer. Math. Soc. 151 (1970), 1-261.

[2] M. Aschbacher, On doubly transitive groups of degree $n \equiv 2$ mod 4, Illinois J. Math. 16 (1972), 87-112.

[3] M. Aschbacher, Doubly transitive groups in which the stabilizer of two points is abelian, J. Alg. 18 (1971), 114-136.

[4] M. Aschbacher, Finite groups with a proper 2-generated core, Trans. Amer. Math. Soc. 197 (1974), 87-112.

[5] M. Aschbacher, F-sets and permutation groups, J. Algebra 30 (1974), 400-416.

[6] M. Aschbacher, 2-transitive groups whose 2-point stabilizer has 2-rank 1, J. Alg. 36 (1975), 98-127.

[7] M. Aschbacher, Finite groups of component type, Illinois J. Math. 19 (1975), 87-115.

[8] H. Bender, Endliche zweifach transitive permutationsgruppen deren Involutionen Keine Fixpunkte haben. Math. Zeit. 104 (1968), 175-204.

[9] H. Bender, Transitive Gruppen gerader ordnung, in denen jede Involutionen genau einen Punkt festlasst, J. Alg. 17 (1971), 527-554.

[10] R. Bose, Strongly regular graphs, partial geometries and partially balanced designs, Pacific J. Math. 13 (1963), 389-419.

[11] F. Buekenhout, Transitive groups in which involutions fix 1 or 3 points, J. Algebra 23 (1972), 438-451.

[12] F. Buekenhout, Doubly transitive groups in which the maximal number of fixed points of involutions is four, Archiv. d. Math. XXIII (1972), 362-369.

[13] F. Buekenhout and P. Rowlinson, On (1,4)-groups I, II, III, J. London Math. Soc., 8 (1974), 493-498, 507-513, 14 (1976), 487-495.

[14] W. Burnside, Theory of groups of finite order, 2nd ed., Cambridge.

[15] P. Cameron and W. Kantor, 2-transitive and anti-flag transitive collineation groups of finite projective spaces, J. Alg. 60 (1979), 384-422.

[16] C. Curtis, W. Kantor and G. Seitz, The 2-transitive permutation representations of the finite Chevalley groups. Trans. Amer. Math. Soc. 218 (1976), 1-59.

[17] E. Dade, Lifting group characters, Annals. Math. 79 (1964), 590-596.

[18] S. Dixmier and F. Zora Essai d'une methode d'etude de certains graphes lies aux groupes classiques, Compte R. Acad. Sc. Paris 282 (1976), 259-262.

[19] W. Feit, On a class of doubly transitive permutation groups, Illinois J. Math. 4 (1960), 170-186.

[20] W. Feit and J. Thompson, Solvability of groups of odd order, Pacific J. Math. 13 (1963), 775-1029.

[21] P. Fong and G. M. Seitz, Groups with a (B,N)-pair of rank 2, I, II, Invent. Math. 21 (1973), 1-57; 24 (1974), 191-239.

[22] G. Glauberman, Central elements in core-free groups, J. Algebra 4 (1966), 403-420.

[23] G. Glauberman, On a class of doubly transitive groups, Illinois J. Math 13 (1969), 394-399.

[24] D. Gluck, A characterization of generalized permutation characters, J. of Alg. 63 (1980), 541-547.

[25] M. Hall Jr., On a theorem of Jordan, Pacific J. Math. 4 (1954), 219-226.

[26] M. Hall Jr., Affine generalized quadrangles, Studies in Pure Mathematics, ed. by L. Mirsky, Academic Press, 1971.

[27] K. Harada, A characterization of the Zassenhaus groups, Nagoya Math. J. 33 (1968), 117-127.

[28] C. Hering, Zweifach transitive permutationsgruppen, in denen zwei maximale Anzahl von Fixpunkten von Involutionen ist, Math. Zeit. 104 (1968), 150-174.

[29] C. Hering, On subgroups with trivial normalizer intersection, J. Alg. 20 (1972), 622-629.

[30] C. Hering, W. Kantor and G. Seitz, Finite groups with a

split (B,N)-pair of rank 1, J. Algebra 20 (1972), 435-475.

[31] G. Higman, Suzuki 2-groups, Illinois J. Math. 7 (1963), 79-96.

[32] D. Holt, Doubly transitive groups with a solvable one-point stabilizer. J. Algebra 44 (1977), 29-92.

[33] D. Holt, Transitive permutation groups in which an involution central in a Sylow 2-subgroup fixes a unique point. To appear in J. London Math. Soc.

[34] N. Ito, On a class of doubly but not triply transitive groups, Arch. Math. (Basel) 18 (1967), 564-570.

[35] W. Kantor, Linear groups containing a Singer cycle, J. Algebra 62 (1980), 232-233.

[36] W. Kantor, Generalized quandrangles associated with $G_2(q)$. To appear.

[37] W. Kantor, M. O'Nan and G. Seitz, 2-transitive groups in which the stabilizer of two points is cyclic, J. Algebra 21 (1972), 17-50.

[38] J. King, A characterization of some doubly transitive groups, Math. Zeit. 107 (1968), 43-48.

[39] J. King, Doubly transitive groups in which involutions fix 1 or 3 points, Math. Zeit. 111 (1969), 311-321.

[40] H. Nagao, On multiply transitive groups V. J. Algebra 9 (1968), 240-248.

[41] R. Noda, Doubly transitive groups in which the maximal number of fixed points of involutions is four, Osaka Math. J. 8 (1971), 77-90.

[42] M. O'Nan, A characterization of $L_n(q)$ as a permutation group, Math. Z. 127 (1972), 301-314.

[43] M. O'Nan, A characterization of $U_3(q)$, J. Algebra 22 (1972), 254-296.

[44] M. O'Nan, The normal structure of the point-stabilizer of a doubly transitive permutation group I, Trans. Amer. Math. Soc. 214 (1975), 1-42.

[45] M. O'Nan, The normal structure of the point stabilizer of a doubly transitive permutation group II, Trans. Amer. Math.

Soc. 214 (1975), 43-74.

[46] M. O'Nan, Doubly transitive groups of odd degree whose one point stabilizers are local, J. Algebra 39 (1976), 440-482.

[47] T. Ostrom and A. Wagner, On projective and affine planes with transitive collineation groups, Math. Zeit. 71 (1959), 186-199.

[48] S. Payne, Quadrangles of order (s-1,s+1), J. Algebra 22 (1972), 97-119.

[49] P. Rowlinson, Simple permutation groups in which an involution fixes a small number of points, J. London Math. Soc. (2) 4 (1972), 655-661.

[50] P. Rowlinson, Simple permutation groups in which an involution fixes a small number of points II. Proc. London Math. Soc. (3), 26 (1973), 463-484.

[51] P. Rowlinson, On (1,6)-groups, J. London Math. Soc. (2) 14 (1976), 481-486.

[52] J. Seidel, A survey of 2-graphs in Teorie Combinatore, Acad. Naz. D. Lincei. Rome 1976, 481-511.

[53] J. Seidel and D. Taylor, A second survey of 2-graphs. Preprint.

[54] E. Shult, On the fusion of an involution in its centralizer. Unpublished.

[55] E. Shult, On a class of doubly transitive groups, Illinois J. Math. 16 (1972), 434-445.

[56] E. Shult, The graph extension theorem, Proc. Amer. Math. Soc. 33 (1972), 278-284.

[57] E. Shult, Certain doubly transitive groups of even degree, (1974). Unpublished.

[58] F. Smith, On transitive permutation groups in which a 2-central involution fixes a unique point, Comm. Algebra 7 (1979), 203-218.

[59] M. Suzuki, On a class of doubly transitive groups, Annals Math. 75 (1962), 105-145.

[60] M. Suzuki, On generalized ZT groups Arch. Math. 13 (1962), 199-202.

[61] M. Suzuki, On a class of doubly transitive groups II, Annals Math. 79 (1964), 514-589.

[62] M. Suzuki, A characterization of the 3-dimensional projective unitary group over a finite field of odd characteristic, J. Algebra 2 (1965), 1-14.

[63] J. Thompson, Finite groups with fixed point free automorphisms of prime order, Proc. Nat. Acad. Sci. 45 (1959), 578-581.

[64] J. Tits, Ovoides à translations, Rendic. Mat. 21 (1962), 37-59.

[65] A. Wagner, On collineation groups of finite projective spaces, I. Math. Z. 76 (1966), 62-69.

[66] R. Weiss, Groups with a (B,N)-pair and locally transitive graphs, Nagoya Math. J. 74 (1979), 1-21.

[67] R. Weiss, Elations of graphs, Acta Math. Acad. Sci. Hungar. 34 (1979), 101-103.

[68] H. Wielandt, Uber die Existenz von Normalteilern in endlichen gruppen. Math. Nachr. 18 (1958), 274-280.

[69] H. Zassenhaus, Kennzeichnung endlicher linearer Gruppen als Permutationsgruppen, Hamb. Abh., 11 (1936), 17-40.

# PROJECTIVITIES AND THE TOPOLOGY OF LINES

Helmut SALZMANN

University of Tübingen

## §1. TOPOLOGICAL PROJECTIVE PLANES

A topological projective plane is a projective plane $\mathcal{P} = (P, \mathcal{L})$ such that $P$ and $\mathcal{L}$ are topological spaces (neither discrete nor indiscrete) and join and intersection are continuous: for any open set $U$ the set of all pairs of distinct lines intersecting in a point of $U$ is open in $\mathcal{L}^2$ and dually. Considering a line $L$ as the set of its points with the induced topology and denoting by $\mathcal{L}_p$ the pencil of lines through $p \notin L$, the perspectivity

$$\pi_{L,p} = (x \mapsto p \vee x) : L \longrightarrow \mathcal{L}_p$$

is obviously a homeomorphism with inverse $\pi_{p,L}$. Consequently, the projectivities of $L$ form a (triply transitive) group $\Pi_L$ of homeomorphisms of $L$ (the von Staudt group). This simple observation, which will



*P. Plaumann and K. Strambach (eds.), Geometry – von Staudt's Point of View, 313–337.*

be referred to as homogeneity of $L$, is at the base of almost all results concerning the topological consequences of the fact that a topological space is a line or the total point set of a topological projective plane. The continuity of the projectivities does not determine the topology of the line, however, as can be seen from the cofinite topology. We begin with a few easy remarks which hold without any additional assumptions; later we shall concentrate on planes with a locally compact and not totally disconnected point space $P$.

In the following, $L$ will always denote a line of a topological projective plane $\mathcal{P} = (P, \mathcal{L})$. The fact that $S$ is open or closed in the topological space $X$ will be indicated by $S \mathrel{\subset\!\!\!\!\circ} X$ or $S \mathrel{\overline{\subset}} X$ respectively, homeomorphism by $\approx$.

(1.1) Any two lines are homeomorphic.

(1.2) Each point is closed in $P$.

Proof. By assumption there is $U \mathrel{\subset\!\!\!\!\circ} P$ with $p \in U$ and $q \notin U$ for some points $p, q$. Homogeneity of $L = p \cup q$ implies that each line is a $T_1$-space. □

(1.3) For $p \notin L$ the projection

$$\rho = (x \longmapsto (p \cup x) \cap L) : P \setminus p \longrightarrow L$$

is continuous.

(1.4) Any line $H$ is closed in $P$.

Proof. For $U = L\setminus x$ and $H = p\cup x$ the continuity of $\rho$ shows $P\setminus H = U^{\overleftarrow{\rho}} \subset P\setminus p \subset P$. □

For $W \in \mathcal{L}$ the affine plane

$\mathfrak{P}^W = (P\setminus W, \mathcal{L}^W = \{L\setminus W ; L \in \mathcal{L}\setminus\{W\}\})$

is a topological geometry in the obvious sense. It is unknown whether in general any topological affine plane can be embedded into a topological projective plane.

(1.5) The point space of an affine plane is homeomorphic to the product of two affine lines.

Proof. Letting $W = u\cup v$, we have

$$\cap : \mathcal{L}_u^W \times \mathcal{L}_v^W \approx P\setminus W . \ \square$$

In the usual coordinatization of $\mathfrak{P}^W$ the coordinate set $K$ can be identified with an affine line and hence is itself a $T_1$-space. The definition of the ternary operation $\tau$ shows immediately

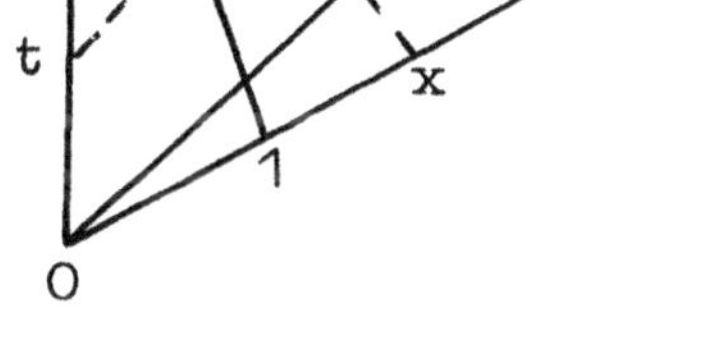


(1.6) $\tau : K^3 \longrightarrow K$ is continuous, and $K^+$ is a topological loop.

Here as customary, addition is given by

$x + t = \tau(1, x, t)$.

Continuity of the addition and its inverses in one variable is a special case of the homogeneity of lines. Instead of $x+t=y$ we write also $x=y-t$. We can now show

(1.7) $P$ and $\mathcal{L}$ are regular topological spaces ($T_3$).

Proof. Let $U \subset K$ be any neighbourhood of the neutral element $0$ of $K^+$. By continuity of "$-$" there is some $V$ such that $0 \in V \subset K$ and $V-V \subseteq U$. If $x \in \overline{V}$, then $(x+V) \cap V \neq \emptyset$ (since $x+V$ is a neighbourhood of $x$), and $x \in V-V$. Because of the homogeneity, $K$ is regular. (1.5) implies that each affine plane $P \setminus L$ is regular, and with (1.4) the regularity of $P$ follows. The second assertion can be obtained dually. □

No example has been found of a plane which is not even completely regular.

Since $K$ has a doubly transitive group of homeomorphisms, the connected component of $x$ in $K$ is either all of $K$ or consists of $x$ alone:

(1.8) An affine line is either connected or totally disconnected.

If $K$ is connected, so is $P$, because any pair of points is contained in a connected subset. On the other hand, if $K = L \setminus a$ is not connected, by homogeneity $\bigcap \{C \mathrel{\overset{\circ}{\subset}} K;\ c \in C\} = \{c\}$, and $c$ has a neighbourhood $A \subset\!\!\!\subset L$ with boundary $\partial A \subseteq \{a\}$ and $a \neq b \notin A$ for some $b$. Similarly, $\partial B \subseteq \{b\}$ and $a \notin B$ for some $B \subset\!\!\!\subset L$ with $c \in B$. Obviously, $A \cap B \overline{\subset\!\!\!\subset} L$, and homogeneity of $L$ implies that $L$ is totally disconnected. Using inverse images for projections from different centres, enough open and closed subsets of $P$ can be found in order to show that also $P$ is totally disconnected:

(1.9) $P$ and all its affine or projective lines are simultaneously connected or totally disconnected.

The arguesian plane $\mathfrak{P}_K$ over a topological (in particular, an ordered) field $K$ is easily seen to be a topological projective plane in the unique topology which induces the given one on $K$ and which is described in (1.5). Each field of cardinality $\mathfrak{m}$ has $2^{2^{\mathfrak{m}}}$ different field topologies (Kiltinen). Hence examples of topological planes abound, and not too much can be said about topological planes in general. In marked contrast, the locally compact commutative fields can be determined completely (Weil): the disconnected ones are the finite extensions of the p-adic fields $\mathbb{Q}_p$ and the fields $\mathbb{F}_q((t))$ of power series over a Galois field; the only connected ones are $\mathbb{R}$ and $\mathbb{C}$.

## § 2. LOCALLY COMPACT PROJECTIVE PLANES

In view of the remarks at the end of § 1, it seems reasonable to restrict attention to locally compact planes. In fact, only locally compact connected planes have been investigated extensively. Again, let $\mathcal{P} = (P, \mathcal{L})$ always denote a topological projective plane over a ternary field $K_\tau = K = L \setminus \infty$. From § 1 follows immediately

(2.1) P (or $\mathcal{L}$) is locally compact if and only if K is.

An important consequence of local compactness is the existence of a countable basis for the open sets.

(2.2) A locally compact projective plane has a separable metric point space P.

The proof will be accomplished in several steps. According to (1.6) the product $s \circ x = \tau(s, x, 0)$ is continuous on $K^2$, in particular at 0. Writing $s = y/x$ for $s \circ x = y$, the right inversion $x \mapsto 1/x$ is a projectivity in $\Pi_L$ which interchanges 0 and $\infty$.

(a) Each element of K has a countable neighbourhood basis, and K is $\sigma$-compact.

Since the topology is not discrete by assumption,

there is an infinite sequence $a_n$ in a compact neighbourhood accumulating at some point $a$, and by homogeneity we may assume $a = 0$. If $V$ is a compact neighbourhood of $0$, the $V_n = a_n \circ V$ form a neighbourhood basis at $0$, and the compact sets $(1/a_n) \circ V$ cover $K$. □

(b) $K$ is metrizable.

More generally, by a criterion of Stone, any topological loop $K^+$ with a countable basis of neighbourhoods $V_n$ of $0$ is metrizable: letting $\{z+V_n\}_{z \in K} = \mathfrak{w}_n$, it suffices to show that the stars

$$(x + V_n) * \mathfrak{w}_n = ((x + V_n) - V_n) + V_n$$

form a neighbourhood basis at $x$, but the latter follows immediately from the continuity of the loop operations. □

Noting that compact metric spaces have a countable basis, (a) and (b) imply

(c) $K$ has a countable basis.

By (1.5) and the fact that $P$ can be covered by 3 affine subplanes, the proof of (2.2) is now complete.

(2.3) $P$ is compact if and only if $L$ is compact.

Proof. If $L$ is compact, so is each pencil $\mathcal{L}_p$. Any infinite discrete pointset contains a sequence $x_n$ such that $p_i \cup x_n \xrightarrow[n]{} L_i$ for finitely many $i$. Choose $p_2 \notin L_1$. Then $x_n \xrightarrow[n]{} L_1 \cap L_2$. □

Presumably, all (non-discrete) locally compact projective planes are even compact, but this has been proved only if P is connected or multiplication is bi-associative or the topology is given by an ordering of K.

(2.4) If P is locally compact and connected, then P is compact.

Proof. Choose $p \notin L$ and a small compact neighbourhood V of p in P. Each line $X \in \mathcal{L}_p$ is connected and hence intersects the compact boundary $\partial V$. By (1.3) the projection $\rho$ from p maps $\partial V$ continuously onto L, so that L is compact. □

## § 3. LOCALLY COMPACT CONNECTED PLANES

By the preceding sections, a topological projective plane has a compact connected pointset if it is locally compact and not totally disconnected.

(3.0) Lemma. In a compact totally disconnected space $\dot{X}$ each point has arbitrarily small compact open neighbourhoods.

Proof. Consider the intersection D of all $S \mathrel{\overline{\circ\!\!\subset}} X$ containing the point a, and let $D \subseteq V \mathrel{\circ\!\!\subset} X$. Then D is connected and $D \subseteq S \mathrel{\circ\!\!\subset} V$ for some compact S:

if $a \in A \subset\!\!\!\subset D$, by normality there is some $U \subset\!\!\!\subset V$ with $A \subseteq U$ and $\overline{U} \cap (D \setminus A) = \emptyset$. Because $X$ is compact and $D \cap \partial U = \emptyset$, there exists $S = S_1 \cap \ldots \cap S_n \subset\!\!\!\subset X$ such that $S \cap \partial U = \emptyset$. But then $S \cap U \subset\!\!\!\subset X$ and $D \subseteq S \cap U$. □

(3.1) The lines of a compact connected plane are locally connected.

Proof. By (3.0) a compact neighbourhood $V$ in a connected ternary field $K = L \setminus \infty$ can not be totally disconnected, and there exists a compact connected subset $C$ of $K$ containing $0$ and $1$. Let $0 \in U \subset\!\!\!\subset K$. Then

$$C \circ U = \bigcup_{c \in C \setminus 0} c \circ U = \bigcup_{u \in U \setminus 0} C \circ u$$

is an arbitrarily small open connected neighbourhood of $0$. □

By the Hahn-Mazurkiewicz theorem (Kuratowski §45 II; Whyburn II 4.1) follows

(3.2) Each line is locally and globally arcwise connected.

In particular, in any locally compact connected ternary field $K$ there is an arc $(t \mapsto e_t) : [0,1] \to K$ joining the elements $0$ and $1$ in $K$, and homogeneity and the homotopy $(t,x) \mapsto e_t \circ x$ shows

(3.3) An affine line is contractible, a projective line is locally contractible.

The latter means that each neighbourhood $V$ of $x$

contains a neighbourhood U of x such that the inclusion $U \hookrightarrow V$ and the constant map $U \to x$ are homotopic within V. Finite dimensional spaces with this condition share important properties with topological manifolds. By definition, $\dim X \leq n$ if every finite open cover of X has a finite open refinement in which at most $n+1$ sets intersect, or equivalently, for normal X, if each continuous $\varphi: A \longrightarrow \mathbb{S}^n$ with $A \subseteq X$ has a continuous extension $\psi: X \longrightarrow \mathbb{S}^n$ into the n-sphere (Pears, Hurewicz-Wallman, Engelking). It seems reasonable to conjecture that all compact projective planes are of finite topological dimension. Examples are known only for $\dim L = 2^m \leq 8$. These things will be discussed in §6.

From lemma (3.0) follows easily that a compact totally disconnected space is 0-dimensional:

(3.4) A compact projective plane is connected if and only if $\dim L > 0$.

Letting $\dim L = \ell > 0$, one can conclude from general results of dimension theory (in particular, Nagami 41-5) and (1.5) that

(3.5) $2\ell - 1 \leq \dim P \leq 2\ell$.

Actually, only the even case is possible, but the proof of this fact uses homology and will be given at the end of this paper. For $\ell = 1$, however, P contains a subset homeomorphic to $[0,1]^2$ by (3.2) and (1.5), and $\dim P = 2$. Planes with this property have been termed *flat*. They will be treated in a separate section because of the special features of low dimensional topology.

The following holds in topological planes in general, but is more useful in the locally compact connected case.

(3.6) The projection $\rho : P \setminus p \longrightarrow L$ is a locally trivial fibering.

Proof. A slicing map (Dugundji XX 4) for $U = L \setminus q$ is given by $(x,u) \mapsto (q \vee x) \wedge (p \vee u)$. □

## §4. FLAT PROJECTIVE PLANES

It will be shown that the point space P of a flat plane is in fact a surface homeomorphic to the real projective plane.

(4.1) A line of a flat projective plane is homeomorphic to a circle $\mathbb{S}$.

Proof. $\dim P = 2$ implies $\dim L = 1$ either by (3.5) or the more elementary fact (Morita) that

$$\dim X \times [0,1] = \dim X + 1 .$$

Now an affine line $K = L \setminus q$ cannot contain a Jordan curve: because $\dim K = 1$, for any embedding $\sigma : \mathbb{S} \hookrightarrow K$ there would exist a retraction $\rho : K \longrightarrow \mathbb{S}$ with $\sigma\rho = 1_{\mathbb{S}} = \varepsilon$. This contradicts the fact that $K$ is contractible but $\varepsilon$ is not homotopic to a constant. Hence any two points $a, b$ are joined by a unique arc in $L \setminus q$ and another one in $L \setminus p$, and $L$ is the union of these arcs which have only the endpoints in common. □

With (1.5) and (2.3) follows immediately

(4.2) The point space $P$ of a flat projective plane is a compact surface.

Since $P \setminus L \approx \mathbb{R}^2$ and $L \approx \mathbb{S}$, the Euler characteristic (and the genus) of $P$ is $1$. This characterizes $P$ among the compact surfaces up to homeomorphism (Seifert-Threlfall):

(4.3) The point spaces of all flat projective planes are homeomorphic.

By duality, the line space $\mathcal{L}$ is also a compact surface of characteristic $1$. Each line $L \in \mathcal{L}$ can be provided with two opposite orientations. The resulting space $\mathcal{L}^+$ is a two-fold covering of $\mathcal{L}$:

(4.4) The space of oriented lines is homeomorphic to a sphere $\mathbb{S}^2$.

Historically, continuity has been introduced into an axiomatic description of classical geometries by means of order. An ordering of a projective plane is a relation $a,b|c,d$ between two point pairs of $L$ which is given by a pair of opposite linear orders on $K = L \setminus \infty$ and which is invariant under $\Pi_L$. The open intervals form a basis for the topology of $K$, and affine planes are topologized as products. The topology of $\mathcal{L}$ is defined dually. Continuity of join and intersection depends on the $\Pi_L$-invariance of the separation $|$. In flat planes let $a,b|c,d$ if $c$ and $d$ are in different components of $L \setminus \{a,b\}$. This relation is obviously invariant under homeomorphisms of $L \approx \mathbb{S}$. Conversely (Bernays),

(4.5) An ordered projective plane with a compact interval is flat.

Proof. If there is a compact closed interval in $K$, then all closed intervals are compact by homogeneity, and $P$ is locally compact by construction. Moreover, $K$ is order dense: by homogeneity, two arbitrary points of $K$ may be taken as $0$ and $1$. Because $z \mapsto 1/z$ is a projectivity, $\infty,1|0,c$ implies $0,1|\infty,1/c$ so that $1/c$ is between $0$ and $1$. A compact order dense interval is connected. □

(4.6) An ordered connected projective plane is flat.

Proof. If B is the set of upper bounds of A in an interval J, then B has a first element or B is open and closed. □

## §5. COMPACT 4-DIMENSIONAL PLANES

In this section $\mathcal{P} = (P, \mathcal{L})$ will always denote a compact projective plane with $\dim P = 4$. For such planes the topological structure of $(P, L)$ has been determined up to homeomorphism, but the proofs cannot be given here in full.

(5.1) A line L of $\mathcal{P}$ is homeomorphic to a sphere $\mathbb{S}^2$.

Proof. $\dim L = 2$ by (3.5). According to Bing-Borsuk (Cor.8.2), a locally compact connected homogeneous locally contractible 2-dimensional space is a topological manifold. Hence L is a compact surface. Since $L \setminus q$ is contractible, $L \approx \mathbb{S}^2$. □

(5.2) Corollary. There is no compact projective plane with a 3-dimensional point space.

(5.3) P is a compact manifold.

Topological projective planes whose lines are manifolds have been investigated by Breitsprecher. Using the theory of vector bundles he proved

(5.4) P is homeomorphic to the point space of the classical complex projective plane.

If $\mathfrak{P}$ carries a complex structure such that all projectivities are holomorphic, then $\Pi_L$ is sharply triply transitive and the theorem of Pappos holds in $\mathfrak{P}$. Consequently (Breitsprecher),

(5.5) A holomorphic projective plane is isomorphic to $P_2\mathbb{C}$.

Projectivities also play an essential role in the following theorem:

(5.6) A closed projective subplane of $\mathfrak{P}$ is flat.

Proof. In coordinates, the assertion says that each proper closed subternary of K is homeomorphic to $\mathbb{R}$ if $K \approx \mathbb{C}$, or in other words, that a locally compact subternary F of K has necessarily positive dimension; in particular, F cannot be finite. Consider the projectivity $\varepsilon = (x \mapsto x+1) : K \circlearrowleft$. Because $K^+$ is a loop, $\varepsilon$ has no fixed point in K. By (3.2) there is an isotopy deforming $\varepsilon$ into the identity. Hence $\varepsilon$ preserves the orientation of $K \approx \mathbb{C}$. Brouwer's translation theorem implies $0^{\varepsilon^n} \xrightarrow[n]{} \infty$ (Sperner, Satz 9). All the elements

$c_n = 1/0^{\varepsilon^n}$ belong to F, and $c_n \xrightarrow[n]{} 0$. By homogeneity, F is dense in itself. If $\dim F = 0$ there would exist a compact $S \subset\subset F$ and a neighbourhood U of 0 in K with $S + U \subseteq S$. (Consider the compact sets $(S+\bar{U}) \cap (F \setminus S)$ and note that S is the intersection of all $S+\bar{U}$.) Let $0 \neq t \in S \cap U$ and $x^\tau = x+t$. Then $s^{\tau^n} \xrightarrow[n]{} \infty$ by the same argument as above, but $S^\tau \subseteq S$, a contradiction. □

Serious obstacles prevent an extension of the results of this section to higher dimensions. Even if P is assumed to be locally euclidean it seems difficult to prove that the lines are manifolds. If L is locally euclidean, however, then L is in fact homeomorphic to $\mathbb{S}^\ell$ with $\ell = 2^m \leq 8$ (Breitsprecher). But also if the latter is true it has been proved only in very special cases that P is then homeomorphic to the point space of the classical quaternion or octave plane (Buchanan). As there is no analogue of Brouwer's translation theorem in higher dimensions, it is still an open question if a compact connected projective plane can have a 0-dimensional closed subplane. By a recent result of Löwen a compact connected plane is classical if it has a point-transitive group of automorphisms.

## §6. FINITE DIMENSIONAL PLANES

Compact connected projective planes with $\dim P \leq 4$ are homeomorphic but in general not isomorphic to classical planes. For $4 < \dim P < \infty$ the spaces $L$ and $P$ have in some cases at least the same homotopy type or the same homology groups as the classical planes over the quaternions or the octaves. This will be sufficient for many applications. A key result is due to Dugundji (unpublished) and cannot be proved here:

(6.0) If $\dim L = \ell$, the integral (singular or Čech) homology group $H_\ell(L) \neq 0$.

It is well known (Kuratowski) that a finite dimensional space is locally contractible if and only if it is an absolute neighbourhood retract (ANR).

A theorem of Łysko implies

(6.1) $L$ is a Cantor manifold. (If $C \subseteq L$, then $L \setminus C$ is connected or $\dim C \geq \ell - 1$).

There is also an easy direct proof using the homogeneity of $L$: if $L \setminus C = U \cup V$ is the disjoint union of open subsets, let $0 \in U$ and $\infty \in V$. Then $\overline{U}$ is a compact neighbourhood of $0$ in $K = L \setminus \infty$, and the sets $a \circ U$ form a basis at $0$ with boundaries

$a \circ \partial U \approx \partial U \subseteq C$ . Hence the inductive dimension $\operatorname{ind} L \leq \operatorname{ind} C + 1$ . For separable metric spaces $\dim = \operatorname{ind}$. □

The fact that $M = K \setminus 0$ is a topological loop with respect to the multiplication in the ternary field $K_\tau$ will play a decisive role in the following. In particular, $M$ is a so-called $H$- space, that is a space with a continuous multiplication having an identity up to homotopy. This has far-reaching consequences for the homotopy and homology of $M$. The homology groups of $L$ and $M$ are related by

(6.2) $H_q(L) \cong H_{q-1}(M)$ for $q > 1$ .

Noting that affine lines are contractible by (3.3) and hence $H_q(K) = 0$ for $q > 0$, the proof follows immediately from the Mayer-Vietoris sequence (Spanier IV 6) applied to $K$ and $K' = L \setminus 0$ . □

As nothing is to be proved in low dimensions, from now on let $\dim P > 4$, or equivalently, $\dim L = \ell > 2$ . Then $L$ and $M$ are simply connected:

(6.3) $\pi_1(L) = H_1(L) = 0$ .

Proof. Because of (3.3) any two curves joining $a$ and $b$ in $L \setminus c$ are homotopic. Hence any continuous map $\alpha : \mathbb{S} \to L$ is homotopic to a piecewise injective map $\omega$. Now $\dim \mathbb{S}^{\omega} = 1 < \ell$ implies that $\mathbb{S}^{\omega}$ is contained in an affine line, and $\omega$ is homotopic to a constant. □

(6.4) $\pi_2(L) \cong H_2(L) \cong H_1(M) \cong \pi_1(M)$.

Proof. $\pi_1(M)$ is commutative because $M$ is a topological loop: let $I = [0,1]$ and $\alpha, \beta : I \longrightarrow M$ with $0^{\alpha} = 1^{\alpha} = 0^{\beta} = 1^{\beta} = 1$. Denote the boundary $\partial I^2$ by $C$ and put $\eta(s,t) = s^{\alpha} \circ t^{\beta}$ (the product in $M$). Then $\eta : I^2 \longrightarrow M$ is a homotopy of $\gamma = \eta | C$ to the constant map $1$. But $\gamma$ represents the element $\alpha\beta\alpha^{-1}\beta^{-1}$ in $\pi_1(M)$. The assertion follows now from (6.3) and the Hurewicz isomorphism theorem (Spanier VII 5) which says, in particular, that for arcwise connected spaces $H_1 \cong \pi_1/\pi_1'$, where $\pi_1'$ is the commutator subgroup, and that $H_2 \cong \pi_2$ for simply connected spaces. □

(6.5) $\pi_1(M) = 0$.

Proof. Because $L$ is locally contractible, each continuous $\alpha : \mathbb{S} \longrightarrow M$ is homotopic to a piecewise injective map $\omega : \mathbb{S} \longrightarrow M \setminus 1$. Let $C = 1/\mathbb{S}^{\omega}$. Then $\dim C = 1 < \ell - 1$, and $L \setminus C$ is connected by (6.1). Hence there is an arc $(t \mapsto e_t) : I \longrightarrow L \setminus C$ joining $0$ and $1$. Put $\eta(s,t) = e_t \circ s^{\omega}$. Then $\eta : \mathbb{S} \times I \longrightarrow K \setminus 1$ is a homotopy between $\omega$ and the constant map 0 . This shows that $K \setminus 1 \approx M$ is simply connected. □

(6.6) $\pi_q(P \setminus p) \cong \pi_q(L)$ for all $q \geq 1$, and $\pi_1(P) = 0$ .

Proof. Since the fibres of the locally trivial fibration $\rho : P \setminus p \longrightarrow L$ are homeomorphic to K , there is an exact sequence

$$\ldots \pi_q(K) \longrightarrow \pi_q(P \setminus p) \longrightarrow \pi_q(L) \longrightarrow \pi_{q-1}(K) \ldots$$

(Spanier VII 2 ) . □

The following proofs make use of a few deeper results from algebraic topology and will only be sketched. Two spaces X and Y are homotopy equivalent ( $X \simeq Y$ ) if there are continuous maps $X \underset{\tau}{\overset{\sigma}{\rightleftarrows}} Y$ such that $\sigma\tau$ and $\tau\sigma$ are homotopic to the identity maps. The homology groups of X and Y are then isomorphic.

(6.7) L is homotopy equivalent to a compact polyhedron N .

By a recent theorem of West this is true whenever L is a compact metric ANR. For the next corollary and applications only the much weaker one-sided "domination" is needed (Dugundji).

(6.7') There exist a finite polyhedron N and maps $L \underset{\tau}{\overset{\sigma}{\rightleftarrows}} N$ such that $\sigma\tau$ is homotopic to the identity on L .

The sum $H_*(L)$ of the homology groups of L is then a direct summand of the finitely generated homology $H_*(N)$ since for the induced maps $\sigma_* \tau_* = 1$. Consequently,

(6.8) $H_*(L)$ is finitely generated, or, in other words, L is of finite type. Because of (6.2) also M is of finite type.

Arcwise connected H-spaces of finite type have been investigated extensively. Noting that M is simply connected, a well known theorem of Hopf implies

(6.9) The rational cohomology $H^*(M;\mathbb{Q}) \cong \mathrm{Hom}(H_*(M),\mathbb{Q})$ is isomorphic to $H^*(S;\mathbb{Q})$ for a product S of spheres of odd dimensions $r_k > 1$.

Moreover, by the Hurewicz isomorphism theorem and a result of Browder

(6.10) $\pi_3(L) \cong H_3(L) \cong H_2(M) \cong \pi_2(M) = 0$.

It follows immediately from the definition of the Čech homology that $H_q(L) = 0$ for all $q > \ell$, but $H_\ell(L) \neq 0$ according to (6.0). In this situation, Browder proved

(6.11) $H_\ell(L) \cong H_{\ell-1}(M) \cong \mathbb{Z}$ and $H^{\ell-q}(M;G) \cong H_{q-1}(M;G)$ for all q and all groups G.

Now $H^r(S) \cong \mathbb{Z}$ for $r = \dim S = \sum_k r_k$. Consequently $r = \ell - 1$, and one has the corollary

(6.12) $\dim L = \ell \neq 3, 5$.

If $\ell = 4$, let $\alpha : \mathbb{S}^4 \longrightarrow L$ represent a generator of $\pi_4(L) \cong H_4(L) \cong \mathbb{Z}$. Then $\alpha$ induces isomorphisms $\alpha_{\#} : \pi_4(\mathbb{S}^4) \cong \pi_4(L)$ and $\alpha_* : H_*(\mathbb{S}^4) \cong H_*(L)$, because $H_q(L) = 0$ for $q \neq 0, 4$. A theorem of Whitehead (Th.3, Cor.1) shows now

(6.13) If $\dim L = 4$, then $\alpha : \mathbb{S}^4 \simeq L$.

Similarly, the homotopy type of M can be determined for $\ell = 4, 6$, or 8, because in these cases $H_q(M) = 0$ for $q \neq 0, \ell-1$ by (6.9-11) and the fact that $\pi_3(M)$ is torsion free (Hubbuck - Kane). Since M is not compact, however, a possibly infinite polyhedron dominating M has to be considered. The technicalities of the proof will be omitted.

(6.14) $M \simeq \mathbb{S}^{\ell-1}$ for $\ell = 4, 6$, or 8.

The homotopy equivalence transforms the multiplication of M into an H - multiplication on $\mathbb{S}^{\ell-1}$. By the theorem of Adams (Husemoller § 14.4) the only spheres admitting an H - multiplication are the $S^{\ell-1}$ with $\ell = 2^m \leq 8$ :

(6.15) $\dim L = 2^m \leq 4$ or $\dim L \geq 7$.

As the possibility $M \simeq S^3 \times S^3$ cannot be excluded by the present arguments, the conjecture $\ell \neq 7$ has not yet been verified. For any $\ell > 1$ finally,

(6.16) $H_{2\ell}(P) \cong \mathbb{Z}$ and $\dim P = 2\ell$ .

Proof. From the Mayer-Vietoris sequence for two affine subplanes follows

$$H_q(P \setminus a) \cong H_{q-1}(P \setminus L' \setminus L'') \cong H_{q-1}(M) \cong H_q(L) .$$

Let $X = P \setminus L \setminus a$ . There is an exact Mayer-Vietoris sequence

$$\ldots H_q(X) \longrightarrow H_q(P \setminus a) \longrightarrow H_q(P) \longrightarrow H_{q-1}(X) \ldots .$$

In particular,

$$H_{2\ell}(P) \cong H_{2\ell-1}(X) \cong H_{2(\ell-1)}(M \times M) \cong$$
$$\cong H_{\ell-1}(M) \otimes H_{\ell-1}(M) \cong \mathbb{Z} \otimes \mathbb{Z} \cong \mathbb{Z} .$$

If $\dim P < 2\ell$, then $H_{2\ell}(P) = 0$ by the definition of the Čech homology. □

REFERENCES

J.F.Adams "On the non-existence of elements of Hopf invariant one"
Ann. Math. 72, 20 - 104 (1960)
Zbl. 96 , 174 ; MRev. 25 , 4530

P.Bernays "Betrachtungen über das Vollständigkeits-axiom und verwandte Axiome"
Math. Z. 63, 219-229 (1955)
Zbl. 68 , 269 ; MRev. 17 , 447

Bing - Borsuk "Some remarks concerning topologically homogeneous spaces"
Ann. Math. 81, 100 - 111 (1965)
Zbl. 127 , 133 ; MRev. 30 , 2475

S.Breitsprecher "Projektive Ebenen, die Mannigfaltigkeiten sind"
Math. Z. 121, 157 - 174 (1971)
Zbl. 229, 50o21 ; MRev. 43, 6935

" "Einzigkeit der reellen und komplexen projektiven Ebene"
Math. Z. 99, 429 - 432 (1967)
Zbl. 147, 389 ; MRev. 35, 48o7

L.E.J.Brouwer "Beweis des ebenen Translationssatzes"
Math. Ann. 72, 37 - 54 (1912) ; FdM 43, 569.

W.Browder "Torsion in H - spaces"
Ann. Math. 74, 24 - 51 (1961)
Zbl. 112, 145 ; MRev. 23, A 22o1

Th. Buchanan "Zur Topologie der projektiven Ebenen über reellen Divisionsalgebren"
Geom. Dedic. 8, 383 - 393 (1979)
Zbl. 414, 51oo8 ;

J.Dugundji " Absolute neighborhood retracts and local connectedness in arbitrary metric spaces"
Compos. Math. 13, 229 - 246 (1958)
Zbl. 89, 389 ; MRev. 22, 4o55

" "Topology" Boston 1966.

R.Engelking "Dimension theory". North Holland 1978.

H.Hopf " Über die Topologie der Gruppenmannigfaltigkeiten und ihre Verallgemeinerungen"
Ann. Math. 42, 22 - 52 (1941)
Zbl. 25, 93 ; MRev. 3, 61

Hubbuck - Kane "On $\pi_3$ of a finite H - space"
Trans.AMS 213, 99 - 1o5 (1975)
Zbl. 312, 55o13 ; MRev. 53, 1582

Hurewicz - Wallman "Dimension theory" Princeton 1948.

D.Husemoller "Fibre bundles"McGraw-Hill 1966, Springer 1975

J.O.Kiltinen "On the number of field topologies on an infinite field"
Proc.AMS 4o, 3o - 36 (1973)
Zbl. 243, 1211o ; MRev. 47, 6667

C.Kuratowski "Topologie II" Warszawa [3]1961.

R.Löwen "Homogeneous compact projective planes"
Journ. reine angew. Math.

J.M.Łysko "Some theorems concerning finite dimensional homogeneous ANR-spaces"
Bull. Acad.Polon. Sci. 24, 491 - 496 (1976)
Zbl. 331 , 54o24 ; MRev. 54 , 1231

K.Morita "On the dimension of product spaces"
Amer. J. Math. 75, 2o5 - 223 (1953)
Zbl. 53 , 124 ; MRev. 14 , 893

K.Nagami "Dimension theory" Academic Press 197o.

A.R.Pears "Dimension theory of general spaces"
Cambridge 1975.

Seifert - Threlfall "Lehrbuch der Topologie"
Leipzig:Teubner 1934.

E.H.Spanier "Algebraic topology" McGraw-Hill 1966.

E.Sperner "Über die fixpunktfreien Abbildungen der Ebene"
Abh.Math.Sem.Univ.Hamburg 1o, 1 - 48 (1934)
Zbl. 7 , 231

P.J.Stone "Sequences of coverings"
Pacif. J. Math. 1o, 689 - 691 (196o)
Zbl. 114 , 14o ; MRev. 22 , 9955

A.Weil "Basic number theory" Springer 1967, [3]1974.

J.E.West "Compact ANR's have finite type"
Bull.AMS 81, 163 - 165 (1975)
Zbl. 297 , 54o15 ; MRev. 5o , 1125o

" "Mapping Hilbert cube manifolds to ANR's"
Ann. Math. 1o6, 1 - 18 (1977)
Zbl. 375 , 57o13 ; MRev. 56 , 9534

J.H.C.Whitehead "On the homotopy type of ANR's"
Bull.AMS 54, 1133 - 1145 (1948)
Zbl. 41 , 319 ; MRev. 1o , 617

G.T.Whyburn "Topological analysis" Princeton [2]1964.

# PROJECTIVITIES AND THE GEOMETRIC STRUCTURE OF TOPOLOGICAL PLANES

Rainer Löwen

Universität Tübingen

ABSTRACT. We prove von Staudt type theorems for projectivities and bundle involutions of topological affine, projective and circle planes (§§5-7). In §1, we discuss some examples. §2 contains background material on topological ovals and topological circle planes. In §3, we prove a crucial auxiliary theorem on connectedness properties of von Staudt groups, and in §4 we present the necessary results on topological transformation groups.

Some of the results have not appeared in the literature (2.1, 2.5, 6.8, 7.6), or have appeared in a different form. In some cases, errors contained in the original papers are corrected (6.6, 7.5, 7.7).

After the investigation, by means of projectivities, of the *topology* underlying a topological projective plane, we shall show in this section how von Staudt's approach can be used in order to study the *geometric* properties of a topological (linear or circle) plane. The presence of a compatible topology has the effect that weaker assumptions on the regularity of projectivities suffice for proving various theorems of von Staudt type. It is possible even to cover the Moufang plane $P_2\mathbb{O}$, which satisfies none of the conditions $P_n$ of von Staudt-Schleiermacher. Moreover, the topology allows to formulate regularity conditions which have



*P. Plaumann and K. Strambach (eds.), Geometry - von Staudt's Point of View, 339–372.*

no analogue at all in abstract geometry.

The previous chapters showed that, in general, the von Staudt group (=group of projectivities) of projective planes is smallest possible in pappian planes and tends to become very big as soon as one gets to nondesarguesian, or even non-Moufang planes. We shall illustrate this abrupt change of size by an example of a compact projective plane (1.4). In general, the von Staudt groups of compact projective planes are too big to be locally compact transformation groups (which would be a kind of smallness and completeness property for a group of homeomorphisms).

It is perhaps interesting to compare this to the behaviour of Klein groups (=automorphism groups) of compact connected projective planes, which is in every way opposite to that of von Staudt groups. Klein groups always are locally compact in some reasonable topology, and they are largest possible in the classical planes. Indeed, the only compact connected projective planes with a point transitive automorphism group are the four classical desarguesian and Moufang planes [21], and the dimension of the automorphism group of a non-Moufang plane homeomorphic to one of the classical planes is never much larger than one half the dimension of the group of that classical plane; see [29,30] and references given there. The contrast in the behaviour of von Staudt and Klein groups is not as marked as that in abstract projective planes; cf. [18]. However, the basic reason for this contrast is independent of the topology: it lies, of course, in the fact that configurational closure properties of a plane may imply existence of collineations on the one hand and existence of relations between projectivities on the other hand.

## § 1 EXAMPLES

We denote by $\mathbb{R}$, $\mathbb{C}$ and $\mathbb{H}$ the classical fields of real and complex numbers and quaternions, and by $\mathbb{O}$ the alternative division algebra of octonions over $\mathbb{R}$.

1.1 *The classical cases.* Let $E = P_2\mathbb{F}$ be the projective plane over $\mathbb{F} = \mathbb{R}, \mathbb{C}$, or $\mathbb{H}$, and let $L$ be a line. Then the group $\Pi = \Pi_L$ of projectivities is the Lie group

$$\mathrm{PGL}_2\mathbb{F}$$

acting on $L = P_1\mathbb{F}$ in the usual way.

If $E = P_2\mathbb{O}$ then the action of $\Pi$ on $L$ is equivalent to the action of the 45-dimensional Lie group

$$\mathrm{PSO}_{10}(\mathbb{R},1)$$

on the 8-sphere of points in $P_9\mathbb{R}$ which are isotropic with respect to a quadratic form of index 1.

If $E$ is the affine plane $A_2\mathbb{F}$ over $\mathbb{F} = \mathbb{R}$, $\mathbb{C}$ or $\mathbb{H}$ then the group $\Pi$ of affine projectivities is

$$\Pi = \mathrm{L}_2\mathbb{F} = \{(x \to ax+b);\ a,b \in \mathbb{F},\ a \neq 0\}.$$

For $E = A_2\mathbb{O}$, finally, $\Pi$ is the group

$$\mathbb{R}^8 \cdot \mathrm{GO}_8^+\mathbb{R}$$

of similitudes of $\mathbb{R}^8$ of positive determinant.

The following considerations show how these facts may be *proved*. In a Moufang plane, the group of affine projectivities is generated by the mappings $x \to ax+b$. If $\mathbb{F}$ is a field, these mappings form a group. Next, observe that in a projective Moufang plane, every perspectivity between lines extends to an elation of the plane. Therefore, $\Pi$ is induced in the projective case by the isotropy group $\Sigma_L$ of the group $\Sigma$ generated by all elations, and in the affine case by $\Gamma_L$, where $\Gamma$ is generated by all elations with centre 'at infinity'. In order to evaluate these groups in the case $\mathbb{F} = \mathbb{O}$, the information provided in [28], (6''), (7) and (5) can be used. – See also Ferrar's contribution to this book for results about projectivities in general Moufang planes.

1.2 REMARKS. a) Observe that the group of projectivities of $A_2\mathbb{O}$ contains elements fixing a nontrivial subspace of $\mathbb{R}^8$ pointwise. Therefore, neither $A_2\mathbb{O}$ nor $P_2\mathbb{O}$ satisfy any one of the regularity conditions $P_n$, $n < \infty$. The same is true for $P_2\mathbb{H}$ (cf. (b) below), but not for $A_2\mathbb{H}$.

b) In $P_2\mathbb{H}$, the isotropy group $\Pi_x$ of $\Pi$ at a point $x \in L$ is $\mathbb{R}^4 \cdot \mathrm{GO}_4^+\mathbb{R}$; cf. [28], (5). Among the examples given above, this is the only case where $\Pi_x$ does not coincide with the group of affine projectivities of the corresponding affine plane.

Going by the size of their automorphism groups, the projective *Moulton planes* $E_k$ (over the real numbers) are the closest relatives of the real projective plane $P_2\mathbb{R}$; see [27], 5.6. Nevertheless, they have an extremely large von Staudt group, as we shall show (1.4).

In affine representation, the plane $E_k$ $(1 < k \in \mathbb{R})$ is obtained from the real affine plane $A_2\mathbb{R}$ as follows. Choose a coordinate system and replace the lines of negative slope $s$ by kinked lines which have slope $s$ in the left half plane defined by the $y$-axis $Y$, and slope $ks$ in the right half plane.

1.3 DEFINITION. A bijection $\pi$ of the real projective line $P_1\mathbb{R}$ is called *piecewise projective linear* if there exists a subdivision

$$P_1\mathbb{R} = I_1 \cup \ldots \cup I_n$$

into closed intervals such that on each $I_j$, the map $\pi$ coincides with some element $\gamma_j \in \mathrm{PGL}_2\mathbb{R}$. The group formed by all these bijections will be denoted by $\mathrm{PPGL}_2\mathbb{R}$.

1.4 THEOREM. [2] *Let* $L$ *be a line of a projective Moulton plane* $E_k$ *and* $\Pi$ *the group of projectivities. There exists an isomorphism of permutation groups,*

$$(\Pi : L) \cong (\mathrm{PPGL}_2\mathbb{R} : P_1\mathbb{R}).$$

1.5 REMARKS. a) The group $\mathrm{PPGL}_2\mathbb{R}$ is easily seen to be transitive on the subsets of any given finite cardinality; in fact, the subgroup fixing some interval pointwise still has this property with respect to the remainder of $P_1$. So the size of this group is, in this respect, comparable to that of the von Staudt group of an existentially closed projective plane (see O.Kegel's contribution to this book).
b) O.Kegel has pointed out that $\mathrm{PPGL}_2\mathbb{R}$ is simple.

c) Observe that $\Pi = \Pi(E_k)$ is independent of $k$.

*Proof of* 1.4. We use the identification between the point sets of the affine Moulton plane and $A_2\mathbb{R}$ arising from the construction of $E_k$. We choose for $L$ the $y$-axis $Y$, and we identify the transformation group $(\mathrm{PGL}_2\mathbb{R}:P_1\mathbb{R})$ with the group $\Gamma$ of ordinary projectivities of $P_2\mathbb{R}$ on the projective line $\overline{Y} = Y \cup \{w\}$. With these identifications, $\Pi = \Pi_{\overline{Y}}(E_k)$ will turn out to be identical with the group $\mathrm{PPGL}_2\mathbb{R} = \mathrm{P}\Gamma$ of 'piecewise-$\Gamma$-bijections'.

First, we indicate why every element $\pi \in \Pi$ lies in $\mathrm{P}\Gamma$. If $Y$ and the line $W$ at infinity neither occur as axes of factor perspectivities of $\pi$, nor contain any of the centres, then $\pi$ can be written as a product of perspectivities $\pi(Y,c,G)^{\pm 1}$, where $c = (x,y) \notin Y$ and $G \neq Y$ is an affine line. Now it suffices to observe that the line pencil $L_c$ of $c$ is the same as the pencil of ordinary straight lines through $c$, except in the quadrant

$$Q = \{(a,b);\ ax < 0,\ b\cdot\mathrm{sgn}x > y\},$$

where $\mathrm{sgn}x = x|x|^{-1}$. In $Q$, on the other hand, $L_c$ is the pencil of ordinary lines passing through the 'virtual centre'

$$c' = (x\cdot k^{\mathrm{sgn}x}, y).$$

So $\pi(Y,c,G)$ is piecewise an ordinary perspectivity of $P_2\mathbb{R}$. If $\pi$ does not satisfy the restrictions made above then $\pi$ can be approximated, using ideas from a later section (3.2b), by a sequence of projectivities $\pi_n$ of constant length (= number of factors) which satisfy those restrictions. The limit of the corresponding sequence $\gamma_n$ in $\mathrm{P}\Gamma$ also belongs to $\mathrm{P}\Gamma$. To see this, note that the number of singularities of $\gamma_n$ is at most three times the number of factors of $\pi_n$ in the standard representation used above. Since $\overline{Y}$ is compact, one can therefore assume that the singularities of $\gamma_n$ converge.

Conversely, we have to show that $\mathrm{P}\Gamma \subseteq \Pi$. First, we show that $\Pi$ contains the isotropy group

$$\Gamma_w = L_2\mathbb{R} = \{y \to ay+b;\ a,b \in \mathbb{R},\ a \neq 0\}.$$

Indeed, the translations $(a = 1)$ can be obtained as composites of parallel projections, first from $Y$ to a parallel $Y'$ contained in the left half plane $L$, and then back to $Y$ in a different direction. The mappings $y \to ay$ are obtained by projecting $Y$ onto $Y' \subseteq L$ from a centre $c \in L$, and back onto $Y$ by parallel projection. Next, it is easy to see that $\Pi$ contains ordinary projectivities $\alpha \in \Gamma$ moving $w$. Thus, $\Pi$ contains $\Gamma_u$ and $\Gamma_v$ for different points $u$, $v$. These generate $\Gamma$, so that

$$\Gamma \subseteq \Pi.$$

To see this, let $\gamma \in \Gamma$. If $u^\gamma \neq v$, choose $\delta \in \Gamma_v$ such that $u^{\gamma\delta} = u$. Then

$$\gamma = (\gamma\delta)\delta^{-1} \in \Gamma_u\Gamma_v.$$

If $u^\gamma = v$, there exists $\sigma \in \Gamma_u$ such that $u^{\gamma\sigma} \neq v$, and $\gamma\sigma \in \Gamma_u\Gamma_v$.

By a *semi dilation*, we mean a mapping $\sigma$ of $\overline{Y}$ which is the identity on some closed interval $I \subseteq \overline{Y}$ and coincides with an element of $\Gamma$ on the complement $J = \overline{Y} \setminus I$. The open interval $J$ is called the *support* of $\sigma$. The semi dilations with support $J$ form a group $S(J)$, which is sharply transitive on $J$. We want to show that $\Pi$ contains the set $S$ of all semi dilations. It suffices to show that $\Pi$ contains an open subset of $S(J)$, where $J = (0,w)$. Project $Y$ from $c = (-1,0)$ onto a parallel $Y_b$ $(x = b > 0)$, and project back horizontally. The resulting projectivity $\pi$ maps $y > 0$ to $y(b+1)$, and $y < 0$ to $y(kb+1)$. Multiplying with

$$\gamma = (y \to y(kb+1)^{-1}) \in \Gamma$$

we get a semi dilation $\sigma = \pi\gamma \in S(J)$ sending $y > 0$ to $y(b+1)(kb+1)^{-1}$.

Finally, we show that $P\Gamma$ is generated by $\Gamma$ and $S$. Let $\pi = \pi_{-1}$ be projective linear on each of the intervals $I_0,\dots,I_n$, where $\overline{Y} = I_0 \cup \dots \cup I_n$. Define $S(\overline{Y}) = \Gamma$. Proceeding inductively, choose, for each $i \geq 0$, elements

$$\rho_i \in S(I_i \cup \dots \cup I_n), \quad \sigma_i \in S(I_i)$$

such that

$$I_i^{\pi_{i-1}\rho_i} = I_i \quad \text{and} \quad \sigma_i | I_i = (\pi_{i-1}\rho_i)^{-1}.$$

Let $\pi_i = \pi_{i-1}\rho_i\sigma_i$. Then $\pi_i \in S(I_{i+1}\cup\ldots\cup I_n)$ for $i < n$, and

$$id = \pi_n = \pi\rho_0\sigma_0\cdot\ldots\cdot\rho_n\sigma_n. \qquad \square$$

1.6 REMARK. I am informed by A.Longwitz that he possesses a general procedure for calculating groups of projectivities, which he intends to publish soon. I understand that he can prove 1.4 for Moulton planes over ordered fields. Note that the present proof, in contrast, uses compactness of $L$ at the end of the second paragraph.

1.7 REMARK. K.Strambach has almost completed a proof of the following conjecture, which is analogous to some results described in H.Lüneburg's contribution to these Proceedings. Let $E$ be a locally compact connected proper translation plane coordinatized by a quasifield $Q$ with kernel $K$. Then the group of affine projectivities of $E$ is $\Pi = GL_nK$, where $n = \dim(Q:K)$.

## § 2 TOPOLOGICAL CIRCLE PLANES AND TOPOLOGICAL OVALS

Throughout the following sections, $E = (P,B)$ denotes a topological plane geometry *with a locally compact connected point set* $P$ and a set $B$ of blocks (lines or circles). If $E$ is projective, the topological axioms for $E$ are the same as in H.Salzmann's contribution. An affine plane is called topological if it can be obtained by removing a line from a topological projective plane. These two types of planes will be referred to as *linear planes*. There,

$$\Pi = \Pi_B(E)$$

denotes the *von Staudt group* of projectivities or affine projectivities of a line $B \in \mathbf{B}$, respectively. We shall tacitly use the fact that $P$, $\mathbf{B}$ and $B \in \mathbf{B}$ are sparable metric, locally connected and even locally contractible spaces; see Salzmann's contribution.

By a *circle plane*, we mean a Möbius, Laguerre, or Minkowski plane. We denote the parallelity relations between points, the parallel projections, and the parallel classes of points $p \in P$ by

$$|| \text{ (Laguerre)}, \quad ||_+, \; ||_- \text{ (Minkowski)},$$

$$pr, \; pr_+, \; pr_-, \quad [p], \; [p]_+, \; [p]_-,$$

respectively. We abbreviate this by $||_{(\pm)}$, etc., when we want to include all possibilites.

In these planes, the von Staudt group $\Pi = \Pi_B(E)$ will be the group of permutations of $B \in \mathcal{B}$ generated by the following types of *perspectivities* $\pi$: For $B, C \in \mathcal{B}$, $b \in B\setminus C$, $c \in C\setminus B$, the map

$$\pi = \pi(B,b,c,C): B \to C$$

is defined by the condition that $x, x^\pi, b, c$ are concircular and different (with the obvious modifications if $x = b$, etc.), unless $b$ and $c$ are parallel. In the latter case, $\pi = pr_{(\pm)}$ if $b||_{(\mp)}c$. This means that we are working with *proper projectivities* only (cf. H.Karzel's contribution).

We denote by

$$E_p = (P_p, \mathcal{B}_p)$$

the affine plane obtained by *localizing* (or *deriving*) $E$ at a point $p$. Here, $P_p$ consists of the points which are not parallel to $p$, and $\mathcal{B}_p$ consists of the circles containing $p$ and the parallel classes not containing $p$. We denote the projective closure of $E_p$ by $\overline{E_p}$.

For a circle $B \in \mathcal{B}\setminus\mathcal{B}_p$, we obtain an oval

$$B_p \subseteq \overline{E_p}$$

by adjoining to $B \cap P_p$ the points $\infty_{(\pm)}$ at infinity incident with the $(\pm)$-parallel classes. By an *oval* in a projective plane, we mean here a set $O$ of points no three of which are collinear, such that each $x \in O$ lies on exactly one *tangent line* (= $T \in \mathcal{B}$ such that $T \cap O = \{x\}$).

We can now formulate the

*Axioms for topological circle planes* $\mathcal{E} = (P, \mathcal{B})$.

(TC 0) $P$ is locally compact and connected.

(TC 1) All geometric operations are continuous: Joining three points by a circle, intersecting two circles, forming tangent circles, and parallel projections onto circles or parallel classes.

(TC 2) For each $p \in P$, the localization $\mathcal{E}_p$ is a topological affine plane with respect to the topology on $P_p$ induced by $P$.

(TC 3) For each $B \in \mathcal{B}\backslash\mathcal{B}_p$, the oval $B_p \subseteq \overline{\mathcal{E}_p}$ corresponding to $B$ is a closed set of points.

(TC 4) The mapping $(x, B, b, c, C) \to x^{\pi(B,b,c,C)}$ is continuous whereever it is defined.

Note that the operations appearing in (TC 2) and (TC 4) are, in general, composed of several ones of the original operations of $\mathcal{E}$. Thus, the axioms (TC 2-4) express that the different operations of $\mathcal{E}$ are compatible at the common boundaries of their domains of definition. This is by no means an obvious consequence of (TC 1), not even of (TC 0,1). However, for Laguerre planes, (TC 2-4) can be deduced from the first two axioms; see [12]. In compact Minkowski planes, (TC 4) is known [25], but (TC 2) still presents considerable difficulties.

In Möbius planes, the axioms that have been considered in the literature are stronger than (TC 1). It is usually required (as in [36]) that the set of pairs of circles satisfying $|B \cap C| = 2$ is open and that for a fixed circle $B$ and a fixed point $p \notin B$, the circle joining $p$, $x_n$ and $x$ tends to the circle containing $p$ and touching $B$ in $x$, whenever $x_n \to x$ on $B$. Even under these assumptions, I could not find a proof of (TC 4) in the literature. However, it is not difficult to prove even more than that.

2.1 PROPOSITION. *In Möbius planes, the axioms* (TC 0) *and* (TC 1) *imply the axioms of Wölk* [36], *as well as our axioms* (TC 2-4).

*Proof.* In each localization $\mathcal{E}_p$, the operations of joining points, intersecting lines and forming parallels are continuous. This implies that $E_p$ can be coordinatized by a topological ternary field as in [27], §7. Therefore, $E_p$ is a topological affine plane ([27], 7.15). The lines of $E_p$ and, hence, also the circles of $E$ are closed sets, so that (TC 2) and (TC 3) are satisfied. By the result 2.3 below (which is proved without using (TC 4)), this implies that the circles are topological 1-spheres. Therefore, $P_p$ is homemorphic to $\mathbb{R}^2$ and $P$ is a 2-sphere. The openness condition of Wölk follows, since $|B \cap C| = |\{x,y\}| = 2$ implies that near $x$, the circles $B$ and $C$ are embedded in $P$ like two intersecting lines in $A_2\mathbb{R}$, so that their intersection must be stable.

It remains to deduce (TC 4). Evidently, it suffices to prove the following stronger form of Wölk's axiom on tangent circles: The circles $T_n$ joining $p_n \to p \notin B$ to $x_n, y_n \in B_n \to B$ tend to the circle $T$ through $p$ touching $B$ in $q$, whenever $x_n, y_n \to q$. We allow that $x_n = y_n$, in which case $T_n$ is understood to be the circle through $p_n$ touching $B_n$ in $x_n$. Now this is easy to see. In $E_r$ for $r \neq p, q$, there exists a line separating $p$ from $q$. Thus, in $E$, there exists a circle $S$ with this property. Each $T_n$ intersects $S$, and by passing to a subsequence, we may arrange that there exists a convergent sequence $s_n \in T_n \cap S$. Then $T_n$ converges to the circle $C$ joining $p, q$ and $s = \lim s_n$. We have to show that $C = T$ for each choice of our subsequence. But if $C \neq T$ then $C$ meets $B$ in two points $q$ and $z$ and, by stability of intersection, $T_n$ meets $B_n$ in a point close to $z$, a contradiction. □

Next, we need a few systematic remarks about ovals. Let $O$ be an oval in a topological projective plane $E = (P,B)$. Denote by $B_i(O)$ the set of all lines meeting $O$ in precisely $i$ points $(0 \leq i \leq 2)$, and let $B(O) = B_1(O) \cup B_2(O)$. The elements of $B_i(O)$ are called *exterior lines* ($i$=0), *tangents* ($i$=1), and *secants* ($i$=2). Let

$$O*O = O\times O/\mathbb{Z}_2$$

be the topological quotient space of the cartesian square modulo the action of $\mathbb{Z}_2$ which interchanges the factors. $O*O$ is called the *symmetric square* of $O$; it can be viewed as the set of all unordered pairs $\{x,y\}$ of (not necessarily different) points of $O$. The oval $O$ will be called a *topological oval* if the following map is a homeomorphism.

$$\varphi_O: O*O \to \mathcal{B}(O): \{x,y\} \to \begin{cases} x \vee y & \text{if } x \neq y \\ T_x \in \mathcal{B}_1(O) \cap \mathcal{B}_x & \text{if } x = y \end{cases}$$

This is a kind of differentiability condition ('tangents are limits of secants'). It implies that $O$ is homeomorphic to a line pencil $\mathcal{B}_x$, $x \in O$ (restrict $\varphi_O$ to $(O\times\{x\})$ mod. $\mathbb{Z}_2$), and, in particular, that $O$ is compact and hence closed in $P$. The converse of the last statement is the first assertion in the following theorem. For the notion of *dimension*, see H. Salzmann's contribution. If $B$ is a manifold then $\dim P = 2\cdot\dim B$.

2.2 THEOREM [3]. *Let* $O$ *be a (topologically) closed oval in a locally compact connected projective plane* $\mathsf{E} = (P,\mathcal{B})$, *where* $\dim P < \infty$.
*a)* $O$ *is a topological oval; in particular,* $O$ *is homeomorphic to a line* $B \in \mathcal{B}$.
*b) Exterior lines exist if and only if* $\dim B = 1$.
*c)* $B \cong O$ *is a sphere* $S_n$ *of dimension* $n = 1$ *or* $2$.

The *proof* of this theorem is too complicated to be included here. We indicate that (a) depends on a topological property of the lines of a projective plane called domain invariance, and that (b) can be proved by defining a continuous commutative multiplication with unit on any exterior line, or by separation arguments. In order to prove (c), one computes the homology in dimension $2l-1$ ($l = \dim B > 2$) of the spaces $\mathcal{B}(O) = \mathcal{B}$ and $O*O \cong B*B$, and shows that they are different. Thus, $l \leq 2$ and $B \cong S_n$; see Salzmann's contribution.

2.3 COROLLARY [3]. *Let* $\mathsf{E} = (P,\mathcal{B})$ *be a topological circle plane with* $\dim P < \infty$, *satisfying* (TC 0-3). *Any circle* $B \in \mathcal{B}$ *is a sphere* $S_n$ *of dimension* $n = 1$ *or* $2$. *If* $\mathsf{E}$ *is a Möbius plane then* $n = 1$.

*Proof*. Axioms (TC 2,3) give us a topological projective plane $\overline{\mathsf{E}_p}$ with a closed oval $B_p$. The lines of $\mathsf{E}_p$ are homeomorphic to circles of $\mathsf{E}$ with one point deleted; so the lines of $\overline{E_p}$, as well as the circles of $E$, are homeomorphic to the one point compactification of a line of $E_p$. It only remains to observe that $B_p$ is entirely contained in $E_p$ if $E$ is a Möbius plane, and hence admits an exterior line in this case.

2.4 THEOREM [4]. *Let $O$ be a closed oval in $P_2\mathbb{C}$. Then $O$ is a conic.*

*Proof*. Take a tangent as the line $W$ at infinity and coordinatize in such a way that the point of tangency lies on the $y$-axis. Then the affine part $O \setminus W$ is the graph of a function $f\colon \mathbb{C} \to \mathbb{C}$, which is holomorphic by 2.2a. Since $O$ is closed in $P_2\mathbb{C}$, $f$ must be continuous at infinity, and the theorem of Casorati-Weierstraß implies that $f$ is a polynomial. It is now easy to deduce from the oval property of $O$ that $f$ must be of degree 2.

2.5 COROLLARY. *Let $E = (P,B)$ be a topological circle plane with* dim $P = 4$, *satisfying* (TC 0–3). *If there exists a point $p$ such that $E_p$ is desarguesian then $E$ is the classical complex Laguerre or Minkowski plane.*

*Proof*. By 2.3, $E$ is a Laguerre or Minkowski plane. The plane $E$ can be redescribed within $E_p$ by the set of ovals $B_p$. By 2.4, each $B_p$ is in fact a conic. If $E$ is a Laguerre plane, the line at infinity is tangent to each $B_p$, with point of tangency $\infty$. In Minkowski planes, each $B_p$ contains $\infty_+$ and $\infty_-$. Thus, the conic $B_p$ is determined as soon as three more points on it are known. On the other hand, $B$ contains a circle joining any three points. This is only possible if $B$ consists of all conics satisfying the conditions menioned above. Then $E$ is the classical complex Laguerre or Minkowski plane.

## § 3 CONNECTEDNESS PROPERTIES OF GROUPS OF PROJECTIVITIES

As we mentioned earlier, the group $\Pi$ of projectivities of a locally compact plane is not, in general, a locally compact group. The only substantial information available on the topology of $\Pi$ concerns connectedness properties; see 3.3 below. This result prepares the way for the application of a theorem of Gleason and Palais ([11]; see 4.4 below) to groups of projectivities.

Let us fix our *notation for projectivities*. It will always be adapted to the case of circle planes; the obvious simplifications are understood in the linear case. $L_\infty$ denotes the line at infinity if $E$ is affine, and we choose a block $B$.

Let $\underline{c} = (c_1,\dots,c_n)$ and $\underline{d} \in P^n$, $\underline{A} \in B^{n-1}$, and assume that for all $k = 1,\dots,n$,

$$(*)\ \begin{cases} d_k = c_k \notin A_{k-1},\ A_k & \text{(linear case) and} \\ c_k \in L_\infty & \text{(affine case), or} \\ c_k \in A_{k-1}\setminus A_k,\quad d_k \in A_k\setminus A_{k-1} & \text{(circle case)} \end{cases}$$

Define $A_0 = A_n = B$, and write

$$\pi = \pi(\underline{c},\underline{d},\underline{A}) = \prod_{k=1}^{n} \pi(A_{k-1},c_k,d_k,A_k)$$

for the projectivity defined by these data; the definition of perspectivities $\pi(X,x,y,Y)$ is given at the beginning of §2. If a fixed representation of $\pi$ as a product of perspectivities is referred to, we may write conversely

$$c_k = c_k(\pi),\ d_k = d_k(\pi),\ A_k = A_k(\pi),\ n = n(\pi).$$

The latter number is called the *length* of $\pi$. Note, however, that none of these last expressions is well defined as a function of $\pi$ alone.

3.1 For a fixed block $B \in \mathcal{B}$, we shall consider the *group of projectivities* (or *von Staudt group*) of $E$ with respect to $B$, consisting of all $\pi(\underline{c},\underline{d},\underline{A})$ as defined above:

$$\Pi = \Pi_B(E).$$

This group will always be endowed with the *compact open topology* $\tau$. This topology has a countable basis in our case, and can therefore be described using sequences. We shall use this description rather than the standard definition of $\tau$; see [8], chapter 12 for this and related questions. We have

$\pi_n \to \pi$ in $(\Pi,\tau)$

$\iff$ $x_n \to x$ in $B$ implies $x_n^{\pi_n} \to x^{\pi}$ ('diagonal convergence')

$\iff$ $\pi_n$ converges to $\pi$ uniformly on each compact subset of $B$.

Recall that $B$ is a metric space, so that the last description makes sense.

3.2 LEMMA. *a)* $(\Pi,\tau)$ *is a topological transformation group of* $B$*; i.e., the group operation* $(\pi,\rho) \to \pi\cdot\rho^{-1}$ *and the evaluation map* $(x,\pi) \to x^{\pi}$ *are continuous. In particular, each* $\pi \in \Pi$ *is a homeomorphism of* $B$.
*b) The map* $(\underline{c},\underline{d},\underline{A}) \to \pi(\underline{c},\underline{d},\underline{A})$ *is continuous where it is defined in* $P^n \times P^n \times B^{n-1}$.

*Proof.* First of all, each $\pi \in \Pi$ is a homeomorphism. This is trivial in the linear case, and follows from (TC 4) in the circle case. Since $B$ is locally compact and locally connected, the rest of assertion (a) follows from a theorem of Arens [1]. Part (b) follows from (TC 4), using induction on the length of $\pi$ and the description of $\tau$-convergence by diagonal convergence. Note that, in linear planes, an assertion analogous to (TC 4) trivially holds.

3.3 THEOREM [20]. *Let* $E$ *be a locally compact connected linear or circle plane and let* $\Sigma$ *be the arcwise connected component of* $1 \in \Pi = \Pi_B(E)$.
*a)* $\Sigma$ *is nontrivial.*
*b) If* $\dim P > 2$, $\Pi = \Sigma$.
*c) If* $\dim P = 2$ *and* $E$ *is projective then* $\Pi/\Sigma$ *is cyclic of order* 2.

REMARK. For the remaining 2-dimensional cases, it is only known that $\Pi/\Sigma$ is at most countable in affine planes and that $\Sigma$ is doubly transitive on $B$ in circle planes; see [20].

*Proof of* 3.3. 1) $(\Pi,\tau)$ is regular and has a countable basis ([8], 12.1.3, 12.5.2) and, hence, is metrizable ([8], 9.9.2). By [35], II,4.1 and 5.3, every continuous image of the unit interval $I \subseteq \mathbb{R}$ in $(\Pi,\tau)$ is arcwise connected. Therefore, a subset of $\Pi$ is pathwise connected (any two points can be joined by a continuous image of $I$) if and only if it is arcwise connected (any two points can be joined by a continuous 1-1 image of $I$).

2) Choose any path $\underline{A}(t)$ in $B^{n-1}$ (e.g., a constant path), and choose paths $\underline{c}(t)$ and $\underline{d}(t)$ in $P^n$ such that (*) is satisfied for each $t \in I$. This is possible because the set of elements satisfying (*) is open. By 3.2b, $\pi(t) := \pi(\underline{c}(t),\underline{d}(t),\underline{A}(t))$ is a path in $\Pi$. It is easy to ensure that this path is nontrivial. The path $\sigma(t) = \pi(t)\cdot\pi(0)^{-1}$ starts from $1 \in \Pi$ and lies in $\Sigma$. This proves (a).

3) Assume that $\dim P > 2$, and let $\pi \in \Pi$. We have to join $\pi$ and $1 \in \Pi$ by a path. We choose $\underline{A}(t) = \underline{A}(\pi)$ constant. In the linear case, we choose a point $c$ ($c \in L_\infty$ in affine planes), and we let $c_k(t)$ vary from $c_k(0) = c_k(\pi)$ to $c_k(1) = c$ within the line $(c_k(\pi) \vee c)\setminus(A_{k-1}\cup A_k)$. This is possible since no finite set can disconnect a line of dimension $l > 1$; indeed otherwise, a coordinatizing ternary, being contractible, would have arbitrarily small neighbourhoods of 0 with finite boundary. The resulting path $\pi(t)$ obviously joins $\pi$ to 1.

For circle planes we start similarly from $\underline{A}(t) = \underline{A}(\pi)$, $\underline{c}(0) = \underline{c}(\pi)$, $\underline{d}(0) = \underline{d}(\pi)$, and we let $c_k(t)$ and $d_k(t)$ vary within $A_{k-1}\setminus A_k$ and $A_k\setminus A_{k-1}$, respectively, such that $c_k(1)||_{(+)}d_k(1)$. Then $\pi(1)$ is a product of (-)-parallel projections, hence is the identity.

4) If $E$ is projective and $\dim P = 2$, we show first that any $\pi \in \Pi$ can be joined by a path to a projectivity $\pi'$ of length $\leq 2$. This will imply that $\pi' \in \pi\Sigma$. So let $n(\pi) > 2$ and consider the pattern formed by $A_i = A_i(\pi)$ and $c_i = c_i(\pi)$ for $0 \leq i \leq 2$. We may perturb the axes $A_i$ slightly, if necessary, so that they are distinct and have no point in common; see Figure 1.

Let $G_i$ be the complementary domains of $A_0\cup A_1\cup A_2$ in $P$, as shown in Figure 1, and let

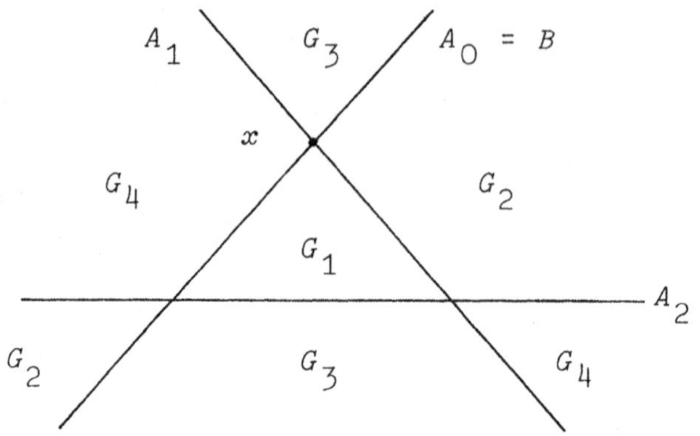


Figure 1.

$G_{ij} = G_i \cup G_j$ $(1 \leq i,j \leq 4)$. The condition (*) allows the centre $c_1$ to cross the line $A_2$, so that $c_1$ moves freely within either $G_{13}$ or $G_{24}$. Likewise, $c_2$ moves freely within $G_{23}$ or $G_{14}$. Thus, it is always possible to move $c_1$ and $c_2$ into the same one of the domains $G_i$. Afterwards, the axis $A_1$ can be rotated about $x = A_1 \wedge A_0$ into $A_0$, sweeping out $G_{13}$ or $G_{24}$. The path in $\Pi$ defined in this way ends at a projectivity of length $< n(\pi)$.

If $\pi'$ is of length 2, $A_0(\pi')$ and $A_1(\pi')$ divide $P$ into two domains $G_1$, $G_2$. If the two centres of $\pi'$ are in the same domain, $\pi'$ can be joined to 1 by a path. This shows that $|\Pi/\Sigma| \leq 3$. If $|\Pi/\Sigma| \neq 2$ then every coset of $\Pi$ mod. $\Sigma$ contains a square, and preserves the orientation of $B$; indeed, if two homeomorphisms of a manifold can be joined by a path of homeomorphisms then they are either both orientation preserving or both orientation reversing. But then $\Pi$ cannot be triply transitive on the 1-sphere $B$. □

3.4 COROLLARY. *In the situation of* 3.3, *let* $\overline{\Pi}$ *be the closure of* $\Pi$ *in the group of all homeomorphisms of* $B$, *with respect to the compact open topology. Let* $\overline{\Pi}^1$ *be the connected component of* $1 \in \overline{\Pi}$.

*a)* $\overline{\Pi}^1$ *is nontrivial.*
*b)* *If* $\dim P > 2$, *then* $\overline{\Pi}^1 = \overline{\Pi}$.
*c)* *If* $\dim P = 2$ *and* $E$ *is projective, then* $\overline{\Pi}/\overline{\Pi}^1$ *is cyclic of order* 2.

*Proof.* It suffices to remark that $\Sigma$ is connected and that, therefore, the closure $\overline{\Sigma}$ also is connected; finally, $\overline{\Pi}$ itself cannot be connected if $\dim P = 2$, since it contains orientation reversing elements.

## § 4 TOPOLOGICAL TRANSFORMATION GROUPS

The special von Staudt theorems for topological planes are all obtained by combining the corresponding theorems of 'abstract' geometry with suitable results on topological transformation groups. In this section, we collect our ingredients from transformation group theory. For proofs, we have to refer to the literature. However, a few explanatory comments will be made. When we say that a group $G$ *acts on a space* $X$, we shall always mean that $G$ is a topological transformation group on $X$, as defined in 3.2a.

4.1 THEOREM [32]. *Let* $G$ *be a locally compact group with a countable basis for its topology, acting transitively on a connected, locally compact and locally contractible space* $X$. *Then* $G$ *is a Lie group, and* $X$ *is a topological manifold.*

A definition of local contractibility can be found in H.Salzmann's contribution. To say that $G$ is a Lie group means that the connected component $G^1$ of the identity is open in $G$ and carries a differentiable structure such that the group operations are differentiable. A topological manifold is a space which is locally homeomorphic to some euclidean space $\mathbb{R}^n$.

Assume for a moment that $\dim G < \infty$ and that $G$ acts on $X = G$ by right translation $x \to xg$. This special case of 4.1 is known as the positive answer to Hilbert's 5th problem; it was first proved in the book of Montgomery and Zippin [22]. In this situation, local connectedness of $G$ suffices instead of local contractibility. The idea of proof is to show that $G$ contains enough homomorphic images of the additive group of real numbers, and to use them in order to build up a local coordinate system around $1 \in G$. In the same book,

p. 243, that theorem was extended to transformation groups on finite dimensional spaces $X$. For spaces of arbitrary dimension, 4.1 was proved by Szenthe [32]. He assumes that $G/G^1$ is compact, but this restriction is inessential since any $G$ contains an open subgroup $H$ with that property ([22], 2.3.1); $H$ acts transitively on the connected space $X$ since all orbits of $H$ are open by [9], (26).

The two notions introduced in the following definition appear in Theorem 4.4 below. In topological planes, the condition of $\omega$-regularity will replace the property $P_n$ of von Staudt groups. Further comments on this condition will be made in §5.

4.2 DEFINITION. Let $X$ be a topological space, and $G$ a group of homeomorphisms of $X$. Let $\tau$ be the compact open topology on $G$ (cf. 3.1).
a) $G$ is said to be *$\omega$-regular* if there exists a finite set $F \subseteq X$ such that the subgroup $G_{[F]}$ fixing $F$ elementwise is discrete with respect to $\tau$.
b) The *modified compact open topology* $\sigma$ for $G$ is defined as the topology generated by the arcwise connected components of all $\tau$-open sets.

Before we state the theorem of Gleason and Palais on the modified compact open topology, we shall illustrate by an example how the modification works. This example will also show that in some relevant cases, this procedure leads to a substantial improvement of the topology.

4.3 EXAMPLE. Let $X$ be the quotient group $\mathbb{R}^2/\mathbb{Z}^2$ (the 2-dimensional torus group), and let $G \cong \mathbb{R}$ be the subgroup $(1,\sqrt{2})\cdot\mathbb{R}$ mod. $\mathbb{Z}^2$; here, $\mathbb{Z}$ denotes the integers. $G$ is obtained by winding the real line round the torus $X$ with no periodicity, in such a way that between any two windings there lie infinitely many other ones. $G$ is a dense subset of $X$. Representing cosets of $\mathbb{Z}^2$ by elements of the unit square in $\mathbb{R}^2$, one obtains the picture of $X$ and $G$ shown in Figure 2.

Let $G$ act on $X$ by right translation, $x^g = x\cdot g$. From the description of the compact open topology by diagonal convergence (3.1), it is easily seen that

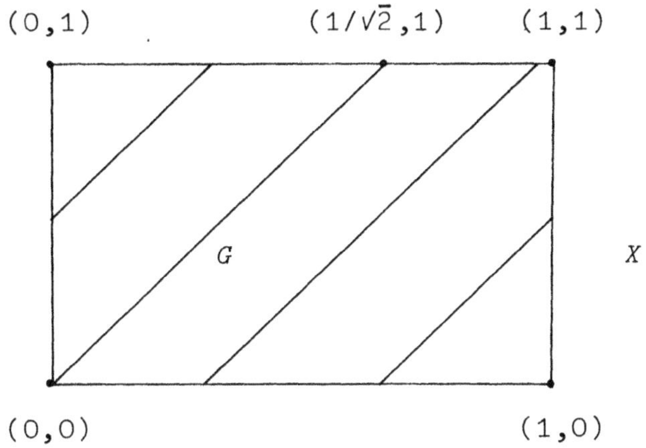


Figure 2.

$\tau$ is the subspace topology of $G \subseteq X$. This topology is rather strange; it is not locally compact, and the complement of any finite set is connected. $(G,\tau)$ is locally homeomorphic to the product of the rational numbers with the real line. This shows that the basic open sets for the modified topology $\sigma$ are open intervals, and it follows that $\sigma$ is the standard topology of $G \cong \mathbb{R}$.

4.4 THEOREM [11]. *Let $X$ be a locally compact, locally connected metric space of finite dimension. Let $G$ be an $\omega$-regular group of homeomorphisms of $X$, endowed with the modified compact open topology $\sigma$.*
*a) $(G,\sigma)$ is a Lie transformation group of $X$.*
*b) As sets, the connected component of $(G,\sigma)$ and the arcwise connected component of $(G,\tau)$ are identical.*

Any discrete group is a Lie group. Without part (b) of the assertion, Theorem 4.4 would therefore be quite useless. We are in position to make use of 4.4 because we possess information on the arcwise connected components of von Staudt groups (3.3).

We give here a rough description of the proof of 4.4. The basic fact about the modification of the topology $\tau$ is that it makes the group $(G,\sigma)$ locally arcwise connected without destroying the transformation group property. The assumption of $\omega$-regularity comes in as follows. Let $F = \{f_1,\ldots,f_n\} \subseteq X$ be such that $(G_{[F]},\tau)$ is discrete. Then

$$g \to (f_i^{\,g})_{1 \leq i \leq n}$$

defines a locally injective continuous mapping of $G$ into $X^n$. This is used to show that the compact connected metric subsets $K$ of $G$ are of bounded dimension. Let $K$ be such a subset which is of maximal dimension. Using local arcwise connectedness of $G$, it is proved next that $K$ has an interior point $k$; otherwise, one finds a small arc $J \subseteq G$ such that the product $G \cdot J$ is of larger dimension. Now $K \cdot k^{-1}$ is a compact connected neighbourhood of $1 \in G$, and $G$ is locally compact, locally connected and finite dimensional. One finally applies the original version of 4.1; see the remarks following that theorem.

The theorems mentioned so far concern the problem of detecting Lie transformation groups. Once we know from these results that a given von Staudt group is a Lie group, the multiple transitivity of that group allows us to determine the group and its action. Actually, all doubly transitive actions of Lie groups have been determined by Tits [33]. We reproduce here only that part of his result which is needed in connection with von Staudt groups, bearing in mind that the blocks of a topological plane belong to a very restricted class of spaces (at least if they are manifolds).

The first steps of Tits's proof are very easy: If a doubly transitive Lie group $G$ contains an abelian normal subgroup $N \neq 1$ then the action of $G$ is determined by the transitive action on $N \setminus \{1\}$, and $N$ has to be a real vector group in order to admit this transitive action. Then $G$ acts as a group of affinities on $N$ considered as a real affine space. If no such $N$ exists, the group is a direct product of simple Lie groups; in fact, $G$ is easily seen to be simple. All simple Lie groups are known, and the difficult part of the proof is to find those subgroups $H \leq G$ which lead to doubly transitive actions on the coset space $G/H$.

4.5 THEOREM [33]. *Let $G$ be a nondiscrete Lie group acting doubly transitively on a space $X$.*
*a) If $X$ is a euclidean space $\mathbb{R}^n$ then $G$ consists of affine maps and contains all translations.*
*b) If $X = S_1$ is a 1-sphere then $G = \mathrm{PSL}_2\mathbb{R}$ or $\mathrm{PGL}_2\mathbb{R}$ acts in the standard way on $X = \mathrm{P}_1\mathbb{R}$.*
*c) If $X = S_2$ then $X = \mathrm{P}_1\mathbb{C}$ and $G = \mathrm{PGL}_2\mathbb{C}$ or*

*$G = \mathrm{PGL}_2\mathbb{C}\cdot\langle\gamma\rangle$, where $\gamma$ is induced by complex conjugation. The action is standard.*

## § 5 VON STAUDT'S THEOREMS FOR TOPOLOGICAL PLANES

As in section 3, we denote by $\Pi = \Pi_B(E)$ the group of projectivities of a block $B \in \mathcal{B}$ of a topological plane $E$. The closure of $\Pi$ with respect to the compact open topology $\tau$ in the group of all homeomorphisms of $B$ is denoted by $\overline{\Pi}$.

5.1 THEOREM [31,20]. *Let $E = (P,\mathcal{B})$ be a locally compact connected linear or circle plane satisfying* (TC 1-4). *Assume either (i) or (ii):*

*i)* $\Pi$ *or* $\overline{\Pi}$ *is locally compact with respect to* $\tau$.
*ii)* $\dim P < \infty$ *and* $\Pi = \Pi_B$ *acts on* $B$ *$\omega$-regularly (see 4.2a).*

*Then 1) If $E$ is affine, $E$ is a translation plane.*
*2) If $E$ is projective, $E$ is the 'classical' desarguesian or Moufang plane $\mathrm{P}_2\mathbb{F}$, $\mathbb{F} = \mathbb{R}, \mathbb{C}, \mathbb{H}$ or $\mathbb{O}$.*
*3) If $E$ is a circle plane, $E$ is the classical geometry of plane sections of a quadric in $\mathrm{P}_3\mathbb{F}$, $\mathbb{F} = \mathbb{R}$ or $\mathbb{C}$.*

REMARKS. a) With its alternative hypotheses, Theorem 5.1 represents the two main types of geometric theorems in topological geometry (there also are topological theorems in topological geometry; see Salzmann's contribution). Indeed, the assertion of 5.1 is primarily geometrical. Hypothesis (i) is mainly topological and cannot be formulated in abstract geometry. Hypothesis (ii), on the other hand, is almost purely geometrical. If one replaces the word 'discrete' in the definition of $\omega$-regularity by 'finite' then condition (ii) becomes entirely geometrical, except that we assume that $E$ admits a topology. In this case, the topology comes in only as a catalyst, as it were, in the proof.
b) Let us compare the condition of $\omega$-regularity (called 'property $Q_\omega$' in [20]) to the von Staudt-Schleiermacher conditions $P_n$ considered in the earlier chapters of this book. We claim that $Q_\omega$ is in several ways weaker than each $P_n$. Indeed, $Q_\omega$ requires *existence* of some

set $F$ of *arbitrary* finite cardinality such that $\Pi_{[F]}$ is *discrete*, while $P_n$ requires that $\Pi_{[F]}$ is *trivial* for *each* set of *fixed* cardinality $n$. This makes it possible to cover the Moufang plane $P_2\mathbb{O}$ and all translation planes in Theorem 5.1, even though the von Staudt groups of these planes all contain nontrivial elements fixing an arc pointwise; cf. 1.2a and 1.7.

We come now to the *proof of* 5.1. If the group $\Pi$ or $\overline{\Pi}$ is locally compact with respect to $\tau$ then 4.1 shows that it is in fact a Lie group. In order to verify the hypotheses of 4.1, recall from 3.1 that $\tau$ has a countable basis, and from Salzmann's contribution that $B$ is locally contractible; cf. also 2.3. If (ii) is assumed, the same result is obtained for $(\Pi,\sigma)$ by applying 4.4. In each case, the Lie group $\Gamma = (\Pi,\tau)$ or $(\overline{\Pi},\tau)$ or $(\Pi,\sigma)$, respectively, is not discrete; this follows from 3.3a and 4.4b.

By 4.1, the line $B$ is a manifold. If $E$ is affine, this implies that $B$ is homeomorphic to some euclidean space $\mathbb{R}^n$; see Salzmann's contribution. In this case, 4.5 asserts that $\Pi$ consists of affine maps of $\mathbb{R}^n$, and the isotropy group $\Pi_{p,q}$ of two points fixes elementwise the line of $\mathbb{R}^n$ joining $p$ and $q$. By a theorem of Schleiermacher ([26], Lemmata 4, 5; cf. Pickert's and Lüneburg's contributions), $E$ is a translation plane.

If $E$ is projective, form an affine plane $E'$ by removing any line, and let $\Pi' = \Pi_B(E')$. Then $\Pi'$ is contained in the Lie group $\Gamma$, and its closure $\Delta$ in $\Gamma$ is locally compact in the topology induced by $\Gamma$. Now the topology $\sigma$ on $\Gamma$ might not induce $\sigma$ on $\Delta$, but $(\Pi',\tau)$ contains arcs; these remain arcs in $(\Gamma,\sigma)$ and hence also in $\Delta$ with the induced topology. So $\Delta$ is certainly not discrete, and we can apply to $\Delta$ the arguments of the first two paragraphs. In this way we see that $E'$ is a translation plane. Since this is true for any affine subplane, $E$ is Moufang and (2) follows; see [27], 7.18 and 7.23.

Finally, if $E$ is a circle plane, $B$ is a sphere $S_n$, $n = 1$ or $2$ (2.3). In that case, 4.5 tells us that $\Gamma$ is sharply triply transitive unless $\Gamma = PSL_2\mathbb{R}$ (which is not even triply transitive and hence is impossible) or $\Gamma = PGL_2\mathbb{C}\cdot\langle\tau\rangle$. The latter group is disconnected contrary to 3.3, 3.4, 4.4. It now follows from results of Freudenthal, Strambach and Kroll [10, 19] that $E$ is the geometry of plane sections of a quadric in $P_3\mathbb{F}$, where $\mathbb{F}$ is a commutative field. $\mathbb{F}$ has to be locally compact and connected, hence $\mathbb{F} = \mathbb{R}$ or $\mathbb{C}$. □

5.2 REMARK. Of the other von Staudt type theorems of abstract geometry, Schleiermacher's result [26] characterizing affine translation planes by the existence of a sharply transitive normal subgroup $\Gamma \triangleleft \Pi = \Pi_B$ has an interesting analogue in locally compact connected affine planes of dimension 2. Instead of $\Gamma \triangleleft \Pi$ sharply transitive on $B$ one needs only $1 \neq \Gamma \triangleleft \Pi_b$ acting freely on $B\setminus\{b\}$ in order to force the plane to be desarguesian (there are no 2-dimensional proper translation planes). This result was proved by K. Strambach, but has not yet been published.

## § 6 THE BUNDLE GROUP OF A TOPOLOGICAL OVAL

Concerning the material presented in the following two sections, the original papers [31, 20] contain several errors, which will be corrected here. In two cases, the errors affect the statement of results. Firstly, 6.6 below cannot be improved in the 2-dimensional case; a counterexample is given in 6.8. Secondly, there exists no such thing as *the* bundle group of a circle plane, cf. 7.3; this forces us to use a somewhat unsatisfactory hypothesis in 7.5. Again, we show by a counterexample (7.6) that the result is best possible. As with 6.6, the original strong form of Theorem 7.5 does hold in dimension 4; see 7.7.

Throughout this section, $O$ denotes a closed oval in a compact connected projective plane $E = (P,\mathcal{B})$, and we assume that $\dim P$ is finite. Recall from 2.2 that $O \cong B \in \mathcal{B}$ is homeomorphic to a sphere of dimension 1 or 2.

6.1 DEFINITION. For $p \in P\setminus O$, the *bundle involution* $\sigma_p : O \to O$ is defined by the condition that

$$y = x^{\sigma_p}$$

if and only if $x$, $y$ and $p$ are collinear and $\{x,y\} = (x \vee p) \cap O$. The *bundle group* $\Sigma = \Sigma_O$ of $O$

is defined as the group generated by all $\sigma_p$, $p \in P\setminus O$. We endow $\Sigma$ with the compact open topology $\tau$.

It is clear that $\sigma_p^2 = 1$, and we shall see later (6.3, 6.4) that actually $\sigma_p \neq 1$ so that $\sigma_p$ deserves its name.

6.2 LEMMA. *a) The map* $(P\setminus O)\times O \to O$: $(p,x) \to x^{\sigma_p}$ *is continuous; in particular, each bundle involution is a homeomorphism.*
*b)* $(\Sigma_O,\tau)$ *is a topological transformation group on* $O$.
*c) The map* $P\setminus O \to \Sigma$: $p \to \sigma_p$ *is continuous.*
*d) The arcwise connected component of* $\Sigma$ *is nontrivial.*

*Proof.* Only part (a) requires a proof; everything else follows as in 3.2. First recall from 2.2 that $O$ is topological. This implies that the map

$$\psi = \varphi_O^{-1}: \mathcal{B}(O) \to O*O: B \to B\cap O$$

is a homeomorphism. ($\mathcal{B}(O)$ is the set of lines meeting $O$.) Now for any Hausdorff space $X$, the map

$$\delta_X: (X*X)\times X \to X: (\{x,y\},x) \to y$$

is continuous. This is easily verified using the basic properties of the quotient topology. Finally, let $\delta = \delta_O$; then we have

$$x^{\sigma_p} = ((x\vee p)^{\psi},x)^{\delta},$$

and the right hand side depends continuously on $(p,x)$. □

Next we show how individual bundle involutions can be used to prove geometric properties of ovals. We denote by

$$\mathcal{B}(p,i) = \mathcal{B}(p,i,O)$$

the set of lines passing through $p$ and meeting $O$ in precisely $i = 0$, 1 or 2 points.

6.3 PROPOSITION [3]. *Assume that* $\dim P = 4$.
*a) Each bundle involution* $\sigma_p$ *is topologically equivalent to a rotation of order* 2 *of the* 2*-sphere* $O$.
*b) Each point* $x \in P\setminus O$ *lies on exactly two tangents:* $|\mathcal{B}(x,1)| = 2$.

*Proof.* It was proved by Brouwer and Kerékjártó (see [17]) that every homeomorphism of order $\leq 2$ of the 2-sphere $S_2$ is topologically equivalent to an orthogonal linear automorphism of $\mathbb{R}^3$, restricted to the unit sphere $S_2 \subseteq \mathbb{R}^3$. We have to exclude the possibilities that $\sigma_p$ is the identity $id$, the antipodal map $-id$, or a reflection at a hyperplane of $\mathbb{R}^3$.

By 2.2b, no exterior lines exist, and the map $\rho_p: O \to B_p: x \to x \vee p$ is a continuous surjection $(p \notin O)$. Since $O$ is compact, $\rho_p$ is a quotient map; i.e., $B_p$ carries the quotient topology with respect to $\rho_p$. Two points $x, y \in O$ are identified by this quotient map if and only if they are interchanged by $\sigma_p$, so that

$$B_p \cong O/\langle\sigma_p\rangle.$$

Next, observe that $\rho_p$ induces a bijection between the set $F(\sigma_p)$ of fixed points of $\sigma_p$ and the set $B(p,1)$ of tangents. We know that the cardinality of $F(\sigma_p)$ is 0, 2, or infinity.

Now assume that $\sigma_p = id$ for some point $p$. Then $F(\sigma_p) = O$ and

$$B(p,1) = B_p = B_1(O)$$

is the set of all tangents of $O$. Then each point $q \neq p$ lies on exactly one tangent, and $|F(\sigma_q)| = 1$, a contradiction. If $\sigma_p = -id$ or if $\sigma_p$ is a reflection at a hyperplane then $O/\langle\sigma_p\rangle$ is homeomorphic to the real projective plane or to a closed disc, respectively, and is certainly not homeomorphic to $B_p$. Again, this is a contradiction. □

Similar considerations work in the 2-dimensional case. In fact, by carrying the argument a little further, one can prove the following theorem, which shows that the topological and geometrical properties of any topological oval resemble those of a conic in a pappian plane as closely as one could wish.

6.4 THEOREM [3]. *Let* $O$ *be a closed oval in a locally compact connected projective plane* $E = (P,B)$ *with* $\dim P < \infty$. *There exists a homeomorphism* $f$ *of* $P$ *onto the point set of* $P_2\mathbb{F}$, $\mathbb{F} = \mathbb{R}$ *or* $\mathbb{C}$, *carrying* $O$ *onto a conic* $C$. *Moreover, for* $p \notin O$ *we have*

$$|B(p,i,O)| = |B(p^f,i,C)|;$$

*i.e.,* $p$ *and* $p^f$ *lie on the same number of secants, tangents, and exterior lines.*

6.5 COROLLARY [3]. *The set* $B_1(O)$ *of tangents is a closed oval in the dual plane* $E^*$.

6.4 and 6.5 are a condensed form of [3], 3.7-3.10. The idea of the *proof* is easiest explained in the 4-dimensional case: One proves first the corollary, which is a direct consequence of 6.3. Then one considers the oval $O^* = B_1(O)$ in $E^*$ and the dual $C^*$ of a conic in $P_2\mathbb{C}$. The map $f$ is obtained as the composite

$$P \xrightarrow{\varphi_{O^*}^{-1}} O^* * O^* \xrightarrow{h*h} C^* * C^* \xrightarrow{\varphi_{C^*}} P_2\mathbb{C},$$

where $h: O^* \cong C^*$ is any homeomorphism. $f$ carries $O$ onto $C$ via the diagonals of the symmetric squares $O^* * O^*$ and $C^* * C^*$, and everything else follows from 6.3 and 2.2b. □

From individual bundle involutions we now turn back to the bundle group. That group was used in Buekenhout's proof of his theorem that any pascalian oval is a conic; cf. Mäurer's contribution. Part of this argument was separated off by Buekenhout [5] and stated like this: Any oval with a sharply triply transitive bundle group is a conic. However, this does not mean that a given plane containing that oval $O$ must be pappian or even contain $O$ embedded as a conic. Indeed, Buekenhout only considers the permutation structure formed by the bundle involutions, and the above statement means that this structure is isomorphic to the corresponding structure of a conic. Now the set of bundle involutions determines only a certain part of the embedding of $O$ into a plane. In topological planes, however, this affects only the low dimensional case. Moreover, we can again replace the regularity condition by a much weaker one.

6.6 THEOREM [31, 20]. *Let* $O$ *be a closed oval with bundle group* $\Sigma$ *in a locally compact connected projective plane* $E = (P,B)$. *Assume either (i) or (ii):*

(*i*) $\Sigma$ *or* $\overline{\Sigma}$ *is locally compact with respect to the compact open topology* $\tau$.

(*ii*) $\dim P < \infty$ *and* $\Sigma$ *is* $\omega$*-regular.*

*Then* $\dim P = 2$ *or* $4$, *and*

*a) If* $\dim P = 4$, *then* $E \cong P_2\mathbb{C}$ *is pappian and* $O$ *is (embedded as) a conic.*

*b) If* $\dim P = 2$, *the partial geometry* $(P,B(O))$ *together with* $O$ *is isomorphic to the corresponding structure defined by a conic in* $P_2\mathbb{R}$.

6.7 REMARK. Stated in a different way, 6.6b says that at most the exterior lines of $O$ differ from those of a conic in $P_2\mathbb{R}$; everything else is classical. In particular, consider the geometry

$$\mathrm{Int}(O)$$

of *interior points* of $P$ with respect to $O$ (a point is called interior if it lies on no tangent) together with the secants of $O$ as lines; 6.6b implies that $\mathrm{Int}(O)$ is isomorphic to the real hyperbolic plane $\mathrm{Int}(C)$.

*Proof of* 6.6. As in the proof of 5.1, case (3), we find that $\Sigma$ is a Lie group and is sharply triply transitive on $O$. We only have to replace 3.2 and 3.3a by 6.2 in that argument; moreover, we do not know that $\Sigma$ is connected if $\dim P = 4$ (3.3b). Instead, we know (6.3a) that then $\Sigma$ is generated by orientation preserving maps of $O$, so that $\Sigma$ cannot be the extension of $PGL_2\mathbb{C}$ by the orientation reversing map induced by complex conjugation. Actually, this proof shows that $\Sigma_O$ is isomorphic as a transformation group to the bundle group $\Sigma_C$ of a conic $C$ in $P_2\mathbb{F}$, $\mathbb{F} = \mathbb{R}$ or $\mathbb{C}$.

Next, we shall establish an isomorphism between $(P,B(O))$ and $(P_2\mathbb{F},B(C))$ carrying $O$ onto $C$. This will prove the theorem since $B = B(O)$ if $\dim P = 4$, see 2.2b. Note first that the set $I_O$ of all bundle involutions of $O$ actually contains all involutions $\sigma \in \Sigma_O$. Indeed, choose $x, y \in O$ such that $x, y, x^\sigma, y^\sigma$ are four distinct points. Let

$$p = (x \vee x^\sigma) \wedge (y \vee y^\sigma).$$

Then $\sigma$ coincides with $\sigma_p$ on those four points, and $\sigma = \sigma_p$ since $\Sigma$ is 3-regular.

Now the geometries $(P, \mathcal{B}(O))$ and $(P_2\mathbb{F}, \mathcal{B}(C))$ can both be reconstructed within the transformation group $(\Sigma, O) \cong (\Sigma, C)$, which shows that they are isomorphic. To do this, map $P$ onto the disjoint union $O \cup I_O$ by sending $p \in O$ to itself and $p \in P \setminus O$ to $\sigma_p$. Next, map $\mathcal{B}(O)$ onto $O*O$ by sending $B$ to $B \cap O$; this is a homeomorphism. The pair consisting of these two mappings is incidence preserving if incidence in the geometry $(O \cup I_O, O*O)$ is defined by

$$p \text{ I } \{x,y\} \iff \{x,y\}^{\sigma_p} = \{x,y\} \quad \text{for } p \in P\setminus O$$

$$z \text{ I } \{x,y\} \iff z \in \{x,y\} \quad \text{for } z \in O.$$

The resulting isomorphism between $(P, \mathcal{B}(O))$ and $(P_2\mathbb{F}, \mathcal{B}(C))$ is a homeomorphism on $\mathcal{B}(O)$. Since convergence of points can be expressed in terms of convergence of lines, it is also a homeomorphism on points. □

The following *example of an oval with a sharply triply transitive bundle group in a nondesarguesian compact connected plane* shows that 6.6b is best possible.

6.8 EXAMPLE. Let $E$ be the nondesarguesian plane described in the 2nd edition of Hilbert's 'Grundlagen' [14]. $E$ is obtained from $P_2\mathbb{R}$ by distorting the lines in the interior of a conic $C$. More precisely, the interior part of a line is replaced by a circular arc with the same end points, such that the corresponding circle passes through a fixed point $x$ in the exterior of $C$. (By a circle, we mean here a circle in the euclidean affine plane.)

$C$ is an oval in $E$. Consider the dual oval $C^*$ in the dual plane $E^*$, as in 6.5. It is easy to see that the construction of the bundle group $\Sigma_{C*}$ involves only the nondistorted part of the geometry. Thus, $\Sigma_{C*}$

coincides with the bundle group of $C^*$ in $(P_2\mathbb{R})^*$ and is sharply triply transitive. Yet, the plane is certainly not pappian; in fact, its automorphism group is of order 2.

## § 7 BUNDLE GROUPS OF TOPOLOGICAL CIRCLE PLANES

7.1 For the considerations in this section, it is necessary to replace our axiom (TC 3) for circle planes by an apparently stronger one. However, any conceivable proof of (TC 3) from (TC 1) should actually prove this stronger axiom.

Consider $B \in \mathcal{B}\setminus\mathcal{B}_p$ and the oval $B_p$ in $\overline{E_p} = \overline{(P_p, \mathcal{B}_p)}$. The identity mapping of $B \cap P_p$ can be extended to a bijection

$$f = f(p,B)\colon B \to B_p$$

by the definition

$$[pr_{(\pm)}(p,B)]^f = \infty_{(\mp)}.$$

Observe that the (+)-parallel projection of $p$ goes to the point at infinity on the (-)-parallel classes; this is suggested by the behaviour of the classical Minkowski plane. Now we require

(TC 3') For each $B \in \mathcal{B}\setminus\mathcal{B}_p$, the oval $B_p \subseteq \overline{E_p}$ is closed, and $f(p,B)\colon B \to B_p$ is a homeomorphism.

If we assume only (TC 0,1,2,3), then (TC 3') is automatically satisfied in Moebius planes and is easy to prove in Laguerre planes since, there, $B$ and $B_p$ are both one point compactifications of $B \cap P_p$. In a Minkowski plane with compact circles, the worst thing that could happen is that $f(p,B)$ interchanges the two end points of $B \cap P_p$.

7.2 DEFINITION. In a circle plane $E$, we consider now two kinds of *bundle involutions*. One kind uses a bundle $\mathcal{B}(p,q)$ of circles defined by $p, q \notin B$, the other uses a pencil $\mathcal{B}(p,C)$ of mutually tangent circles defined by $p \in C \in \mathcal{B}$, $p \notin B$. In both cases, the intersection points of $B$ with an element of

$B(p,q)$ or $B(p,C)$ are interchanged by the bundle involution $\sigma(p,q)$ or $\sigma(p,C)$, respectively. Moreover, we define

$$[pr_{(\pm)}(p,B)]^{\sigma(p,q)} = pr_{(\mp)}(q,B)$$
$$[pr_{(\pm)}(p,B)]^{\sigma(p,C)} = pr_{(\mp)}(p,B),$$

as it is suggested by the classical case. Note that, again, the two parallel projections are interchanged in the Minkowski case. Finally, we define the *bundle group*

$$\Sigma = \Sigma_B(E)$$

as the group generated by all bundle involutions of $B$.

7.3 REMARK. The groups of projectivities $\Pi_B(E)$ are isomorphic as permutation groups for all choices of $B \in \mathcal{B}$. *This need not be true for bundle groups*; see 7.6 below. Thus, we can not speak of *the* bundle group of $E$. The reason is that all perspectivities of $E$ generate a groupoid $\Gamma(E)$, in which the groups $\Pi_B(E)$ appear as object groups; thus, they are conjugate in $\Gamma(E)$. No such groupoid exists in the case of bundle groups. (A groupoid $\Gamma$ is a category in which all morphisms are invertible; an object group is the group of all morphisms of one object to itself.)

The axiom (TC 3') was devised so that bundle involutions can be transferred into the localizations of $E$:

7.4 LEMMA. *a) All bundle involutions are homeomorphisms.*
*b) For $p \in P \setminus B$, the bundle group $\Sigma_B(E)$ contains the permutation group*

$$f(p,B)\Sigma_{B_p}(\overline{E_p})f(p,B)^{-1}$$

*as a subgroup.*

For the *proof*, it suffices to check that

$$f(p,B)^{-1}\sigma(p,q)f(p,B) = \sigma_q \in \Sigma_{B_p}(\overline{E_p})$$
$$f(p,B)^{-1}\sigma(p,C)f(b,B) = \sigma_c \in \Sigma_{B_p}(\overline{E_p}),$$

where $c$ is the point at infinity of $E_p$ corresponding to the pencil $\mathcal{B}(p,C)$. This proves (a) and (b).

7.5 THEOREM [31, 20]. *Let* $E = (P,\mathcal{B})$ *be a topological circle plane with* dim $P = 2$, *satisfying* (TC 0-2) *and* (TC 3'). *If* $E$ *is a Möbius plane, assume that it satisfies the bundle theorem. Assume that the bundle group of* e a c h *circle* $B$ *satisfies (i) or (ii):*

*(i)* $\Sigma = \Sigma_B(E)$ *or* $\overline{\Sigma}$ *is locally compact with respect to the compact open topology.*

*(ii)* $\Sigma$ *is* $\omega$*-regular.*

*Then* $E$ *is the classical real Möbius or Laguerre or Minkowski plane.*

PROBLEM. Is 7.5 true for Möbius planes without assuming the bundle theorem?

*Proof of* 7.5. Let $p \notin B \in \mathcal{B}$. First of all, the usual proof shows that $\Sigma_B(E)$ or $\overline{\Sigma}$ is a sharply triply transitive Lie group. By 7.4, it contains a triply transitive subgroup permutation isomorphic to $\Sigma_{B_p}(\overline{E_p})$. Therefore, $B_p$ satisfies the hypotheses of 6.6. We want to show that $\overline{E_p}$ is pappian and $B_p$ is a conic. We know from 6.7 that $\mathrm{Int}(B_p)$ is isomorphic to the real hyperbolic plane. For a fixed $p$, we show next that each point $x \in P_p$ lies on one of the hyperbolic planes $\mathrm{Int}(B_p)$. To see this, first choose a circle $B$ such that $x \in B_p$ and then move it slightly to the 'exterior' side, keeping it within some bundle $\mathcal{B}(q,r)$. This shows that $E_p$ is a locally hyperbolic plane in the sense of Polley [23] and, hence, is pappian. That each $B_p$ has to be a conic now follows from the fact that all embeddings of the hyperbolic plane into $P_2\mathbb{R}$ are projectively equivalent; see [7], § 29. (In fact, a far more general proposition about embeddings of stable planes into desarguesian projective planes is true, which I hope to publish soon.)

If $E$ is a Laguerre or Minkowski plane, $E$ can now be redescribed within any of the localizations $E_p$, as in the proof of 2.5. This completes the present proof in that case.

If $E$ is a Möbius plane satisfying the bundle theorem then $E$ is ovoidal [13, 24, 16]; i.e., $E = E(O)$ is the geometry of plane sections of an ovoid in $\mathbb{R}^3$. We know that all plane sections of $O$ are conics, and $O$ must be a quadric ([6]; cf. Mäurer's contribution). Therefore, $E$ is classical. □

The following example shows that 7.5 *fails if regularity is assumed only for one bundle group* $\Sigma_B(E)$. At the same time, it is an example of *a Möbius plane with different bundle groups for different circles.*

7.6 EXAMPLE. Let $S$ be the unit sphere in $\mathbb{R}^3$, and let $O$ be an ovoid contained in the convex hull of $S$. Assume that $O$ and $S$ have a circle $B$ in common. Then the bundle groups $\Sigma_B(E(O))$ and $\Sigma_B(E(S))$ of $B$ with respect to the Möbius planes defined by $O$ and $S$ coincide.

Indeed, each bundle involution of $B$ in $E(O)$ can be described as follows. Take a line $l$ of $\mathbb{R}^3$ meetig $O$ but not $B$; for each plane $E$ containing $l$, interchange the points of $E \cap B$. If $l$ meets $O$ then it meets $S$ also, and $\Sigma_B(E(O))$ is contained in $\Sigma_B(E(S))$. The smaller group is triply transitive, and the larger one satisfies $P_3$, so they coincide. Clearly, $E(O)$ can be chosen nonmiquelian; then $E(O)$ must contain some other circle with a nonregular bundle group.

Nothing similar can happen in dimension 4:

7.7 THEOREM. *Let* $E = (P, \mathcal{B})$ *be a topological circle plane with* dim $P = 4$, *satisfying* (TC 0-2) *and* (TC 3'). *If there exists a circle* $B \in \mathcal{B}$ *such that the bundle group* $\Sigma_B(E)$ *satisfies the condition (i) or (ii) of* 7.5, *then* $E$ *is the classical complex Laguerre or Minkowski plane.*

*Proof.* By 2.5, it suffices to find a desarguesian localization of $E$. This can be done as in the first steps of the proof of 7.5.

## REFERENCES


1. Arens, R.: *Topologies for homeomorphism groups*, Amer. J. Math. 68, 1964, pp. 593-610.
2. Betten, D.: *Die Projektivitätengruppe der Moulton-Ebenen*, J. Geom. 13, 1979, pp. 197-209.
3. Buchanan, T., Hähl, H., and Löwen, R.: *Topologische Ovale*, Geom. Dedic. 9, 1980, pp. 401-424.
4. Buchanan, T.: *Ovale und Kegelschnitte in der komplexen projektiven Ebene*, Math. Phys. Sem. Ber. 26, 1979, pp. 244-260.
5. Buekenhout, F.: *Etude intrinsèque des ovales*, Rend. Math. V. Ser. 25, 1966, pp. 333-393.
6. Buekenhout, F.: *Ensembles quadratiques des espaces projectifs*, Math. Zeit. 110, 1969, pp. 306-318.
7. Busemann, H., Kelley, P.J.: *Projective geometry and projective metrics*, New York: Academic Press, 1953.
8. Dugundji, J.: *Topology*, 5th ed., Boston: Allyn & Bacon, 1970.
9. Freudenthal, H.: *Einige Sätze über topologische Gruppen*, Ann. Math. 37, 1936, pp. 46-56.
10. Freudenthal, H., Strambach, K.: *Schließungssätze und Projektivitäten in der Möbius- und Laguerregeometrie*, Math. Zeit. 143, 1975, pp. 213-234.
11. Gleason, A.M., Palais, R.S.: *On a class of transformation groups*, Amer. J. Math. 79, 1957, pp. 631-648.
12. Groh, H.: *Topologische Laguerreebenen, I, II*, Abh. math. Sem. Univ. Hamb. 32, 1968, pp. 216-231 and 34, 1969/70, pp. 11-21.
13. Hesselbach, B.: *Über zwei Vierecksätze in der Kreisgeometrie*, Abh. math. Sem. Univ. Hamb. 9, 1933, pp. 265-271.
14. Hilbert, D.: *Grundlagen der Geometrie*, 2nd. ed., Leipzig: Teubner, 1903.
16. Kahn, J.: (to appear in Math. Zeit.)
17. Kerékjártó, B. von: *Über die periodischen Abbildungen der Kreisscheibe und der Kugelfläche*, Math. Ann. 80, 1919, pp.36-38.
18. Kegel, O.H., Schleiermacher, A.: *Amalgams and embeddings of projective planes*, Geom. Dedic. 2, 1973, pp. 379-395.
19. Kroll, H.: *Die Gruppe der eigentlichen Projektivitäten in Benz-Ebenen*, Geom. Dedic. 6, 1977, pp. 407-413.
20. Löwen, R.: *Schleiermachers Starrheitsbedingung für Projektivitäten in der topologischen Geometrie*, Math. Zeit. 155, 1977, pp. 23-28.

21. Löwen, R.: *Homogeneous compact projective planes*, J. reine u. angew. Math., to appear.
22. Montgomery, D., Zippin, L.: *Topological transformation groups*, New York: Interscience, 1955.
23. Polley, C.: *Lokal desarguessche Salzmann-Ebenen*, Arch. Math. 19, 1968, pp. 553-557.
24. Reidemeister, K.: *Eine Kennzeichnung der Kugel nach W. Blaschke*, J. reine u. angew. Math. 154, 1925, pp. 8-15.
25. Schenkel, A.: *Topologische Minkowskiebenen*, Dissertation, Erlangen-Nürnberg 1980.
26. Schleiermacher, A.: *Reguläre Normalteiler in der Gruppe der Projektivitäten bei projektiven und affinen Ebenen*, Math. Zeit. 114, 1970, pp. 313-320.
27. Salzmann, H.: *Topological planes*, Adv. in Math. 2, 1967, pp. 1-60.
28. Salzmann, H.: *Compact planes of Lenz type III*, Geom. Dedic. 3, 1974, pp. 399-403.
29. Salzmann, H.: *Compact 8-dimensional projective planes with large collineation groups*, Geom. Dedic. 8, 1979, pp. 139-161.
30. Salzmann, H.: *Automorphismengruppen 8-dimensionaler Ternärkörper*, Math. Zeit. 166, 1979, pp. 265-275.
31. Strambach, K.: *Der von Staudtsche Standpunkt in lokal kompakten Geometrien*, Math. Zeit. 155, 1977, pp. 11-21.
32. Szenthe, J.: *On the topological characterization of transitive Lie group actions*, Acta Sci. Math. Szeged. 36, 1974, pp. 323-344.
33. Tits, J.: *Sur certaines classes d'espaces homogènes de groupes de Lie*, Acad. roy. Belgique, Cl. Sci., Mem., Coll. $8^o$, 29 No. 3, 1955.
35. Whyburn, G.T.: *Analytic Topology*, Providence: American Mathematical Society, 1942.
36. Wölk, D.: *Topologische Möbiusebenen*, Math. Zeit. 93, 1966, pp. 311-333.

# SEMIMODULAR LOCALLY PROJECTIVE LATTICES OF RANK 4 FROM v.STAUDT'S POINT OF VIEW

Armin Herzer

Johannes Gutenberg-Universität Mainz

We consider groups of projectivities in a certain kind of lattices called "Spaces", also comprising the circle planes, and give theorems of v.Staudtian type, which characterize those Spaces which can be represented by a sublattice of a projective geometry of rank 4.

## 1. BASIC DEFINITIONS

Let L be a lattice with least element 0 and greatest element 1 and partial order $\leq$, where for $a,b \in L$ the <u>join</u> (lowest upper bound) of a and b is denoted by $a \vee b$, and the <u>intersection</u> (greatest lower bound) of a and b is denoted by $a \wedge b$.

We say that b <u>covers</u> a ($a \lessdot b$), if $a<b$ holds but $a<c<b$ is not satisfied for any $c \in L$. A <u>rank function</u> is a map r of L into the set of the natural numbers together with 0, such that $r0 = 0$ and $a \lessdot b \Leftrightarrow rb = ra+1 \ \forall a,b \in L$. We say that L is <u>semimodular</u> of rank n, if L admits a rank function r with $r1 = n$, for which the following "rank formula" holds:

$$r(a \vee b)+r(a \wedge b) \leqq ra+rb \qquad \forall a,b \in L.$$

The latter condition is equivalent to the following:

$$a \wedge b \lessdot a \Rightarrow b \lessdot a \vee b \qquad \forall a,b \in L.$$



*P. Plaumann and K. Strambach (eds.), Geometry – von Staudt's Point of View, 373–400.*

In any case, for $c \in L$ the rank of c is the length of a maximal chain from 0 to c. The set of all elements of L of rank i we denote by $L_i$.

The atoms (the elements of $L_1$) we call <u>points</u>, the elements of $L_2$ are called <u>lines</u> and the elements of $L_3$ <u>planes</u>. In general the elements of L also are called <u>subspaces</u> (of the <u>whole space</u> 1).

In fact, a semimodular lattice of rank 4 behaves rather as a nice geometry: To any two points there is exactly one line containing these points; to any three non-collinear points (or a line and a point outside of this line) there is exactly one plane containing these subspaces; if a line l is contained in the different planes e and f, then $e \wedge f = l$ and so on. The only strange things which can occure, are lines which contain only one point; such lines are called <u>tangent lines</u>, whereas the lines containing at least two points are called <u>proper lines</u>. More general, a subspace is called <u>proper</u> if it is the <u>span</u> (i.e. the join) of its points.
A <u>geometric lattice</u> is a semimodular lattice (of finite rank) all subspaces of which are proper.

For $c \in L$ let $L(c) = \{x \in L \mid c \leqq x\}$ be the lattice of all subspaces of L containing c (with the induced lattice operations). Then L(c) is a semimodular lattice too; his least element is c. If for instance c is a point, the lines of L containing c are the points of L(c) and so on.

Let L be a semimodular lattice of rank n. We call L <u>locally projective</u>, if for every $p \in L_1$ the lattice L(p) is a projective geometry. (We consider a projective geometry as the lattice of its subspaces.) The projective geometries L(p) have the same order for all $p \in L_1$, called the <u>order of</u> L.

EXAMPLE. (1) Let L be a projective geometry of rank n (considered as lattice of its subspaces), and let $\mathcal{Q}$ be a set of points of L spanning L. Define

$$L(\mathcal{Q}) = \{x \in L \mid \text{there is a } p \in \mathcal{Q} \text{ with } p \leqq x\} \cup \{0\}$$

and consider $L(\mathcal{Q})$ as lattice induced by the partial order of L. Then $L(\mathcal{Q})$ is a semimodular locally projective lattice of rank n. For we can use the rank in L as rank function for $L(\mathcal{Q})$ and also the join is the same in L and in $L(\mathcal{Q})$, whereas the intersection in $L(\mathcal{Q})$ is contained in the intersection of L. Because the rank formula is valid in L (with equality sign), the rank formula also holds in L($\mathcal{Q}$). So L($\mathcal{Q}$) is semimodular. Furthermore the lattice of all subspaces of L($\mathcal{Q}$) containing the point p of $\mathcal{Q}$ is identical with L(p); thus L($\mathcal{Q}$) is locally projective.

(2) Let L be a projective geometry of rank 4 and $\mathcal{Q}$ consist of all points of four planes of L in general position. Then L($\mathcal{Q}$) contains all lines and planes of L and moreover every line of L possesses at least two points of $\mathcal{Q}$. So L($\mathcal{Q}$) is even geometric.

(3) Let L be as in (2) and $\mathcal{Q}$ be an ovoid in L. Then the planes of L($\mathcal{Q}$) containing more than one point represent the circles of the corresponding inversive plane. Through every point p of L($\mathcal{Q}$) (= point of $\mathcal{Q}$) goes exactly one plane containing all tangent lines through p (the tangent plane).

Such a semimodular lattice of rank 4 in which - as in example (3) - all the tangent lines through any common point lie in a common plane, shall be called <u>smooth</u>. (Also the lattice of example (2) is smooth, because there are no tangent lines.) Now we are ready to define as our main object the concept of "Space", namely a smooth locally projective semimodular lattice of rank 4 and order >4. More exactly:

DEFINITION. A <u>Space</u> is a lattice L with 0 and 1 satisfying the following conditions (L1)-(L3):

(L1) L is a semimodular lattice of rank 4.

(L2) For every $p \in L_1$ the lattice L(p) is a projective plane of order >4.

(L3) There exists a mapping $L_1 \to L_3$; $p \mapsto \langle p \rangle$, such that every tangent line through p lies in $\langle p \rangle$.

The planes of the form $\langle p \rangle$ sometimes we call <u>tangent planes</u>. The condition (L2) has the consequence that in L every two different planes which have a point in common already intersect in a line. This property is very important for our definition of perspectivity in section 4, - and important at all. The excluded cases of order $\leq 4$ are not without interest, but must be omitted here. An interisting example of a geometric locally projective lattice of rank 4 which is excluded from Spaces by the order condition in (L2) comes from the Steiner system S(22,6,3) (cfr.[12],pp.424 Example, [5],pp.481, Beispiel 3.3).

There arises the question which Spaces are of the form $L(\mathcal{Q})$, or, as we say, which Spaces are "projectively embeddable". I think, this is an important and nice property which should be characterized by an appropriate group of projectivities, and this might be the real v.Staudt's point of view in this matter. We give a more exact definition:

Let L be a semimodular lattice and $\bar{L}$ a projective geometry, both of rank n. A mapping $\alpha: L \to \bar{L}$ is called a strong embedding, if $\alpha$ is a rank and join-preserving monomorphism; so it satisfies

(1) $\alpha$ is injective,

(2) $ra = ra\alpha \quad \forall a \in L$,

(3) $(a \vee b)\alpha = a\alpha \vee b\alpha$.

We call L projectively embeddable, if there exists a strong embedding of L into a projective geometry $\bar{L}$. Furthermore we may choose $\bar{L}$ minimal; in the case of locally projective L this means $L(p) \cong L(p\alpha)$ for $p \in L_1$.

Now it seems that the concept of "Space" is a rather restrictive definition. Why we do not consider more general structures? The first reason is the following theorem due to Wille ([12], Proposition 3): If L is a semimodular locally projective lattice of rank $n>4$, such that the whole space is spanned by the set $L_1$ of its points, then L is projectively embeddable. (This is a generalisation of a theorem of Mäurer [10], which states that Möbius geometries of dimension at least 3 are projectively embeddable.) Wille's theorem is false for L of rank 4; therefore this is the most interesting case. Here is the task to characterize the projectively embeddable lattices by incidence propositions (see section 3, specially Kahn's theorem (3.1)), or - which is our main aim - by properties of the group of projectivities (section 4).

Secondly, if one looks for a generalisation of the concept "locally projective" and corresponding theorems, it comes out that there is essentially nothing new with respect to the group of projectivities, but the same theorems can be applied (cfr. section 5.4, which comprises for instance n-affine geometries (Möbius-m-structures) etc.).

## 2. THE CIRCLE PLANES AS SPACES

We consider lattices not only for themselves but also as tools for better describing other geometric structures. So circle planes can be considered as Spaces, and it seems to me, that one then better can understand the definition of perspectivity in circle planes where one has to add ecceptional elements (cfr. H.KARZEL, these Proceedings). Also it might be fruitful to introduce the general concept of "topological Space" as a first step before going in details for topological circle planes (cfr.R.LÖWEN, these Proceedings). -
In the sequel we consider only circle planes of order >4; but this is really no restriction, because the circle planes of order $\leqq 4$ are wellknown: They all are miquelian and therefore are projectively embeddable.

A circle plane (Benz plane) of order >4 is a triplet $\mathcal{B}=(P,\mathcal{G},\mathcal{C})$, where $\mathcal{G}$ and $\mathcal{C}$ are sets of subsets of P. The elements of P are called points, the elements of $\mathcal{G}$ are called generators, and the elements of $\mathcal{C}$ are called circles. Two points are called parallel, if they lie in a common generator. The following conditions are satisfied:

1. If $\mathcal{B}$ is an inversive plane, then $\mathcal{G}=\emptyset$.
   If $\mathcal{B}$ is a Laguerre plane, then through every point goes exactly one generator.
   If $\mathcal{B}$ is a Minkowski plane, then $\mathcal{G}$ is the distinct union of two sets $\mathcal{G}_1$ and $\mathcal{G}_2$ and through every point goes exactly one generator from each of the two sets.
2. Every generator intersects every circle in exactly one point.
3. Through every three pairwise non-parallel points goes exactly one circle.
4. If c is a circle and p is a point of c and q is a point outside of c and non-parallel to p, then there is exactly one circle through q which intersects c exactly in p.(We say that c and this circle touch each other in p.)
5. There exists a circle c with $c\neq P$, and c contains at least n points, where n is 6 minus the number of different generators through one point.

Let $p\in P$. A pencil through p is a set of circles which

touch each other in p, such that every point q, non-parallel to p, is contained in some circle of this pencil. By 4. to every circle c and point p of c there is exactly one pencil through p containing c. We call p the carrier of this pencil.

Now we will construct the Space L corresponding to the circle plane $\mathcal{B}$. Put $L_1$=P and let $L_2$ consist of all two-sets of non-parallel points, all generators and all pencils. The partial order relation is the natural one. The pencils are the tangent lines: The only point a pencil contains is its carrier.

To construct $L_3$, we look at first for the proper planes which are no tangent planes. Naturally all circles in L belong to this class. In the Laguerre-case there is still another kind of such planes consisting of the union of two different generators. For all these planes the partial order is the natural one given by elements and set-theoretic inclusion.

Now we have still the tangent planes (planes of the form $\langle p\rangle$ for all $p \in P$). In the case of inversive planes the only point the plane $\langle p\rangle$ contains is p, and the only lines which belong to $\langle p\rangle$ are the pencils with carrier p. In the case of Minkowski planes the points of $\langle p\rangle$ are exactly the points of the two generators through p (so in this case $\langle p\rangle$ also is a proper plane), and the lines of $\langle p\rangle$ are the two generators through p, the two-sets of non-parallel points of $\langle p\rangle$ and the pencils with carrier p. In the case of Laguerre planes the points of $\langle p\rangle$ are exactly the points of the generator through p, and the lines of $\langle p\rangle$ are the generator through p and all pencils with carrier some point of $\langle p\rangle$. (In the Laguerre-case for parallel points p and q we have $\langle p\rangle=\langle q\rangle$.)

To prove that L is semimodular of rank 4, it suffices to show: Through any two different points goes exactly one line, and through three non-collinear points there goes exactly one plane. Moreover, if two lines have a point in common, then they lie in a common plane. All these things are easily verified; one has to distinguish the proper and the tangent elements.

To prove (L2) consider two circles c,d intersecting in one point p. Either they have a second point in common - then we are done - or not: then there is a pencil with carrier p containing c and d. This pencil then is a common line to c and d. In a similar manner we can handle the combinations of different kinds of planes. For instance in the case of the intersection of a plane <p> with a circle c containing p, both have in common the line which is the pencil with carrier p containing c, etc.

Clearly L(p) is a projective plane: By the former things it suffices to find four lines through p, no three of which are complanar. You can find them easily by considering two circles touching each other in p. That the order of L(p) is >4, comes from 5. - Finally (L3) comes from the definition of L by means of $\mathcal{B}$.

Conversely one can characterize those Spaces which arise from circle planes by this proceeding. For instance the Space L corresponds to an inversive plane if and only if (1) every line of L has at most 2 points and (2) <p> has p as its only point for all $p \in L_1$.

## 3. INCIDENCE PROPOSITIONS IN SPACES

In this section we study incidence propositions characterizing those Spaces which are projectively embeddable.

We say that the lines $l_1, l_2, \ldots$ lie <u>in a bundle</u>, if they are pairwise complanar.

Let $A$ be a set of quadruplets of lines and n a natural number, $n \leq 4$. We say that the <u>bundle theorem for n tangents</u> $(b_n)$ holds for $A$, if the following condition is satisfied:

$(b_n)$ Let $(l_1, l_2, l_3, l_4) \in A$, such that no two of the lines $l_1, l_2, l_3, l_4$ have a point in common and no three lie in a common plane, but $l_i$ and $l_k$ are complanar for five pairs $(i,k), 1 \leq i < k \leq 4$, and there are at most n tangent lines among the lines $l_1, l_2, l_3, l_4$. Then $l_1, l_2, l_3, l_4$ lie in a bundle.

We say that $(b_n)$ holds in L, if it holds for the set of all quadruplets of lines of L. The <u>Bundle Theorem</u> (b) is the bundle theorem for 4 tangents $(b_4)$. (In fact, (b) corresponds to the bundle theorem in inversive planes , German "Büschelsatz")

(3.1) THEOREM (J.Kahn 1979,[8]). The Space L is projectively embeddable if and only if (b) holds in L.

(Really Kahn uses a weaker condition instead of (L3), which we cannot handle here.)

The lines $l_1,l_2,l_3,l_4$ are in <u>standard position</u>, if no two of them have a point in common and no three lie in a common plane, but $l_i$ and $l_k$ are complanar for i=1,2 and k=2,3,4. Clearly the concept of standard position is symmetric in {1,2} and {3,4}. We also use the notion of $p_i$ and $q_i$ as <u>different</u> points of $l_i$. (If $l_i$ is a tangent line, then $p_i$ is its only point.)

(3.2) LEMMA. For $n \leq 3$ let $p_i \in L$ be fixed points and $(b_n)$ hold for the set A of all quadruplets of lines $l_1,l_2,l_3,l_4$ such that $p_i \leq l_i$. Let $l_1,l_2,l_3,l_4$ be in standard position such that $p_i \leq l_i$ and at most n tangent lines are among the lines $l_1,l_2,l_3$. Then $l_1,l_2,l_3,l_4$ lie in a bundle.

<u>Proof</u>: We may assume $l_4$ to be a tangent line and $l_3,l_4$ not complanar. Define

$$l_4' = (l_1 \vee p_4) \wedge (l_3 \vee p_4), \qquad l_4'' = (l_2 \vee p_4) \wedge (l_3 \vee p_4).$$

Then the lines $l_4, l_4', l_4''$ do not lie in a common plane, and so, by (L3), at least one of $l_4', l_4''$, say l, is a proper line. Then $(l_1,l_2,l_3,l)$ belong to A and thus $l_1,l_2,l_3,l$ lie in a bundle, because we can apply $(b_n)$. Therefore

$$l = (l_1 \vee p_4) \wedge (l_2 \vee p_4),$$

a contradiction. □

We say that the <u>Central Bundle Proposition</u> (CB) holds in L, if the following condition is satisfied:

(CB) Let be $p_i \in L_1$, $h_{ik} \in L_2$, $e_i \in L_3$,

with $p_i \leq h_{ik} \leq e_i$, $h_{1k} \vee h_{2k} \vee h_{3k} = 1$,

the $h_{ik}$ 9 different lines for i,k=1,2,3

but $p_i \nleq e_k$ for $i \neq k$.

If the lines $h_{ik}$, $h_{jk}$ for i<j are complanar in eight cases, then also in the ninth.

(3.3) THEOREM. The Space L is projectively embeddable if and only if (CB) holds in L.

The proof uses (3.2) and the following lemmas. (We prove the equivalence of (b) and (CB). For a direct proof see [7]. But (3.3) is an improvement; so we give the proofs in detail.)

(3.4) LEMMA. If the Space L is projectively embeddable then (CB) holds in L.

Proof: Let L be a projective geometry of rank 4 and $p_i$, $h_{ik}$, $e_i$ satisfy the hypotheses of (CB), say $h_{ik}, h_{jk}$ are all complanar with exception of $h_{23}$, $h_{33}$. We have planes $e_{jk}=h_{sk}\vee h_{tk}$ with $\{j,s,t\}=\{1,2,3\}$, excepted $e_{13}$.

For k=1,2 we have the point

$$z_k = e_{1k}\wedge e_{2k}\wedge e_{3k} = h_{1k}\wedge h_{2k}\wedge h_{3k} \leqq e_1\wedge e_2\wedge e_3,$$

and since $z_1\neq z_2$, the tree planes $e_i$ intersect in a line $l = z_1\vee z_2$. Furthermore

$$0 \neq h_{13}\wedge h_{23} \leqq e_1\wedge e_2 = l,$$

$$0 \neq h_{13}\wedge h_{33} \leqq e_1\wedge e_3 = l,$$

thus

$$0 \neq h_{13}\wedge h_{23} = l\wedge h_{13} = h_{13}\wedge h_{33} =: s,$$

$$h_{23}\wedge h_{33} = s \neq 0.$$

So also $e_{13}$ is a plane. □

(3.5) LEMMA. Let (CB) hold in the Space L. Let $l_1, l_2, l_3, l_4$ be in standard position, such that both $p_1, p_2, p_3, p_4$ and $p_1, p_2, p_3, q_4$ span the whole space. Then $l_1, l_2, l_3, l_4$ lie in a bundle.

Proof: For i=1,2 define (cfr.Fig.1)

$$h_{i1} = l_i, \quad h_{i2} = p_i\vee p_4, \quad h_{i3} = p_i\vee q_4.$$

Then $h_{1k}, h_{2k}$ are complanar for k=1,2,3. Now let be

$$h_{3k} = (h_{1k}\vee p_3)\wedge(h_{2k}\vee p_3), \quad k = 1,2,3.$$

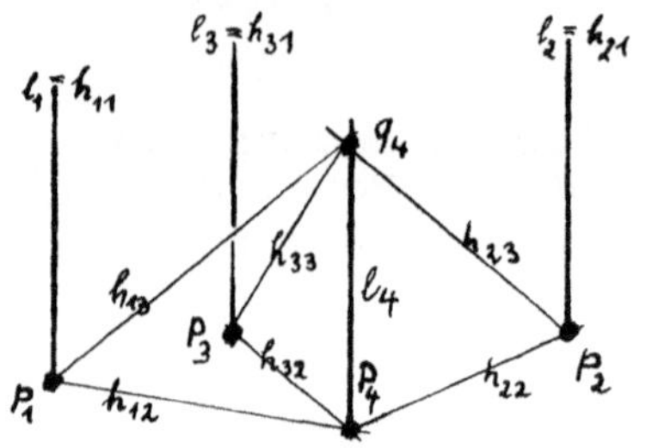

Fig.1

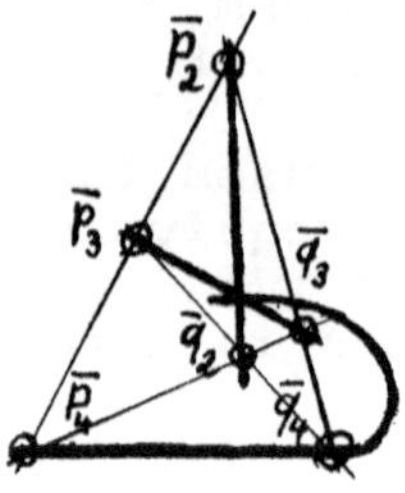

Fig.2

Then $h_{3k}$ is the unique line through $p_3$, which is complanar to $h_{1k}$ and $h_{2k}$. So by (CB) the three lines $h_{3k}$, $k=1,2,3$ lie in a common plane. But $h_{31}=l_3$, $h_{32}=p_3 \vee p_4$, $h_{33}=p_3 \vee q_4$; thus $l_3$ and $l_4=p_4 \vee q_4$ are complanar. □

(3.6) LEMMA. In a Space L the condition (CB) implies $(b_o)$.

Proof: Let (CB) hold for the Space L, and let $l_1, l_2, l_3, l_4$ be proper lines of L in standard position. Assume that $l_3, l_4$ are not complanar. Then the hypotheses of (3.2) are not satisfied.

We consider this situation in $L(p_1)$, writing $\bar{x}$ instead of $x \vee p_1$ for all $x \in L$: In $L(p_1)$ there is no line joining a point of $\bar{l}_2$ and a point of $\bar{l}_3$ such that at least two points of $\bar{l}_4$ are outside of this line. So $l_4$ has exactly two points and also the lines $l_2$ and $l_3$ have exactly two points.

Then in $L(p_1)$ the lines $\bar{l}_2, \bar{l}_3, \bar{l}_4$ necessarily are the diagonal lines of a complete quadrilateral with vertices $\bar{p}_2, \bar{p}_3, \bar{p}_4, \bar{q}_2, \bar{q}_3, \bar{q}_4$, cfr. fig.2. Since moreover $\bar{l}_2, \bar{l}_3, \bar{l}_4$ are copunctal in the point $l_1$ of $L(p_1)$, there exists a Fano-configuration in $L(p_1)$. By symmetry also $l_1$ has exactly two points, and the statements for $L(p_1)$ analogously hold for $L(q_1)$.

Now assume at first, that there exists a point s outside the five planes $e_{ik}=l_i \vee l_k$ for $i<k$ and $i=1,2$ and $k=2,3,4$. Let be

$$l_5 = (s \vee l_1) \wedge (s \vee l_2),$$

then $l_1, l_2, l_3, l_5$ are in standard position, and in $L(p_1)$

the lines $\bar{l}_2, \bar{l}_3, \bar{l}_5$ cannot form a configuration as mentioned before. So by (3.5) and (3.2) the lines $l_1, l_2, l_3, l_5$ lie in a bundle. In the same way we may conclude that also $l_1, l_2, l_4, l_5$ lie in a bundle. So also $l_5, l_1, l_3, l_4$ are in standard position, and by the same conclusion we get $l_5, l_1, l_3, l_4$ lying in a bundle. So we have the contradiction of complanar lines $l_3, l_4$. Therefore every point of L lies on some plane $e_{ik}$ mentioned before.

Now we can label the points of the lines $l_i$, i=1,2,3,4 in such a way, that the configurations of $L(p_1)$, $L(q_1)$ give us quadruplets of points lying in planes $e_i$ as follows (see fig.3):

$p_1, p_2, p_3, p_4 < e_0$

$p_1, p_4, q_2, q_3 < e_1$

$p_1, p_3, q_2, q_4 < e_2$

$q_1, q_4, p_2, p_3 < e_3$

$q_1, q_3, p_2, p_4 < e_4$

Then with $e_{12} = l_1 \vee l_2 = e_5$ we have lines $l_i$ lying in planes $e_k$ for k=5,6,7,8:

$l_5 = p_1 \vee q_2 < e_1, e_2, e_5$

$l_6 = q_1 \vee p_2 < e_3, e_4, e_5$

$l_7 = p_3 \vee q_4 < e_2, e_3$

$l_8 = q_3 \vee p_4 < e_1, e_4$.

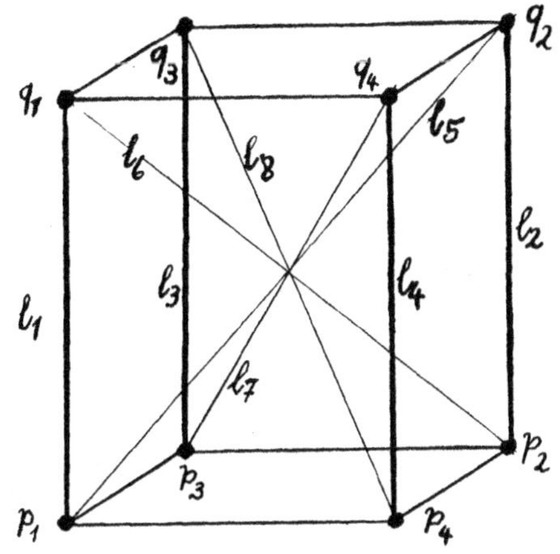


Fig.3

So the lines $l_5, l_6, l_7, l_8$ are in standard position. Since $l_7$ and $l_8$ by assumption not are complanar (else also $l_3$ and $l_4$ were complanar), every point of L also lies on some plane $e_i$ for i=1,...,5.

Finally consider a plane f through $l_1$ different from $\langle p_1 \rangle$ and $e_{1k}$, k=1,2,3. We will count the lines of f through $p_1$. At first we see the lines $f \wedge e_k$ for k=0,1,2 and the line $l_1 = f \wedge e_5$. Points of f outside of these lines only can arise by intersection of f with the lines $p_2 \vee q_4 = e_{24} \wedge e_3$ and $p_2 \vee q_3 = e_{23} \wedge e_4$. Adding one tangent line through $p_1$ in f we see that in $L(p_1)$ the line f has at most 7 points. So $L(p_1)$ has order 5, and therefore $L(p_1)$ is argesian and does not admit any Fano-configuration - in contradiction to our original assumption, that $l_3, l_4$ are not complanar. □

(3.7) LEMMA. In a Space L the conditions (CB) and $(b_0)$ imply $(b_2)$.

Proof: Let the Space L satisfy (CB) and $(b_0)$, let the lines $l_1, l_2, l_3, l_4$ be in standard position and among them at most 2 tangent lines. Using (3.2) and symmetry of the standard position in $\{1,2\}$ and $\{3,4\}$ we may assume, that at most $l_1$ or (in a second step) $l_1$ and $l_2$ are tangent lines.

Now assume $l_3$ and $l_4$ not to be complanar. Then we can label the points in such a way that $e_3 := l_3 \vee p_4 \neq \langle p_3 \rangle$ holds and moreover the points $p_1, p_2, p_3, p_4$ lie in a common plane $f_0$ (otherwise we could apply (3.5) to get a contradiction). Furthermore in the case of order 5 of L we may assume that both $l_1$ and $l_2$ are tangent lines. For, if $l_2$ were a proper line we had the situation of (3.5) too, because $L(p_1)$ does not admit any Fano-configuration. In either case we may choose a plane $e_1$ through $l_1$ different from $\langle p_1 \rangle$ and planes $f_2, f_3$ through $p_1 \vee p_2$ such that defining

$$\begin{aligned} h_{i1} &= l_i, && i = 1,2,3 \\ h_{1k} &= e_1 \wedge f_k, && \\ h_{3k} &= e_3 \wedge (p_3 \vee h_{1k}), && k = 2,3 \\ h_{2k} &= f_k \wedge (p_2 \vee h_{3k}), && \\ e_2 &= h_{21} \vee h_{22} && \end{aligned}$$

the plane $e_2$ is different from $\langle p_2 \rangle$, the lines $h_{ik}$ are proper lines for $i=1,2$ and $k=2,3$ and are outside of $f_0$. So the hypotheses of (CB) are satisfied and therefore $h_{23} \leqq e_2$ holds. At next we define

$$\begin{aligned} h_{41} &= l_4, && \\ h_{4k} &= (h_{1k} \vee p_4) \wedge (h_{2k} \vee p_4), && k = 2,3 \\ e_4 &= h_{42} \vee h_{43}. && \end{aligned}$$

We can still choose $f_2, f_3$ in such a way that moreover $h_{3k}, h_{4k}$ not at the same time are tangent lines for some k with $k=2,3$. So by $(b_0)$ and (3.2) the lines $h_{1k}, h_{2k}, h_{3k}, h_{4k}$ lie in a bundle for $k=2,3$. Therefore

$$\begin{aligned} h_{4k} &\leqq h_{3k} \vee p_4 = e_3, && k = 2,3 \\ e_4 &= h_{42} \vee h_{43} = e_3. && \end{aligned}$$

Now at last we get by (CB) applied to $h_{1k}, h_{2k}, h_{4k}$ that

$$l_4 = h_{41} \leqq e_4 = e_3$$

holds, and thus $l_3$ and $l_4$ are complanar - a contradiction. □

(3.8) FINAL PROOF OF THE THEOREM (3.3): Let the Space L satisfy the condition (CB). Then by (3.6) and (3.7) in L the bundle theorem for two tangents ($b_2$) is valid. Let $l_1, l_2, l_3, l_4$ be in standard position and exactly three tangent lines among them. Using the symmetry of the standard position in {3,4} we may apply (3.2) to get $l_1, l_2, l_3, l_4$ lying in a bundle. So ($b_3$) holds in L. Now apply (3.2) once more to the tangent lines $l_1, l_2, l_3, l_4$ to get the validity of (b) in L.

Then by Kahn's theorem (3.1) we conclude, that L is projectively embeddable.

The converse direction of the theorem is stated in (3.4). □

## 4. GROUPS OF PROJECTIVITIES IN SPACES

For $p \in L_1$ and $e \in L_3$ with $p \leqq e$ let $(p,e) = \{x \in L \mid p < x < e\}$ be the bundle of all lines through p in e. For $p,q \in L_1$ and $e,f \in L_3$ with $p \leqq e \neq \langle p \rangle$ and $q \leqq f \neq \langle q \rangle$, but $p \nleqq f$ and $q \nleqq e$ define $\sigma = [(p,e),(q,f)]$ to be the map

$$\sigma: (p,e) \to (q,f); \; x \mapsto (x \vee q) \wedge f,$$

and call $\sigma$ a <u>proper perspectivity</u>.

If $\sigma_i = [(p_{i-1}, e_{i-1}), (p_i, e_i)]$ for $i=1,\dots,n$ are proper perspectivities, then we call $\pi = \sigma_1 \sigma_2 \dots \sigma_n$ a <u>proper projectivity</u> from $(p_0, e_0)$ onto $(p_n, e_n)$. The elements $p_i, e_i$, $i=0,\dots,n$, form the <u>support of</u> $\pi$. If $p_0 = p = p_n$ and $e_0 = e = e_n$ is valid, $\pi$ is called a proper projectivity from (p,e) onto itself. Clearly the set of projectivities from (p,e) onto itself forms a group with composition as multiplication, which we denote by $\Pi(p,e)$ and call it the <u>group of proper projectivities from</u> (p,e) <u>onto itself</u>.

A more general concept is the following. Let be $e,f \in L_3$ and $l \in L_2$ such that $p = l \wedge e$ and $q = l \wedge f$ are points. Then define $\tau = [e,l,f]$ as the map

$$\tau\colon (p,e) \to (q,f);\ x \mapsto (x \vee l) \wedge f,$$

and call $\tau$ a <u>perspectivity</u> (in the general sense). We construct analogously the projectivities (in general sense) $\pi$ of $(p,e)$ onto itself by $\pi = \tau_1\tau_2\ldots\tau_n$, where $\tau_i = [e_{i-1}, l_i, e_i]$ is a perspectivity for $i=1,2,\ldots,n$ and both $e_0 \wedge l_0 = p = e_n \wedge l_n$ and $e_0 = e_n$ hold.

The set of all projectivities from $(p,e)$ onto itself again form a group in the natural way denoted by $\Pi^*(p,e)$ and called <u>General Group of Projectivities from</u> $(p,e)$ <u>onto itself</u>.

Naturally a perspectivity $\tau = [e,l,f]$ may be a proper perspectivity. But if $e \wedge l = f \wedge l$, we call it a <u>free perspectivity</u> in case $l$ is a proper line, and an <u>affine perspectivity</u> in case $l$ is a tangent line. Moreover now also $e = \langle p \rangle$ or $f = \langle q \rangle$ (or both) can happen, in which case the perspectivity shall be called <u>tangential</u>.

One sees at once the correspondence of these definitions to the different kinds of perspectivities in circle planes (cfr.H.KARZEL, these Proceedings). But there are differences in the definition of projectivities. For in circle planes (considered as Spaces as in section 2) for a projectivity $\pi = [e_0, l_1, e_1]\ldots$ $\ldots[e_{n-1}, l_n, e_n]$ nevermore the condition $e_0 \wedge l_1 = e_n \wedge l_n$ is necessary for $\pi$ to be a projectivity from $(e,p)$ onto itself. On the other hand in a Space there are projectivities which are not used in circle planes, because they contain tangential perspectivities. By example, let L be a Space, which belongs to an inversive plane and consider the projectivity $\pi = \sigma_1\sigma_2$, where $\sigma_i = [(p_{i-1}, e_{i-1}), (p_i, e_i)]$ and $e_1$ is a tangent plane, $e_1 = \langle p_1 \rangle$, whereas $e_0$ and $e_2$ are proper planes. Then in terms of the inversive plane we can describe the map $\pi$ as follows (fig.4): Any point $x$ of the circle $e_0$ determines uniquely a circle $f$ through the points $x, p_0, p_1$. (For $x = p_0$ this is the circle through $p_1$ touching the circle $e_0$ in $p_0$.) Now there exists a unique circle $g$ through $p_2$ tuoching the circle $f$ in $p_1$. Then $y = x\pi$ is defined by the set $\{y, p_2\}$ of common points of the circles $g$ and $e_2$. - The different concept of projectivity in Spaces and in circle planes also explains the different results.

As in other geometric structures groups $\Pi(p,e)$ and

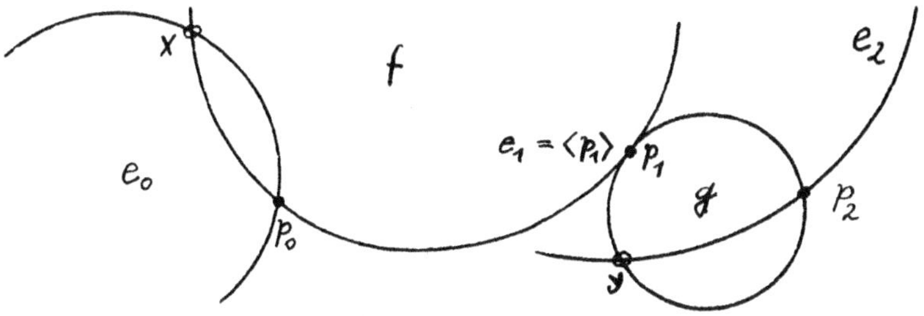


Fig.4

$\Pi(q,f)$ ($\Pi^*(p,e)$ and $\Pi^*(q,f)$) are similar to each other as permutation groups. So in the sequel we shall omit the arguments and simply write $\Pi$ and $\Pi^*$. We denote by $\Pi^*_p$ the group of all projectivities $\pi$ of some $(p,e)$ onto itself, where $\pi=\tau_1\tau_2\ldots\tau_n$, $\tau_i=[e_{i-1},l_i,e_i]$, but now $e_{i-1}\wedge l_i=p=e_i\wedge l_i$, $i=1,\ldots n$.

So $\Pi^*_p$ simply is the group of projectivities of the projective plane $L(p)$. Clearly $\Pi^*$ possesses a subgroup similar to $\Pi^*_p$ for every $p \in L_1$.

Now, after some more or less trivial observations, we shall turn to the main theorems of this section.

(4.1) Every free perspectivity which is not tangential is the product of two proper perspectivities.
<u>Proof</u>: Let $\tau=[e,l,f]$ be a free perspectivity with $e\wedge l=p=f\wedge l$ and $e\neq\langle p\rangle\neq f$. There is a point q of l different from p and a plane g through q with $l\nleq g\neq\langle q\rangle$. Then

$$\tau = [(p,e),(q,g)][(q,g),(p,f)]$$

is as desired.

(4.2) If L is geometric, it makes no real sense to speak of tangential perspectivities. So in the definition of proper perspectivity in this case we omit the restriction $e\neq\langle p\rangle$ and $f\neq\langle q\rangle$. Then we get $\Pi = \Pi^*$, because every perspectivity is either proper or free now.

(4.3) $\Pi$ is 3-transitive: If we restrict ourselves to free perspectivities of the form $[f,l,g]$ with $f\wedge l=p=g\wedge l$ for some fixed p, the proof reduces to the task to construct in an affine plane of order >4 a projectivity $\pi$ which maps three given collinear points onto three given others, such that the support of $\pi$ contains only proper elements of the affine plane, and this task may be given as an exercise to the reader.

Let $\Sigma$ be a group of projectivities. We say, that $(P_{33})$ holds for $\Sigma$, if every projectivity of $\Sigma$ which is the product of three perspectivities of $\Sigma$ and has three fixed elements is the identity.

$(P_n)$ holds for $\Sigma$, if every projectivity of $\Sigma$ with n fixed elements is the identity. Instead of $(P_3)$ we will write (P); this is the original v.Staudt's "Fundamental Theorem". The following statement is a direct transfer of the analogous statement for projective planes (cfr. G.PICKERT, these Proceedings).

(4.4) The following statements are equivalent:
1) For $\Pi$ the Fundamental Theorem (P) is valid,
2) $\Pi$ is sharply 3-transitive,
3) if $\pi$ is a proper projectivity from (p,e) onto (q,f) with $p \nleq f$ and $q \nleq e$ such that for three different $x_i$ of (p,e) the lines $x_i$ and $x_i\pi$ are complanar, then $\pi$ is a proper perspectivity.

(4.5) THEOREM. The Space L is projectively embeddable if and only if $(P_{33})$ holds for $\Pi$.

<u>Proof</u>: By theorem (3.3) it suffices to show that $(P_{33})$ holds for $\Pi$ if and only if (CB) is valid in L.

1\. Let $(P_{33})$ hold for $\Pi$. Let $p_i, h_{ik}, e_i$ satisfy the hypotheses of (CB) such that $h_{ik}, h_{jk}$ are complanar with the eventual exception of $h_{13}$ and $h_{33}$

$$\pi = [(p_1,e_1), (p_2,e_2)][(p_2,e_2),(p_3,e_3)][(p_3,e_3),(p_1,e_1)] ,$$

$$x_0 = (p_1 \vee p_2 \vee p_3) \wedge e_1, \qquad x_k = h_{1k}, \quad k=1,2,3.$$

Then $x_i\pi = x_i$ for i=0,1,2, and from $(P_{33})$ follows $x_3\pi = x_3$, so necessarily $h_{13}$ and $h_{33}$ are complanar too.

At the next step assume that $e_1 = \langle p_1 \rangle$ holds and $h_{13}$ and $h_{23}$ are not complanar. Define $h_{13}' = (h_{23} \vee p_1) \wedge (h_{33} \vee p_1)$. Since $h_{11}, h_{12}, h_{13}'$ are not complanar, it follows that $e_1' := h_{11} \vee h_{13}'$ is different from $\langle p_1 \rangle$. By the part already proved we conclude $h_{12} < e_1'$ - a contradiction.

As a second possibility with $e_1=<p_1>$ assume that $h_{23}$ and $h_{33}$ are not complanar. To reduce this to the first case replace $h_{23}$ by $h'_{23}=(h_{33}\vee p_2)\wedge e_2$. By the preceding considerations we gain the contradiction of $h_{13}$ and $h'_{23}$ complanar and thus $h_{23}=h'_{23}$.

All cases $e_i=<p_i>$ for <u>one</u> i by relabelling can be brought to one of the two previous forms. Now in a second step in the same way one proves the case $e_i=<p_i>$ for <u>two</u> i, - and at last in a third step for three i.

2. Let (CB) be valid in L, let $\pi\in\Pi$, $\pi=\sigma_{12}\sigma_{23}\sigma_{31}$, where

$$\sigma_{ij} = [(p_i,e_i),(p_j,e_j)],$$

let $x_i\in(p_1,e_1)$ with $x_i\pi=x_i$ for i=0,1,2 be different elements. Let $x_3$ be another element of $(p_1,e_1)$. We will show $x_3\pi=x_3$. we can label the xs in such a way that with

$$h_{1k} = x_k, \quad h_{2k} = x_k\sigma_{12}, \quad h_{3k} = x_k\sigma_{12}\sigma_{23},$$

for i,k=1,2,3 the elements $p_i, h_{ik}, e_i$ fulfill the hypotheses of (CB), and we conclude that also $h_{13}$ and $h_{33}$ are complanar. Now we see immediately $x_3\pi=x_3$. □

(4.6) THEOREM. Let L be a Space. Then the following statements are equivalent:

1) $(P_{33})$ holds for $\Pi^*$,
2) there exists a strong embedding of L into a Pappian projective geometry,
3) (P) holds for $\Pi^*$,
4) (P) holds for $\Pi$.

<u>Proof</u>: 1)⇒2). Since $(P_{33})$ holds for $\Pi$, by (4.5) there exists a strong embedding of L into a projective geometry $\bar{L}$ such that L(p) for $p\in L_1$ is isomorphic to a projective plane of $\bar{L}$. On the other hand $(P_{33})$ also holds for $\Pi^*_p$, which means that $(P_{33})$ holds for the group of projectivities of the projective plane L(p). As a wellknown fact (cfr. G.PICKERT, these Proceedings) this implies that L(p) is Pappian. So we get, that also $\bar{L}$ is a Pappian projective geometry.

2)⇒3). In a Pappian projective geometry the perspectivities preserve ratios; so the Fundamental Theorem holds for $\Pi^*$.

The implications 3)⇒1) and 3)⇒4) are trivial.

4)⇒2). Since ($P_{33}$) holds for $\Pi$, by (4.5) there exists a strong embedding of L into a projective geometry $\bar{L}$ such that L(p) for $p \in L_1$ is isomorphic to a projective plane of $\bar{L}$. That (P) holds for $\Pi$ implies, that - according to products of three free perspectivities - in L(p) every Pappos-configuration exists, which has no points or lines in <p>, i.e.which consists of proper points and lines of an affine plane. Because L(p) is desarguesian and therefore transitive on quadrangles, we conclude that $\bar{L}$ is Pappian.

(4.7) REMARK. By (4.6) it makes sense to say, that the Fundamental Theorem holds (doesn't hold) in L. Secondly if L is projectively embeddable so clearly $\Pi^* = \Pi^*_p$ holds.

Let L be projectively embeddable but let the Fundamental Theorem not hold for L. Then it would be of interest to give a method for representing any projectivity of (p,e) onto itself as an element of $\Pi^*_q$ for en appropriate $q \in L_1$.

To this purpose let L be a projective geometry over an infinite field and let $\pi \in \Pi^*(p,e)$. (We can still omit a finite number of subspaces to get a more general Space.) Since now $\Pi = \Pi^*$, we may assume $\pi = \sigma_1\sigma_2 \ldots \sigma_n$, where

$$\sigma_i = [(p_{i-1}, e_{i-1}), (p_i, e_i)], \qquad i=1,\ldots,n.$$

$$p_0 = p = p_n, \qquad e_0 = e = e_n.$$

There exists a point q with $q \not\leq e_i$, $p_{i-1} \vee p_i$ for i=1,...n. By ($P_{33}$) there exists $f_i \in L(q) \cap L_3$, such that for

$$\rho_i = [(p_{i-1}, e_{i-1}), (q, f_i)],$$

$$\tau_i = [(p_i, e_i), (q, f_i)], \qquad i=1,\ldots,n,$$

the projectivity $\rho_i^{-1}\sigma_i\tau_i$ is the identity on $(q, f_i)$. Now with $\tau_0 = \tau_n =: \tau$, $f_0 = f_n =: f$ define $\phi_i = \tau_i^{-1}\rho_{i+1}$ for i=0,...,n-1. Then we get $\phi_i = [f_i, q \vee p_i, f_{i+1}]$, and furthermore we have

$$\pi^* = \phi_0\phi_1 \ldots \phi_{n-1} \in \Pi^*_q, \quad \pi^*: (q,f) \to (q,f),$$

$$\pi^* = \phi_0(\rho_1^{-1}\sigma_1\tau_1)\phi_1(\rho_2^{-1}\sigma_2\tau_2)\phi_2\ldots\phi_{n-1}(\rho_n^{-1}\sigma_n\tau_n)$$
$$= \tau_0^{-1}\sigma_1\sigma_2\ldots\sigma_n\tau_n = \tau^{-1}\pi\tau. \qquad \square$$

## 5. SPECIAL KINDS OF SPACES

### 5.1 Free extensions

We proceed very similar to the construction of free planes (cfr. A.BARLOTTI, these Proceedings); we consider only the simplest possible construction (see [9], 5. Example 5). Call partial (geometric) Space an incidence structure where the blocks are called planes - the lines simply are all two-sets of points and need no extra attention - such that every three points are incident with at most one plane and every two planes are incident with at most two points. Construct a sequence of partial Spaces $S_i$ as follows. Let $S_0$ consist of a point-set of cardinality at least 5, and empty sets of planes and of incidence. Let these points have stage 0.

If $S_{n-1}$ is already given, $S_n$ is constructed as follows: To any three points which are still not incident with a common plane add a new plane, and let it be incident with exactly these three points. And to any two planes which are incident with exactly one point add a new point and let it be incident with exactly these two planes. The new planes and points constructed in this way are by definition of stage n.

It is easy to see that the union over all these partial Spaces $S_i$, i=0,1,2,... is a geometric Space L, called a Free Space. Moreover for $n \geq 1$ every plane of stage n is incident with exactly 3 points of lower stage, at least one of these beeing of stage n-1; and every point of stage n is incident with exactly two planes of lower stage, at least one of these beeing of stage n-1.

A confined configuration of L is a finite set $F$ of points and planes of L with the property, that every point of $F$ is incident with at least three planes of $F$ and every plane of $F$ is incident with at least four points of $F$. Looking at an element of $F$ of largest stage we see at once that in a Free Space the only confined configuration is the empty set.

The representation of a projectivity $\pi=\tau_1\tau_2\ldots\tau_n$ is reduced, if with $\tau_i = [e_{i-1}, l_i, e_i]$ never occurs $e_{i-1} = e_i$ or $l_i = l_{i+1}$. Let L be a Free Space. If $\pi \in \Pi = \Pi(p,e)$ and $\pi = \tau_1\tau_2\ldots\tau_n$ is reduced and $n \geq 1$, then $\pi$ never is the identity; else one could construct a non- empty confined configuration.

In a Free Space ($P_6$) holds. The proof runs in the same way as for free planes (cfr. A.BARLOTTI, these Proceedings or [2]): Assume $\pi = \tau_1\tau_2\ldots\tau_n$ for $n \geqq 1$ has the 6 fixed lines $x_1, \ldots, x_6$. Then with

$$p_k = l_k \wedge e_k, \quad x_i\tau_1\tau_2\ldots\tau_k = \{p_k, r_i^k\} := x_i^k, \quad l_k = \{p_k, q_k\}$$

one can construct a confined configuration $F$ consisting of $p_k, q_k, e_k$ for $k=1,2,\ldots,n$ (the support of $\pi$) and a certain subst of the points $r_i^k$ and the planes $x_i^{k-1} \vee x_i^k$ (at least three for every k, $k=1,2,\ldots,n$). It is important for the proof that every line of L contains exactly two points.

On the other hand $\Pi_p^*$ contains projectivities with 5 fixed elements which are not the identity; for one can construct in open projective planes (planes with-out confined configurations) projectivities with 5 fixed points which are not the identity. The group $\Pi=\Pi^*$ of a Free Space L is a free group of rank the order of L; the same is true for the stabilizers in $\Pi$ of at most 5 elements. (For the statements of this last paragraph see [11].)

## 5.2 Translation Spaces

A dilatation $\alpha$ of the Space L is an automorphism of L such that for every $p \in L_1$, $x \in L$

$$p, p^\alpha \leqq x \quad \text{implies} \quad x = x^\alpha.$$

A translation of L is the identity or a dilatation without fixed points. The Space L is called a translation Space if it admits a point-transitive group T of translations. (Then T acts regularly on $L_1$.) Clearly the projective planes L(p) in a translation Space all are isomorphic, and so are all the groups $\Pi_p^*$. We need the following

LEMMA. In a translation Space $\Pi^* = \Pi_p^*$ holds.

Proof: For every automorphism $\alpha$ of a Space L we have

$$\alpha^{-1}[e,l,f]\alpha = [e^{\alpha},l^{\alpha},f^{\alpha}].$$

Furthermore, let $\tau$ be a translation of L with $l^{\tau} = l$. Then

$$[e,l,f]\tau = [e,l,f^{\tau}]$$

holds. Now let L be a translation Space and

$$\pi = [e_0,l_1,e_1]\ldots[e_{n-1},l_n,e_n]$$

with

$$e_i \wedge l_i = p_i = e_i \wedge l_{i+1} \qquad \text{for } i=0,\ldots,n$$

and

$$l_0=l_n, \quad l_1=l_{n+1}, \quad e_0=e_n, \quad p_0=p_n:=p.$$

Let T be a point-transitive translation group of L and let $\tau_i \in T$ satisfy

$$p_i^{\tau_i} = p_{i-1}.$$

Then

$$\tau_n\tau_{n-1}\cdots\tau_1 = 1,$$

$$([e_{i-1},l_i,e_i]\tau_i)\tau_{i-1}\cdots\tau_1$$

$$= [e_{i-1},l_i,e_i^{\tau_i}]\tau_{i-1}\cdots\tau_1$$

$$= \tau_{i-1}\cdots\tau_1[e_{i-1}^{\tau_{i-1}\cdots\tau_1},l_i^{\tau_{i-1}\cdots\tau_1},e_i^{\tau_i\cdots\tau_1}],$$

therefore

$$\pi = [e_0,l_1,e_1]\ldots[e_{n-1},l_n,e_n]\tau_n\tau_{n-1}\cdots\tau_1$$

$$= [e_0,l_1,e_1^{\tau_1}][e_1^{\tau_1},l_2^{\tau_1},e_2^{\tau_2\tau_1}]\ldots$$

$$\ldots[e_{n-1}^{\tau_{n-1}\cdots\tau_1},l_n^{\tau_{n-1}\cdots\tau_1},e_0],$$

i.e.

$$\pi \in \Pi_p^*.$$ □

COROLLARY 1. A translation Space L is projectively embeddable, if and only if L(p) is desarguesian for $p \in L_1$.

Proof: To show $(P_{33})$ let

$$\pi = [e_0,l_1,e_1][e_1,l_2,e_2][e_2,l_3,e_3]$$

with $e_0=e_3$ and $p_0=p=p_3$ have three fixed elements. Let $p_0,p_1,p_2$ span a plane f. Then $\tau_1,\tau_2,\tau_3$ all fix f, and therefore $l_1$, $l_2^{\tau_1}$, $l_3^{\tau_1\tau_2}$ in L(p) are collinear and are the centers of the three corresponding perspectivities in L(p). Because $\pi$ has 3 fixed elements, by the dual of Desargues'Theorem the three lines

$e_0$, $e_1^{\tau_1}$, $e_2^{\tau_1\tau_2}$ of L(p) are confluent and now in a

desarguesian plane $\pi$ (by the methods of the proof of the Lemma considered as an element of $\Pi_p^*$) is the identity. □

(If L(p) is even Pappian, the Corollary 1 follows directly from $\Pi^* = \Pi_p^*$ and (4.6).)

COROLLARY 2. Let L be a translation Space. Then the Fundamental Theorem in L is valid if and only if $(P_5)$ holds for $\Pi^*$.
Proof: By the Lemma $\Pi^*=\Pi_p^*$ holds. But in a projective plane (P) and $(P_5)$ are equivalent (see G.PICKERT, these Proceedings). □

REMARK. One can gain similar results by the following concept. Let L be a Space and $e \in L_3$. Call $\gamma$ an elation of L with axis e if $\gamma$ is an automorphism of L with the following properties:
(1) $\forall p \in L_1 \ \forall x \in L: \quad p, p^\gamma \leqq x \Rightarrow x = x^\gamma$,

(2) $\gamma$ fixes e and every subspace of e,

(3) $\forall p \in L_1: (p \nleqq e \ \& \ p = p^\gamma) \Rightarrow \gamma = 1$.

Now let the Space L have the property, that for any points $p_0, p_1, \ldots, p_n$ (n a natural number) there exists a plane e and a group $\Gamma(e)$ of elations with axis e and

$\gamma_i \in \Gamma(e)$ **with** $p_i^{\gamma_i} = p_{i-1}, \quad i=1,\ldots,n.$

Then also $\Pi^*=\Pi_p^*$ is valid, which can be proved in the same way as in the Lemma, and therefore the Corollaries mutatis mutandis also hold for L. (For Corollary 1 one needs only n=3.)

(If in section 1, example (3), $\mathcal{Q}$ is (a quadric and) a translation ovoid in the sense of Tits, then for every $p \in L_1$ the elations with axis $\langle p \rangle$ form a group $\Gamma(\langle p \rangle)$ acting transitively on $L_1 \setminus \{p\}$.)

CONSTRUCTION OF TRANSLATION SPACES WHICH ARE NOT PROJECTIVELY EMBEDDABLE (cf.[5], 3.5 pp.482f, [6],(2.6) pp.33ff). Let V be a vector space and H a hyperplane of V. Let $F'$ be a spread of H and let $F$ be the set of subspaces consisting of $F'$ and all one-dimensional subspaces outside of H. Furthermore, let $S$ be the set of subspaces consisting of H and all subspaces not contained in H and intersecting H in a member of $F$.

Now put $L_1=V$, let $L_2$ consist of all cosets of members of $F$ and $L_3$ consist of all cosets of members of $S$.Then $L = \{\emptyset\} \cup L_1 \cup L_2 \cup L_2 \cup \{V\}$ with the set-theoretical order ("subset" and "element") is a translation Space, the translations of L beeing just the translations of the affine space corresponding to V, and L(p) for $p \in L_1$ is a translation plane belonging to the spread $F'$.

So we have at once a lot of examples (also finite ones) of translation Spaces which are not projectively embeddable. Moreover, because there are translation planes where the group of projectivities is symmetric (alternating), we now have examples of Spaces where (by the Lemma) the group of projectivities of (p,e) onto itself is the symmetric (alternating) group of (p,e) (cfr. H.LÜNEBURG, these Proceedings). Examples of Spaces where the group of projectivities is alternating but not symmetric, occur in the case of characteristic 2 of the kernel of the translation plane (and therefore the coordinatizing field of V). -It should be mentioned that the translation Spaces constructed in this manner all are geometric.

## 5.3 A General Construction Principle

A more general construction due to J.Kahn is the following. Let $A_1, A_2$ be two distinct affine planes which possess a common line at infinity (so their improper points are identical, say they are identified by a bijection). The points of L are 1. the points of $A_1$ and $A_2$ and 2. isomorphisms $A_1 \to A_2$ fixing all improper points. The line through the point p of $A_1$ and q of $A_2$ contains as further points exactly those isomorphisms $\alpha$ of L which satisfy $p^\alpha = q$. In general the isomorphisms $\alpha, \beta, \gamma$ of L are collinear if and only if

$$\alpha\beta^{-1} \text{ and } \beta\gamma^{-1} \text{ have the same center}$$

(in the projective closure of $A_1$). Now it is clear what the planes are. (The points of those planes which do not intersect the $A_i$ are a maximal set of isomorphisms $A_1 \to A_2$ belonging to L, such that for any two isomorphisms $\alpha, \beta$ of this set $\alpha\beta^{-1}$ is a translation of $A_1$. This definition works because the product of two translations of $A_1$ is a translation of $A_1$ too.)

One can gain these examples also from the duals of the spaces constructed by Ewald in [3]. They can be characterized as geometric Spaces furnished with two distinguished planes e and f with $e \wedge f = 0$, such that holds

$$\forall l \in L_2: \quad l \wedge e \in L_1 \Leftrightarrow l \wedge f \in L_1 .$$

## 5.4 A Generalization of the Concept of Space

Let $\mathcal{W}$ be the class of semimodular lattices of finite rank $\geqq 4$ with the following property:

> Let L be a lattice of the class $\mathcal{W}$ and let L be of rank n. Then L is projectively embeddable if and only if for every $a \in L_{n-4}$ the lattice L(a) is projectively embeddable.

Then a theorem of Kantor ([9], theorem 2) says, that $\mathcal{W}$ contains at least the class of geometric lattices.

Now let us consider semimodular lattices L of finite rank $n \geqq 4$, such that for every $a \in L_{n-4}$ the lattice L(a) is a Space. Then in L(a) we can define perspectivities as before. More general and more exactly: For i=1,2,

$$p_i \in L_{n-3}, \quad e_i \in L_{n-1}, \quad p_i \leqq e_i, \quad p_1 \vee e_2 = 1 = p_2 \vee e_1,$$
$$r(p_1 \wedge p_2) = n-4,$$

define the proper perspectivity $\sigma = [(p_1,e_1),(p_2,e_2)]$ by the map

$$\sigma: (p_1,e_1) \rightarrow (p_2,e_2); \; x \mapsto (x \vee p_2) \wedge e_2 .$$

(Naturally $(p,e) = \{x \in L \mid p < x < e\}$). Now we define the groups $\Pi$ and $\Pi^*$ as in section 4.

Let $\pi$ be the product of three perspectivities which is the identity,

$$\pi: (p_1,e_1) \rightarrow (p_2,e_2) \rightarrow (p_3,e_3) \rightarrow (p_1,e_1).$$

We restrict ourselves to the case where all these perspectivities are proper and for $1 \leqq i < k \leqq 3$ we put

$$l_{ik} = p_i \vee p_k .$$

We have $p_i \wedge p_k \lessdot p_i$; because L is semimodular it follows $p_k \lessdot p_i \vee p_k = l_{ik}$, therefore $r(l_{ik}) = n-2$.

Assume at first that two of the subspaces $l_{ik}$ are equal, say $l_{12} = l_{23}$. Then from $p_3 < l_{12}$ follows $l_{13} \leqq l_{12}$, and $r(l_{12}) = r(l_{13})$ now implies $l_{12} = l_{13} = l_{23}$; this is a trivial case for $\pi$ to be the identity. So we may assume $l_{12} \neq l_{23}$ and therefore $p_2 \leqq l_{12} \wedge l_{23} < l_{12}$, thus

$$l_{12} \wedge l_{23} = p_2, \quad \text{and similar}$$
$$l_{12} \wedge l_{13} = p_1,$$
$$l_{13} \wedge l_{23} = p_3.$$

Finally

$$p_1 \wedge p_2 = (l_{12} \wedge l_{13}) \wedge (l_{12} \wedge l_{23}) = l_{12} \wedge l_{13} \wedge l_{23}, \quad \text{so by symmetry}$$

$$p_1 \wedge p_2 = p_1 \wedge p_3 = p_2 \wedge p_3 := a.$$

We have shown: If $\pi$ is proper and is the identity in a non-trivial way, then there is $a \in L_{n-4}$, such that the whole support of $\pi$ lies in L(a).
So theorem (4.5) generalizes to the following

THEOREM. Let L be a lattice of the class $\mathcal{W}$ and let L have rank n and the property that for all $a \in L_{n-4}$ the lattice L(a) is a Space.
Then L is projectively embeddable if and only if $(P_{33})$ holds for $\Pi$.
Furthermore for L the four statements 1)-4) of theorem (4.6) also are equivalent.

To find examples for lattices L of the theorem, one can look for the incidence geometries of grade n and rank n+3 and the n-affine geometries of rank n+2 (n>1), both in the sense of Wille [12]. The latter are - in other terms - also called Möbius-m-structures or $\kappa$-affine spaces for appropriate $m,\kappa$. Also one may consider the Minkowski- and Laguerre-m-structures (cf.[4],p.235,268); these are related to sharply (m+2)-fold transitive permutation sets of degree m+n and optimal (m+n,m+2)-codes respectively. If L is the lattice of one of these geometries, L of rank s, then the subspaces of rank $t \leqq s-4$ are exactly the t-sets of points of L and for $a \in L_{s-4}$ the lattice L(a) is a Space (if the order of L is >4). Moreover, in the case of an incidence geometry, L is geometric, so by Kantor's theorem belongs to the class $\mathcal{W}$. In the other cases previously we have to be satisfied with some conjecture until a real proof is known.

## 5.5 Sperner-spaces which are restricted dual Spaces

We call Sperner-space (S-space, generalized affine space) an incidence structure of points and lines, where every line is incident with the same (cardinal) number of points and through any two points there goes exactly one line. Moreover we have a parallelism, i.e. an equivalence relation on the line-set, such that for any line l and point s there is exactly one line m, which is parallel to l and incident with s. Given an S-space *S*, we call (affine) plane of *S* every set *A* of lines of *S*, such that *A* together with all points incident with some line of *A* (under the induced incidence relation) is an affine plane. Now we add in the usual way improper points represented by the equivalence classes of mutual parallel lines and call the set of all improper points the improper plane. Furthermore the improper lines belonging to the affine planes of *S* are introduced. The latter are welldefined by an additional property (3) we will ask for. Namely, in addition we ask for the following properties:

(1) Every line lies in at least one plane and contains at least five points,
(2) to every plane p of *S* there is a unique point <p> of p, such that every line of p which does not lie in any other plane goes through <p>,
(3) for every plane p and every line l which does not meet p, there is a line l' of p parallel to l.

By dualizing we get a semimodular lattice L of rank 4 with a distinguished point o such that: if two planes of L do not intersect they both contain o. All other pairs of planes of L intersect in lines. So omitting the point o we get a Space. Conversely the dual of any Space of this kind leads to an S-space as described above. If for such a Space we dualize the concept of perspectivity we get simply the usual definition of perspectivity in the projective closure of the affine planes of *S* together with the analogue of theorems (4.5),(4.6). So for certain classes of S-spaces it might be useful to define perspectivities in this manner.

proper planar
EXAMPLE (cf.[1]). Let F be any nearfield with distributive law a(b+c)=ab+ac. In the abelian group $F^3$ (with + as composition) consider the partition *P* of the subspaces $\{(xa,xb,xc)\,|\,x\in F\}$ for all $(a,b,c)\neq(0,0,0)$, with $a,b,c\in F$. Then the translation structure of all cosets of *P* is an S-space with the properties mentioned above (every line is contained in at least two different planes). The Space L connected with this S-space can be described as follows: All the points of L are contained in any of three distinguished planes e,f,g; they forme projective dual nearfield-planes and by pairs are

glued together along one of their distinguished lines - and their (common) distinguished point o is removed. The translations of the S-Space now become elations of L, the axes of which run over all planes through o, and every such elation fixes all subspaces through o.

References

1 Arnold,H.J.: Algebraische und geometrische Kennzeichnung der schwach affinen Vektorräume über Fastkörpern 1968, Abh.Math.Sem.Hamburg 32, pp.73-88

2 Barlotti,A.: Sul gruppo delle proiettività di una retta in se nei piani liberi e nei piani aperti 1964, Rendic.Sem.Mat.Padova 34, pp. 135-159

3 Ewald,G.: Kennzeichnungen der projektiven dreidimensionalen Räume und nichtdesarguessche räumliche Strukturen über beliebigen Ternärkörpern 1961, Math.Z. 75, pp.395-418

4 Halder,H.-R. and Heise,W.: "Kombinatorik", München Wien 1976

5 Herzer,A.: Projektiv darstellbare stark planare Geometrien vom Rang 4 1976, Geom.Dedic. 5, pp.467-484

6 Herzer,A.: "Halbprojektive Translationsgeometrien", Gießen 1977 (Mitt.Math.Sem.Gießen Heft 127)

7 Herzer,A.: Büschelsätze zur Charakterisierung projektiv darstellbarer Zykelebenen 1979, Math.Z.164 pp.215-238

8 Kahn,J.: Locally projective-planar lattices which satisfy the Bundle Theorem, to appear in Math.Z.

9 Kantor, W.M.: Dimension and Embedding Theorems for Geometric lattices 1974, J.Comb.Th.(A) 17 pp.173-195

10 Mäurer,H.: Ein axiomatischer Aufbau der mindestens 3-dimensionalen Möbiusgeometrie 1968, Math.Z. 103, pp.282-305

11 Schleiermacher,A. and Strambach,K.: Über die Gruppe der Projektivitäten in nichtgeschlossenen Ebenen 1967, Arch.Math.18, pp.299-307

12 Wille,R.: On Incidence Geometries of Grade n, 1971 Atti Conv.Geom.Comb.Appl.Perugia, pp.421-426

# THE IMPACT OF VON STAUDT'S FOUNDATIONS OF GEOMETRY

Hans Freudenthal

Mathematisch Instituut
University of Utrecht

What did Projective Geometry mean before von Staudt? It owed much to Monge, but its true founder was J.V. Poncelet. He invented the so-called continuity principle (Traité des propriétés projectives des figures, 1822, p. XIII:)

Let us consider an arbitrary figure, in a general position which in a certain sense is indeterminate among all positions it can assume without violating the laws, the conditions, the bonds that exist between the different parts of the system; let us suppose according to these data one has found one or more relations or properties, which may be metric or descriptive, belonging to the figure, by way of ordinary explicit reasoning, that is, the procedure which is in certain cases considered as the only rigorous one. Is



*P. Plaumann and K. Strambach (eds.), Geometry - von Staudt's Point of View, 401–425.*

it not obvious that if while preserving those data one undertakes to vary the original figure ever so slightly and subjects parts of it to an arbitrary but continuous motion - is it not obvious that the properties and relations, found in the first system, remain valid in its successive stages, provided that due account is attributed to the particular modifications that may arise, for example if certain magnitudes vanish or change their direction or sign, and so on, modifications that can easily be recognized a priori and by sure rules?

Thus, what can be asserted of a figure that contains two intersecting lines must remain valid if the lines become parallel - this is to legalize the infinite, which in geometry has been a heuristic tool, certainly since antiquity. But there is more to it than that, and this is easily forgotten today: the same must be allowed to be said about the situation of a straight line with respect to a curve whether - in the real plane - both of them intersect or not - and this is to grant civil rights in the realm of geometry to the imaginary. Of both aspects von Staudt will face the consequences, but beyond this the continuity principle will, later on, give rise to the principle of the conservation of number. Is Poncelet's principle only a heuristic one? (See p. XIV:)

...would it not be legal to accept the continuity principle in its total generality in Theoretical Geometry, such as one has first accepted it in Algebra, and then in the application of this calculus to Geometry, as a means of discovery and invention if not of proof? As a teaching subject are not the expedients used at different periods by ingenious people to find out truth at least as important as the painstaking efforts they were afterwards obliged to make to prove it to the satisfaction of the timid ones who could not grasp it otherwise?

To be sure, the situation in geometry was not too much different from that in Calculus. Infinitely small and imaginary numbers were as unreal as infinite and imaginary points, though in the course of time they had become more familiar and developed beyond the stage of mere heuristic tools. In Geometry the development would start later but proceed faster and the stage of elucidation would be arrived at in analysis and geometry simultaneously, though finally geometry with its pretended reality of geometric space would be a tougher case than analysis.

The next name I have to recall ist that of

J. D. Gergonne (from 1824 onwards), because it is related of course to the discovery of duality, or rather, since dualities had been known earlier, because he discovered the fundamental meaning of duality, though Gergonne must also be remembered as the geometer who introduced pencils and sheaves of curves. Collineations, too, were instrumental from the start of projective geometry, and in his maiden paper von Staudt used them systematically. The concept of collineations, however, was first grasped and formulated by A. F. Möbius (Der baryzentrische Calcul, 1927). Möbius is also to be credited with the cross ratio, the Möbius grid and of course with the barycentric coordinates, which were followed after a short delay by J. Plücker's general homogeneous coordinates (Crelle 6 (1829), 107 - 146).

The preceding sketch does not at all account for the great variety of special results that were obtained in projective geometry up to about 1830. The first who tried to arrange them systematically was J. Steiner in his work with the most characteristic title "Systematische Entwicklung der Abhängigkeit geometrischer Gestalten voneinander" (1832). Steiner put the cross ratio and its perspectivity invariance in front, generated conics and ruled surfaces projectively, defined, and systematically used, general correspondences and arrived at a great variety of new results. Priority for many among them was also claimed by Chasles, who had progressed in the same direction in his Aperçu Historique, which though dated 1830 was published not earlier than 1837.

Von Staudt started something that was essentially new. Ernst Kötter's "Die Entwicklung der synthetischen Geometrie von Monge bis auf Staudt" (1847) has a subtitle: "Erster Band eines Berichtes, erstattet der Deutschen Mathematiker-Vereinigung."[1] The second part seems never to have appeared, but the subtitle suggests that von Staudt's "Geometrie der Lage" marks the beginning of a new period. It is the aim of this lecture to show you that this assertion is, indeed, justified.

In his address to the International Congress of Mathematicians at Chicago in 1893, Hilbert asserted that the history of a mathematical theory could be seen to develop in three stages: the naive, the formal, and the critical one. In the theory of invariants the naive period is represented by Cayley and Sylvester, the formal by Clebsch and Gordon, whereas Hilbert's finiteness proof and its consequences comprise the critical period. Though Hilbert's triad of reshuffling history is more to the point than Hegel's, it is a deplorable fact that history is never as simple as people would like it to be. If, however, one wants to play the game of dividing the 19th century history of geometry according to this pattern, it would not be too farfetched to count the naive period up to von Staudt, to have him open the formal one, and to start the critical period with Pasch, or the Italians, or with Hilbert. Von Staudt's work is characterized by the fact that, rather than contributing to the contents of projective geometry, he made an attempt at its formal recasting, though even in this respect his work was exemplary rather than permanently fundamental.

To appreciate some formal progress in mathematics one has first to find out what is the thing its author put upside down. This, indeed, has been the favorite business of formalizers from antiquity onwards. It is not as striking in Euclid's work because the tradition of his predecessors has almost been lost, but even so Euclid's taste for putting the cart before the horse has troubled people for centuries and has not been genuinely understood until more recent times. We can, however, still see how the chapter of mathematics which is now called conics has been put upside down in history; even if we did not know the details, it should be clear from the terminology. Conics have not, indeed, been invented as sections of a cone but as curves of the second degree, as solutions of the quadratic equation such as formulated by the Greeks in geometric language: the application of an area $y^2$ as a rectangle to a side $a$ exactly (parabolic), or with a square $x^2$ exceeding (hyperbolic) or falling short (elliptic), thus

$$y^2 = ax$$
$$y^2 = ax + x^2$$
$$y^2 = ax - x^2$$

The curve described by this relation between $x$ and $y$ was consequently named a parabola, hyperbola, and ellipse. Only afterwards were these curves identified as conic sections. It was Apollonius' eccentricity, a historically important and exemplary eccentricity, to redefine parabolas, hyperbolas and ellipses as sections of a cone and then conversely, to derive their old equations from this new definition. Today such inversions in mathematics are our daily bread, but if they are particularly ingenious ones they still excite our admiration. In geometry, which after a long sleep since antiquity had just been flourishing again for half a century, von Staudt's approach was a striking new example of that old strategy, an example that was bound more and more to determine the spirit of geometry in the 19th century.

Von Staudt's first inverse approach concerns the redefinition of harmonicity of a point quadruple, which before - one may even say since the days of antiquity - was defined by the cross ratio -1; likewise it had been known, essentially from antiquity, that such cross ratios are found in the figure of the complete quadrilateral. Now von Staudt who tried to avoid any appeal to metric relations, started from this incidence figure to define by it what he would call a harmonic quadruple. Of course, he had now to show that the fourth member of a harmonic quadruple is determined by three of them, but this follows from Desargues' theorem or the quadrilateral theorem (properly said, even from weakened versions of these theorems).

Von Staudt's next inversive definition is that of projectivities of point series (or pencils): a mapping is called projective if it preserves harmonicity. As a freshman at university, after having become acquainted with projective geometry at high school through Reye's book, I learned von Staudt's approach in a course of von Mises, who followed the book of Enriques. It made a strong impression on me then, and it still does, though for different reasons. Of course, I had learned at school that in mathematics you are allowed to make definitions as you want of what you want. The difference is at university it is not only said but even done. Von Staudt's definition, however, was an example of a liberty of forging definitions, on the verge of impudence. No doubt we are still more impudent today but we can do so only because so many before us presumed so big audacities.

Today, hardboiled, a Jack of all axiomatic trades, but also anointed with the oil of history, I feel still charmed by von Staudt's definition. I will soon explain why.

When organizing some field of mathematics one again and again faces the problem which properties of an object should be used to define it. To von Staudt's predecessors the projectivities were given explicitly, though separately defined for point series, line pencils, plane pencils. Clearly von Staudt looked for one definition which, first, showed perspicuously the group property of projectivities; second, applied uniformly to all one-dimensional entities, and third, led in a straight way to the fundamental theorem of projective geometry. This excluded a constructive definition of projectivities; another solution would have been to define them as restrictions of collineations, but this probably was not elegant enough. In the historical framework of that period it was an astonishing occurrence that he hit upon the utterly implicit definition by invariance of harmonicity, and this event has had exemplary consequences in algebra and geometry. It is true that we now know that something was wrong in von Staudt's approach. In fact, we have to distinguish three groups on the line, which are different in principle:

$G_1$: the products of perspectivities (the projectivities).

$G_2$: the restriction of the plane collineation group (the collineations).

$G_3$: the mappings preserving harmonicity (von Staudt's projectivities).

In the algebraization over a skew field of characteristic $\neq 2$ these groups are algebraically induced as follows:

$G_1$: by the linear mappings,

$G_2$: by the linear mappings together with

automorphisms of the underlying field,

$G_3$: by the linear mappings together with the automorphisms and antiautomorphisms of the underlying field.

This leads to the (in general, strict) inclusions $G_1 \subset G_2 \subset G_3$ . Even with order axioms, which are always explicitly presupposed by von Staudt, the equality of these groups cannot be enforced, and the Pappus theorem grants only the equality $G_2 = G_3$. If von Staudt believes he can prove $G_1 = G_2 = G_3$, he is making tacit topological assumptions, which were quite natural at that time. Not until a quarter of a century later had people become critical enough to notice them. I shall come back to this point in just a minute.

It was von Staudt's aim in his "Geometrie der Lage" to found projective geometry independently of any metric assumptions. How far did he succeed? When answering this question, one should not bother too much about the lack of axiomatic form. What we would call axioms is in von Staudt's approach copied from "real" space though it is not yet concentrated in a few "Grundsätze", as it would be in Pasch's work. It is circumstantially formulated in the first paragraph, where it can easily be identified by the characteristic fact that it is neither proved nor introduced as a definition. Anyhow though not as axioms, incidence axioms are explicitly formulated and - a rather important fact - duality is already axiomatically rooted. Even the order axioms are somehow satisfactory. To be able to consider von Staudt's approach as a rigorous foundation of projective geometry, one need only add explicitly the topological axioms which are tacitly used by von Staudt. But this would be a serious historical mistake as becomes clear from that happened a quarter of a century later. The problems that cropped up then show that von Staudt had dealt too lightheartedly with topology, or as they called it a century ago, continuity. Historically-psychologically the order properties were a much simpler thing than the topology. The cyclic order in one dimension and its invariance under perspectivities introduces in a straight way relations, independent of metric. But how can one formulate the topology of the projective space without the support of a metric? Von Staudt was still far from raising this question, which a quarter of a century later would become urgent. In Cayley's work F. Klein discovered a model of non-Euclidean geometry, constructed within the frame of projective geometry. To Cayley's view geometry was only a kind of algebra; to

Klein, who was nothing less than a modern axiomatician, this model was the proof that non-Euclidean geometry could be founded independently of the Euclidean geometry. F. Klein noticed the gap in von Staudt's approach; he was aware of the need to formulate the topology of projective space independently of that of Euclidean space. Not before the 20th century is this detachment consciously performed. In the 19th century it was generally considered impossible, and the few who possessed a more profound insight, lacked the techniques to rationalize their intuition. People suspected a vicious circle in Riemann's and Helmholtz's concepts: they did not believe that the concept of "n-dimensional number manifold" would be made independent of a metric, and even Hilbert's "Grundlagen der Geometrie" were met with suspicion by people who firmly believed that a topology could only be defined through a metric. Whereas von Staudt did not even raise the question of a metric-free definition of the topology of projective space, it remained a vague intuition in F. Klein's thought; all he attempted in this respect was so confused that it tended to convince people rather of the impossibility of detaching projective space from Euclidean geometry. Even Pasch introduced the topology of the projective line via Euclidean geometry; the Italians were the first to find truly satisfactory solutions for the problem of a purely projective foundation of projective geometry, which von Staudt had tried to solve.

In my previous inquiries into the history of the foundations of geometry I appreciated von Staudt's contribution only within a broader context, so it was a matter of course that I did not fully do justice to von Staudt himself since I cut off all that seemed to be less important for the further development of geometry. When preparing the present lecture and focussing more closely on von Staudt, I arrived at attributing more historical meaning to some of von Staudt's ideas which I had been inclined before to consider as mere whims. I should say that in this course my high estimation for von Staudt only increased.

The subject I mean is the main subject of the "Beiträge", namely the foundation of the imaginary in geometry. Viewing Hilbert's "Grundlagen" as the goal of the development, one easily forgets that in the middle of the 19th century the imaginary rather than the infinite in geometry needed explanation. There can be little doubt that from the start onwards, von

Staudt aimed at this goal. Unfortunately, the "Geometrie der Lage" of 1847 was founded on the concept of harmonic separation, which is just good for real geometry. Probably in 1847 von Staudt did not yet have any idea how to tackle the imaginary, and still less what kind of difficulties he would meet when solving this problem in the "Beiträge" of 1856, 1857, and 1860. Fortunately, I would say, because if he had guessed it, he maybe would never have written the work of 1847. But once the A of real geometry was said, he could not avoid any more the B of reducing imaginary to real geometry.

The idea to represent a conjugate pair of complex points of a real line by an elliptic involution was not new; it can be found in the literature before von Staudt. The big problem was to find a postulate that achieves splitting this pair of points into its components. Some illumination between 1847 and 1856 must have shown von Staudt the way: a complex point on a real straight line is an elliptic involution combined with one of the two cyclic orientations of the line. In the midst of the 19th century this was an unheard of abstraction, and certainly too much for most of von Staudt's contemporaries, even for those who held the author of the "Geometrie der Lage" in high esteem. Of course, today such an eccentricity would not impress us much, but after von Staudt is lasted many years until in formulating mathematical definitions people dared to take liberties such as von Staudt's. However, von Staudt's trick did not solve all problems. On the contrary, it was rather the first step on a long and thorny path, on which von Staudt persevered to the bitter end - it looks like a crime story where every foul deed commits the wrongdoer to seven fresh ones. Also in this respect von Staudt's work has become exemplary. Of course it was not enough to define complex points on one real line and complex planes in one real pencil. For example one should also tell under which conditions such a point should be incident with such a plane (namely if the determining involutions and orientations were perspective). Among the imaginary straight lines one should distinguish two kinds, the first kind that intersect their conjugates and consequently have a real point and a real plane in common, and the second kind that form with their conjugate a skew pair. The first kind is defined by elliptic involutions in an oriented plane pencil; the second by involutions in an oriented ruled surface pencil (von Staudt's followers chose instead an oriented linear congruence with no real directrices). But on the ground of these definitions one had also to prove anew all elementrary incidence properties. Collineations and correlations must be investigated and conics had to be redefined. And even this was not yet enough. In the second issue of "Beiträge" von Staudt had to suffer for the wrong he had done in "Geometrie der Lage": the definition of projectivities by the invariance of the harmonic separation. This definition is obviously wrong in the complex domain if the fundamental theorem of projective geometry is to be saved. To give some idea of how von Staudt slips his neck out of the noose I will show the heuristic way which von Staudt must have followed because, as it has been the habit among mathematicians from olden times, he keeps silent about it - and therefore there were hardly any contemporaries who tried to understand von Staudt's method at this point.

On the complex line such as represented by the Riemann sphere some real figures, the real lines and circles, are characterized by the reality of the cross ratio of any point quadruple - von Staudt calls such figures chains. On the oriented complex line every oriented chain shows a left and a right side, depending on the sign of the imaginary part of the cross ratio. Projectivities of the complex line are to be defined as transformations that map chains into chains and preserve the orientation, that is, carry the left side of a chain into the left side of its image. Chains and orientations must now be redefined while departing from the real geometry. A chain is determined by three of its points. Let us take three points $\ulcorner P_i, I_i \urcorner$ on a line of the second kind, that is, determined by the real oriented lines $P_i$ and the elliptic involutions $I_i (i = 1,2,3)$. The three lines are lying in one pencil $R$ of a ruled surface, with the transversal pencil $R'$. The chain $\mathfrak{K}$ through the $\ulcorner P_i, I_i \urcorner$ consists of all $\ulcorner P, I \urcorner$ with $P$ running through $R$ and the different $I$ mapped upon each other by the pencil $R'$. The line of the second kind containing the $P_i$ is determined by the pencil $R'$ provided with an orientation and an involution $J$. Let now $\ulcorner P^*, I^* \urcorner$ be some point of the line $\ulcorner R', J \urcorner$. If it does not belong to the chain $\mathfrak{K}$, then $P^*$ does not intersect the ruled surface bearing $R$ and $R'$. An oriented line that is skew to $P^*$ determines an orientation in the pencil of $P^*$, and so does $R'$ in the pencil of $P^*$. By orienting the chain $\mathfrak{K}$ of the $\ulcorner P, I \urcorner$ one gets a new orientation in the pencil of $P^*$. If both of them coincide, then one can say that $\ulcorner P^*, I^* \urcorner$ is lying on the positive side of the oriented chain, otherwise on the negative one.

This is the heuristic background of von Staudt's definition, which more precisely should run as follows: A chain $\mathcal{K}$ on a line of the second kind consists of a line of the oriented pencil R of a ruled surface with orientations and involutions, connected to each other by the transversal pencil R' ; such a line is represented by $\ulcorner R', J \urcorner$ . A point $\ulcorner P^*, I^* \urcorner$ outside the chain $\mathcal{K}$ is lying on the positive or negative side of $\mathcal{K}$ according to whether the two orientations determined by R and R' or P coincide or not.

Of course, something like this should also be done for the lines of the first kind. It is done in an indirect way. The original definition is invariant under perspectivities and is extended by postulating invariance. Finally projectivities have to be re-defined, as indicated earlier, by the invariance property for chains and orientations.

It is easy to imagine how tedious such a foundation is with its numerous case distinctions and with de-finitions which without the heuristic background are incomprehensible. Certainly von Staudt has spent an enormous labour on it, and though he was a careful man, I would not go through fire and water for the rigor of his approach. J. Lüroth, who in 1875 (Math. Ann. 8) dealt with this subject anew, complains that "this ingenious investigation has not drawn the attention it would have merited". Previous to Lüroth only F. August[2] (1872) can be mentioned, and after him only R. Sturm[3]. This subject has never become popular, which is not to be wondered at. Scrutinizing the result, everybody would ask: Is this worth the formidable effort? Anyone who has understood the idea and grasped the plan will admit that it works, and perhaps may give this problem to an assiduous student to work it out in detail - today such techniques are very common. But from a historical point of view this would be a wrong assessment. It is not true that von Staudt's "Beiträge" could be erased from the course of history of mathematics. Whether something was operative in the past is not necessarily shown by its straight continuations. The exemplary impact, though less easily proved, is not less important. In von Staudt's case this influence must have been significant. Definitions like von Staudt's are quite common now, for instance if we define a real number as a cut in the set of rational numbers, or an algebraic number as a coset in a polynomial ring, or a spot of an algebraic variety as a maximal idea, and so on. Von

Staudt's bold abstraction, when he defines the imaginary in geometry in terms of real figures, sounds like an anachronism, like a premature chiming in, if one remembers that according to his age von Staudt was halfway Cauchy and Kummer, older than Dirichlet, and a whole generation previous to Kronecker and Dedekind, and that it would not be until twenty years after von Staudt that a wave of abstractions as bold as von Staudt's sets in. It strikes one that this step was done in geometry earlier than in algebra. Was not geometry more closely tied to the concrete than algebra? Or was it just easier in geometry because there the abstraction could safely be guided by intuition?

Till now I have dealt with two accomplishments of von Staudt: his attempt at a metric-free foundation of projective geometry in "Geometrie der Lage" and his reduction of complex to real geometry in "Beiträge zur Geometrie der Lage". A third accomplishment should still be mentioned: the calculus of throws, likewise in "Beiträge". It is again a most modern looking idea: the direct algebraization of geometry by means of geometrical definitions of the algebraic operations.

A throw is a class of projectively equivalent

quadruples on a one dimensional projective entity (with certain obvious conventions if some elements of the quadruple coincide). Departing from algebra, a throw is defined by its cross ratio - if $0, 1, \infty$ are the coordinates of the first three points, the cross ratio is the coordinate of the fourth. The throws are now to be added and multiplied as their cross ratios:

The quadruples $\ulcorner p_0, p_1, p_\infty, x \urcorner$, $\ulcorner p_0, p_1, p_\infty, y \urcorner$, $\ulcorner p_0, p_1, p_\infty, z \urcorner$ determine three throws. The third is called the sum of the first two, if there is an involution fixing $p_\infty$ and taking $x$ into $y$ and $0$ into $z$.

The quadruples $\ulcorner p_0, p_1, p_\infty, x \urcorner$, $\ulcorner p_0, x, p_\infty, y \urcorner$, $\ulcorner p_0, y, p_\infty, z \urcorner$ determine three throws, the third being the product of the first and the second.

These are von Staudt's definitions. Carefully von Staudt proved that with these operations the throws form a field, and he ascertained the laws that in our terminology characterize a field[4]. Von Staudt, in fact, does not go as far as to show that it is an ordered field, and to connect it to the real and complex field - this was finally done by Lüroth. It should be stressed that up to Hilbert there is no other example for such a direct derivation of the algebraic laws from the geometric axioms as is found in von Staudt's "Beiträge". There can be little doubt that Hilbert's segments calculus depended on von Staudt's throws calculus, though as usual quotations are lacking in Hilbert's work.

Let us summarize: von Staudt hardly contributed to the contents of projective geometry. He has been the first to raise the foundational question and to aspire to purety of methods in projective geometry. His definition of projectivities by the invariance of harmonicity and his definition of complex points by involutions on oriented lines are almost anachronistic examples of bold abstraction; by his calculus of throws he has outlined the modern method of algebraization of an axiomatic theory.

NOTES.

This article appeared originally R.S. Cohen et al. (eds.), For Dirk Struik, 189 - 200.  It is the translation of the non-biographical part of the commemoration lecture on the centenary of von Staudt's death, held on June 20, 1967, in Erlangen. Biographical details on von Staudt can be found in: Archiv der Mathematik und

Physik 49 (1869), Literarischer Bericht CLXXXXIII, and further in : M. Noether, "Zur Erinnerung an Karl Georg Christian von Staudt, Festschrift der Universität Erlangen zur Feier des 80. Geburtstages Sr. Kgl. Hoheit des Prinzregenten Luitpold von Bayern", Erlangen-Leipzig, A. Deichert 1901. Carl Georg Christian von Staudt (January 24, 1798 Rothenburg ob der Tauber - June 1, 1867 Erlangen) was a student of Gauss; got a Ph.D. degree at Erlangen; was a professor at the Gymnasium in Würzburg (1822) and Nürnberg (1827); and a university professor at Erlangen (1835). Descendants of von Staudt are still living in the house in Rothenburg where he was born.

Von Staudt's principal works are: "Geometrie der Lage" 1847, Nürnberg, Bauer & Raspe. "Beiträge zur Geometrie der Lage", 1856, 1857, 1860, idem.

[1] Jahresbericht d. Deutschen Math. Ver. 5 (1901).

[2] Untersuchungen über das Imaginäre in der Geometrie. Pr. Berlin (quoted after Jahrbuch Fortschr. Math. 4 (1872), 242).

[3] Math. Ann. 9 (1876), 333 - 346.

[4] Elsewhere (Nieuw Archief Wiskunde (4) 5, 105 - 142 (1957) I claimed that von Staudt derived these laws from the theory of conics. This is true of Lüroth, but misleading as regards von Staudt. It is not

absolutely clear what intervenes if all inferences are traced back ad ovum, and I am not sure whether then the conics would play the part I ascribed to them.

INDEX OF SUBJECTS

GPSR Compliance

*The European Union's (EU) General Product Safety Regulation (GPSR) is a set of rules that requires consumer products to be safe and our obligations to ensure this.*

*If you have any concerns about our products, you can contact us on ProductSafety@springernature.com*

In case Publisher is established outside the EU, the EU authorized representative is:

Springer Nature Customer Service Center GmbH
Europaplatz 3
69115 Heidelberg, Germany

**Batch number: 10761599**

Printed by Printforce, the Netherlands